An Introduction to

Electrospinning
and Nanofibers

An Introduction to

Electrospinning
and Nanofibers

Seeram Ramakrishna, Kazutoshi Fujihara,

Wee-Eong Teo,Teik-Cheng Lim & Zuwei Ma

National University of Singapore

World Scientific

NEW JERSEY · LONDON · SINGAPORE · BEIJING · SHANGHAI · HONG KONG · TAIPEI · CHENNAI

Published by

World Scientific Publishing Co. Pte. Ltd.

5 Toh Tuck Link, Singapore 596224

USA office: 27 Warren Street, Suite 401-402, Hackensack, NJ 07601

UK office: 57 Shelton Street, Covent Garden, London WC2H 9HE

British Library Cataloguing-in-Publication Data
A catalogue record for this book is available from the British Library.

AN INTRODUCTION TO ELECTROSPINNING AND NANOFIBERS

ISBN-13 978-981-256-415-3
ISBN-10 981-256-415-2
ISBN-13 978-981-256-454-2 (pbk)
ISBN-10 981-256-454-3 (pbk)

Foreword

Even though research and development related to the electrospinning process and the electrospun nanofibers has increased in recent years, the availability of the corresponding literature is mainly confined to research journals. As a consequence, information on electrospinning process and electrospun nanofibers is comprehensible only for the highly specialized readers. The situation is further compounded as a result of differing approaches and perspectives from various interdisciplinary backgrounds. In spite of the numerous groups throughout the world investigating on electrospinning of nanofibers, the lack of introductory reading materials in this field is felt.

The appearance of this book is timely in view of this rapidly expanding field of science and technology. There exists a critical mass of information in this area for a book to be written. In order to attain a balanced perspective, this book was written by a group of researchers from various backgrounds – mechanical and chemical engineering, materials science, chemistry.

The contents page shows that this book covers a wide spectrum, which includes the basic materials used for manufacturing nanofibers, processing techniques and parameters, various characterization methods, various ways to produce different types of nanofibers, the surface modification & functionalization, theoretical understanding and modeling approaches, and finally the potential applications.

The book is intended for the use of practicing engineers and scientists, as well as the students interested in electrospinning process and applications of the electrospun mats and nanofibers.

A.L. Yarin
Technion -Israel Institute of Technology
Haifa

Contents

Chapter 1

Introduction

1.1. Preface of Nanofibers

It is essential in the beginning of this book to firstly define what a nanofiber is. To do so, we split the term into two parts, namely "nano" and "fiber". As the latter term is more common, we begin by considering the latter from various professional viewpoints. Botanists identify this term with elongated, thick-walled cells that give strength and support to plant tissue. Anatomists understand "fibers" as any of the filaments constituting the extracelullar matrix of connective tissue, or any of various elongated cells or threadlike structures, especially muscle fiber or nerve fiber. The textile industry views fibers as natural or synthetic filament, such as cotton or nylon, capable of being spun into yarn, or simply as material made of such filaments. Physiologists and biochemists use the term "fiber" to refer to coarse, indigestible plant matter, consisting primarily of polysaccharides such as cellulose, that when eaten stimulates intestinal peristalsis. Historically, the term "fiber", or "fibre" in British English, comes from Latin "fibra". In this book we define a "fiber" from a geometrical standpoint – a slender, elongated, threadlike object or structure.

The term "nano" is historically interesting. Like many prefixes used in conjunction with Système International d'unités (SI units), "nano" comes from a language other than English (see Table 1.1). Originating from the Greek word "nannas" for "uncle", the Greek word "nanos" or "nannos" refer to "little old man" or "dwarf". Before the term nanotechnology was coined and became popular, the prefixes "nanno-" or "nano-" were used in equal frequency, although not always technically correct. For example the term nannoplanktons or nanoplanktons have been used, and has since been socially accepted, for describing very small planktons measuring 2 to 20 micrometers – a technical error which should be more correctly termed as microplanktons.

Table 1.1. Modern definition of "nano" and other prefixes.

Prefixes	Meaning	Original language
Yokto	10^{-24}	Latin
Femto	10^{-15}	Danish Norwegian
Pico	10^{-12}	Spanish
Nano	10^{-9}	Greek (dwarf)
Micro	10^{-6}	Greek (small)
Hecto	10^{+2}	French Greek (hundred)
Giga	10^{+9}	Greek (giant)
Yotta	10^{+24}	Latin

With the modern definition, we use "nano" to technically refer to physical quantities within the scale of a billionth of the reference unit – hence nanometer, nanosecond, nanogram and nanofarad for describing a billionth of a meter (length), second (time), gram (weight) and farad (charge) respectively. At the point of writing, nanotechnology refers to the science and engineering concerning materials, structures and devices which at least one of the dimension is 100nm or less. This term also refers to a fabrication technology in which objects are designed and built by the specification and placement of individual atoms or molecules or where at least one dimension is on a scale of nanometers.

At this juncture, we wish to point out that whilst the academic community has somewhat agreed to the <100nm criterion as the benchmark for the nanotechnology classification, the commercial sector has allowed broader flexibility – such as 300nm or even up to 500nm – which some academics would classify as sub-microtechnology. The authors are of the opinion that both benchmarks have their own merits. Whilst the imposition of a strict guideline is essential for maintaining some form of standard, the loose definition would be beneficial for the industry – for which the product quality and dimension is ultimately determined by the consumers and not mere measurements. Having pointed out the merit of the loose definition, the importance of the strict definition becomes evident in the light of the loose definition – a strict standard inhibits the loose definition from getting out of hand.

Ever since the the term "nanotechnology" was coined by K. Eric Drexler in his book "Engines of Creation", the field of nanotechnology has been a hot topic both in the academia and the industry. Although the positional manipulation of xenon atoms on a nickel substrate in 1990 (to spell the logo of a very large computer company) was hailed by some as the "first unequivocal" nanofabrication experiment, it should be borne in mind that the growth of nanowires and nanorods by vapor-liquid-solid method was reported in the beginning of 1960s and the spontaneous growth of nanowires and nanorods was in the 1950s. The scientific know-how of gold nanoparticle synthesis was, in fact, performed by Faraday. Perhaps the first nanotechnologists can be attributed to medieval stained-glass makers who, by prescribing varying amount of gold particles, produced gold with colors other than gold color. Unbeknown to the medieval stained-glass makers, these tiny gold spheres, which absorbed and reflected sunlight in differing frequency, will forever be part of the history of size effects in nano-scale object.

1.2. Nanotechnology and Nanofibers

In this sub-chapter an attempt is made to classify nanofibers into one or more sub-category of nanotechnology.

To do so we briefly review some common sub-fields of nanotechnology itself. As far as "nanostructures" are concerned, one can view this as objects or structures whereby at least one of its dimensions is within nano-scale. A "nanoparticle" can be considered as a zero-dimensional nano-element, which is the simplest form of nanostructure. It follows that a "nanotube" or a nanorod" is a one-dimensional nano-element from which slightly more complex nanostructure can be constructed of. Following this train of thought, a "nanoplatelet" or a "nanodisk" is a two-dimensional element which, along with its one-dimensional counterpart, are useful in the construction of nanodevices.

The difference between a nanostructure and a nanodevice can be viewed upon as the analogy between a building and a machine (whether mechanical, electrical or both). It goes without saying that as far as nano-scale is concerned, one should not pigeon-hole these nano-elements – for an element that is considered a structure can at times be used as a significant part of a device. For example, the use of carbon nanotube as the tip of an Atomic Force Microscope (AFM) would have it classified as a nanostructure. The same nanotube, however, can be used as a single-molecule circuit, or as part of a miniaturized electronic component, thereby appearing as a nanodevice. Hence the function, along with the structure, is essential in classifying which nanotechnology sub-area it belongs to.

Whilst nanostructures clearly define the solids' overall dimensions, the same cannot be said so for nanomaterials. In some instances a nanomaterial refers to a nano-sized material while in other instances a nanomaterial is a bulk material with nano-scaled structure. Nanocrystals appear to be a misnomer. It is understood that a crystal is highly structured and that the repetitive unit is indeed small enough. Hence a nanocrystal refers to the size of the entire crystal itself being nano-sized, but not of the repetitive unit.

Nanophotonics refers to the study, research, development and/or applications of nano-scale object that emit light and its corresponding light. These objects are normaly quantum gots. Whilst the emission of photon is largest for bulk (3-dimensional), followed by quantum well (2-dimensional) and finally quantum dot (0-dimensional), the ranking is reversed in terms of efficiency.

Although the term nanomagnetics is self expanatory, we wish to view it in terms of highly miniaturized magnetic data storage materials with very high memory. This can be attained by taking advantage of the electron spin for memory storage – hence the term "spin-electronics", which has since been more popularly and more conveniently known as "spintronics".

In nanobioengineering, the novel properties at nano-scale are taken advantage of for bioengineering applications. The many naturally occurring nanofibrous and nanoporous structure in the human body further adds to the impetus for research and development in this sub-area. Closely related to this is molecular functionalization whereby the surface of an object is modified by attaching certain molecules to enable desired functions to be carried out – such as for sensing and/or filtering chemicals based on molecular affinity.

With the rapid growth of nanotechnology, nanomechanics is no longer the narrow field it used to be. This field can be broadly categorized into the molecular mechanics and the continuum mechanics approaches – which view objects as consisting of discrete many-body system and continuous media respectively. Whilst the former inherently includes the size-effect, it is a requirement for the latter to factor in the influence of increasing surface-to-volume ratio, molecular reorientation and other novelties as the size shrinks.

As with many other fields, nanotechnology includes nanoprocessing – novel materials processing techniques by which nano-scale structures and devices are designed and constructed.

Depending upon the final size and shape, a nanostructure or nanodevice can be produced by the top-down or the bottom up approach. The former refers to the act of removal or cutting down a bulk to the desired size whilst the latter takes on the philosophy of using fundamental building blocks – such as atoms and molecules – to build up nanostructures in the same manner as one would towards lego sets. It is obvious that the top-down and the botton-up nanoprocessing methodologies are suitable for the larger and to smaller ends respectively in the spectrum of nano-scale construction. The effort of nanopatterning – or patterning at the nano-scale – would hence fall into nanoprocessing.

So where does all these descriptions point nanofibers to? It is obvious that nanofibers would geometrically fall into the category of 1-dimensional nano-scale elements that includes nanotubes and nanorods. However, the flexible nature of nanofibers would align it along with other highly flexible nano-elements such as globular molecules (assumed as 0-dimensional soft matter), as well as solid and liquid films of nano-thickness (2-dimensional). A nanofiber is a nanomaterial in view of its diameter, and can be considered a nanostructured material material if filled with nanoparticles to form composite nanofibers.

Where application to bioengineering is concerned, such as the use of nanofibrous network to tissue engineering scaffolds, these nanofibers play significant roles in nanobioengineering [Lim and Ramakrishna (2005)]. The study on the nanofiber mechanical properties as a result of manufacturing techniques, constituent materials, processing parameters and other factors would fall into the category of nanomechanics. Indeed, while the primary classification of nanofibers is that of nanostructure or nanomaterial, other aspects of nanofibers such as its characteristics, modeling, application and processing would enable nanofibers to penetrate into many subfields of nanotechnology. Finally the processing techniques of nanofibers are diverse, and include both the top-down and the bottom-up approaches as we shall see in the next sub-chapter.

1.3. Various Ways to Make Nanofibers

Polymeric nanofibers can be processed by a number of techniques such as Drawing, Template Synthesis, Phase Separation, Self-Assembly and Electrospinning, which are briefly reviewed in this section. A comparison of the various issues relating to these processing methods and some of the polymers that can be converted into nanofibers can be found in Tables 1.2 and 1.3, respectively.

Table 1.2(a).　Comparison of processing techniques for obtaining nanofibers.

Process	Technological advances	Can the process be scaled?	Repeatability	Convenient to process?	Control on fiber dimensions
Drawing	Laboratory	X	√	√	X
Template Synthesis	Laboratory	X	√	√	√
Phase Separation	Laboratory	X	√	√	X
Self-Assembly	Laboratory	X	√	X	X
Electro-spinning	Laboratory (with potential for industrial processing	√	√	√	√

Table 1.2(b).　Advantages and disadvantages of various processing techniques.

Process	Advantages	Disadvantages
Drawing	Minimum equipment requirement.	Discontinuous process
Template Synthesis	Fibers of different diameters can be easily achieved by using different templates.	
Phase Separation	Minimum equipment requirement. Process can directly fabricate a nanofiber matrix. Batch-to-batch consistency is achieved easily. Mechanical properties of the matrix can be tailored by adjusting polymer concentration.	Limited to specific polymers
Self-Assembly	Good for obtaining smaller nanofibers.	Complex process
Electrospinning	Cost effective. Long, continuous nanofibers can be produced	Jet instability

Table 1.3(a). Effect of processing method, material and solvent on the nanofiber dimension.

Process	Material	Solvent	Fiber diameter	Fiber length
Drawing	Sodium Citrate	Chloroauric acid	2 nm to 100 nm	10 microns to mms
Template Synthesis	Polyacrylonitrile	Dimethylformamide	100 nm	10 microns
Phase Separation	PLLA PLLA-PCL blends	Tetrahydrofuran	50 nm to 500 nm	Porous structure or continuous network
Self-Assembly	PCEMA core - PS shell	Tetrahydrofuran	100 nm	20 microns
	PAA/γ- Fe_2O_3 core – PCEMA middle layer – PS corona	Tetrahydrofuran	100 nm	20 microns
	PS core - P4VP corona	Chloroform	25 nm – 28 nm	up to 1 micron
	Peptide-Amphiphile	Chloroform	7 to 8 nm	several microns

Table 1.3(b).   Effect of processing method, material and solvent on the nanofiber dimension.

Process	Material	Solvent	Fiber diameter	Fiber length
Electro-spinning	Polyimides Polyamic acid Polyetherimide	Phenol m-cresol Methylene chloride	3 nm to 1000 nm	several cms to several meters
	Polyaramid Poly-gamma-benzyl-glutamate	Sulphuric acid Dimethylformamide		
	Poly (p-phenylene terephthalamide) Nylon 6-polyimide	Sulphuric acid Formic acid		
	Polyacrylonitrile Polyethylene-terephthalate Nylon	Dimethylformamide Trifluoroacetic acid Dichloromethane		
	Polyaniline	Sulphuric acid		
	DNA Polyhydroxybutyrate-valerate PLLA	Water Chloroform Chloroform or Mixed Methlyene chloride and Dimethylformamide		
	Poly (D,L-lactic acid)	Dimethylformamide		
	PEO	Water		
	PMMA	Toluene		
	PU	Dimethylformamide		

1.3.1. Drawing

Nanofibers have been fabricated with citrate molecules through the process of drawing [Ondarcuhu and Joachim (1998)]. A micropipette with a diameter of a few micrometers was dipped into the droplet near the contact line using a micromanipulator (see Fig. 1.1).

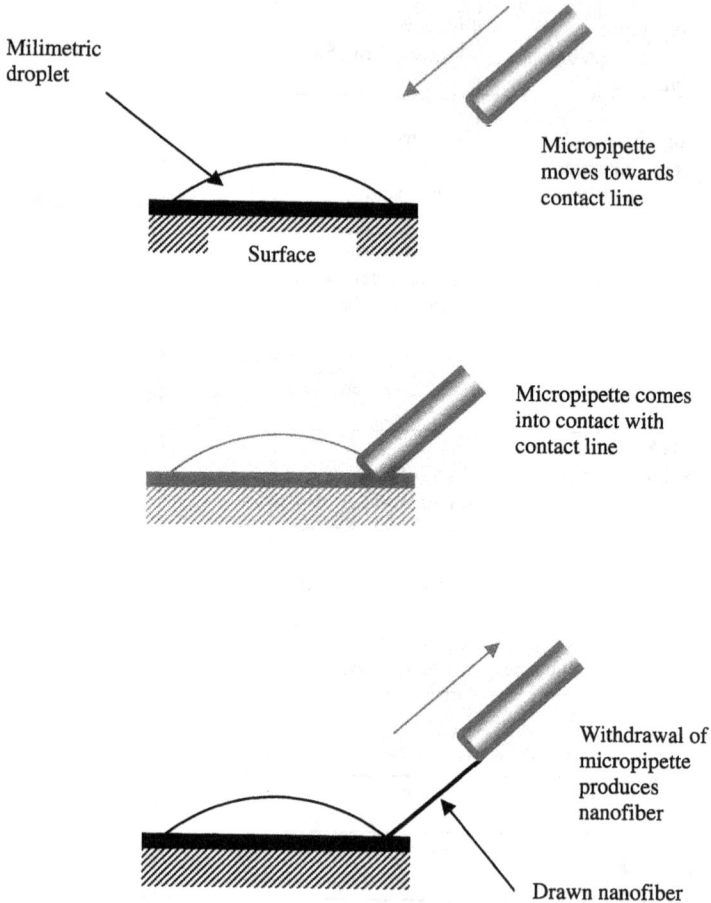

Milimetric droplet

Micropipette moves towards contact line

Surface

Micropipette comes into contact with contact line

Withdrawal of micropipette produces nanofiber

Drawn nanofiber

Fig. 1.1. Obtaining nanofiber by drawing.

The micropipette was then withdrawn from the liquid and moved at a speed of approximately 1×10^{-4} ms^{-1}, resulting in a nanofiber being pulled. The pulled fiber was deposited on the surface by touching it with the end of the micropipette. The drawing of nanofibers was repeated several times on every droplet. The viscosity of the material at the edge of the droplet increased with evaporation. At the beginning of evaporation corresponding to part X of the curve in Fig. 1.2, the drawn fiber broke due to Rayleigh instability. During the second stage of evaporation corresponding to part Y of the curve, nanofibers were successfully drawn. In the final stage of evaporation of the droplet corresponding to part Z of the curve, the solution was concentrated at the edge of the droplet and broke in a cohesive manner. Thus, drawing a fiber requires a viscoelastic material that can undergo strong deformations while being cohesive enough to support the stresses developed during pulling. The drawing process can be considered as dry spinning at a molecular level.

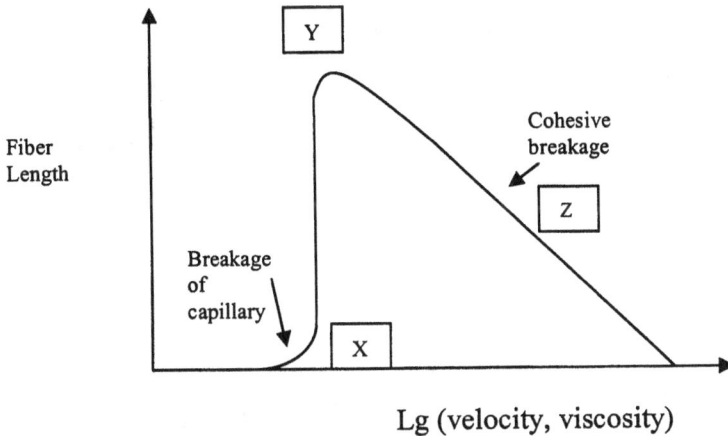

Fig. 1.2. Schematic representation of the length of the drawn nanofiber as a function of the drawing velocity and viscosity of the material (Adapted from [Ondarcuhu and Joachim (1998)]).

1.3.2. Template Synthesis

Template synthesis implies the use of a template or mold to obtain a desired material or structure (see Fig. 1.3).

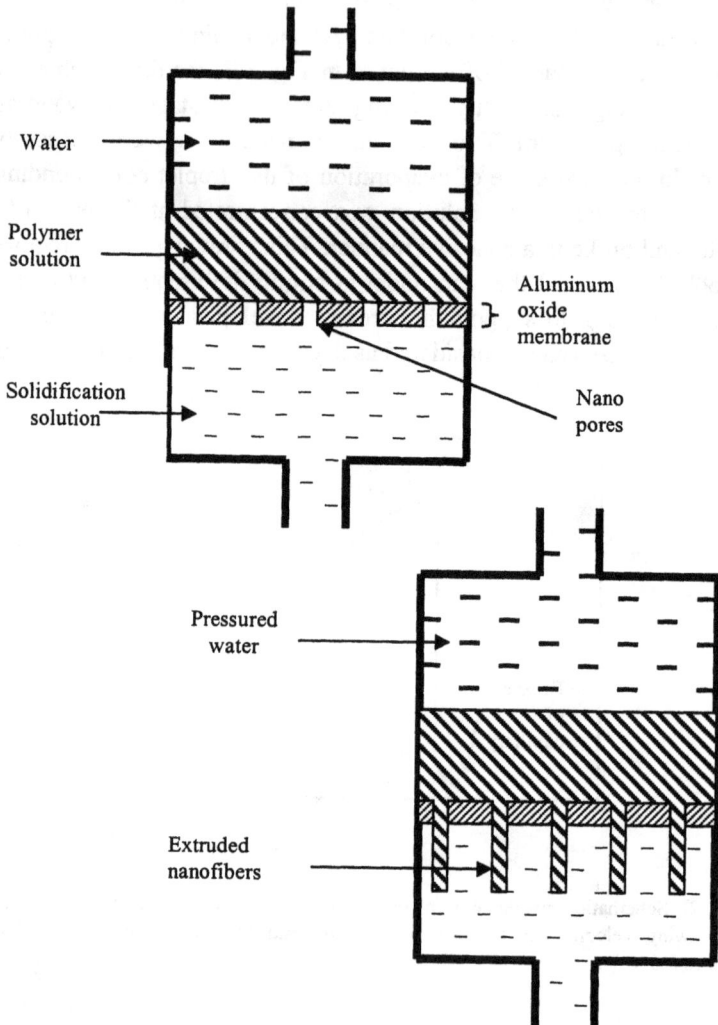

Fig. 1.3. Obtaining nanofibers by template synthesis.

Hence the casting method and DNA replication can be considered as template-based synthesis. For the case of nanofiber creation by [Feng et al. (2002)], the template refers to a metal oxide membrane with through-thickness pores of nano-scale diameter. Under the application of water pressure on one side and restrain from the porous membrane causes extrusion of the polymer which, upon coming into contact with a solidifying solution, gives rise to nanofibers whose diameters are determined by the pores.

1.3.3. Phase Separation

In phase separation, a polymer is firstly mixed with a solvent before undergoing gelation. The main mechanism in this process is – as the name suggests – the separation of phases due to physical incompatibility. One of the phase – which is that of the solvent – is then extracted, leaving behind the other remaining phase. A detailed procedure for producing nanofibrous poly(L-lactic) acid (PLLA) has been described by [Ma and Zhang (1999)], which consists of 5 major steps: (i) polymer dissolution, (ii) gelation, (iii) solvent extraction, (iv) freezing and (v) freeze-drying, as follows:

(i)   Tetrahydrofuran (THF) was added to PLLA for making a solution with the required concentration (1% w/v to 15% w/v). The solution was stirred at 60°C for two hours to produce a homogeneous solution.

(ii)   Two mililiters of the solution at 50°C was poured into a Teflon vial and then transferred to a refrigerator set to gelation temperature (-18°C to 45°C) which was chosen based on the PLLA concentration.Upon formation of the gel, it was kept at the gelation temperature for two hours.

(iii)   The vial that contains the gel was immersed in distilled water to allow solvent exchange and the water was changed three times a day for two days.

(iv)   The gel was then removed from water, blotted with filter paper and then transferred to a freezer at −18°C and kept for two hours.

(v)   Finally, the frozen gel was transferred into a freeze-drying vessel and freeze-dried at –55°C under a vacuum of 0.5 mm of Hg for a week.

A simplified generic representation of phase separation is shown in Fig. 1.4.

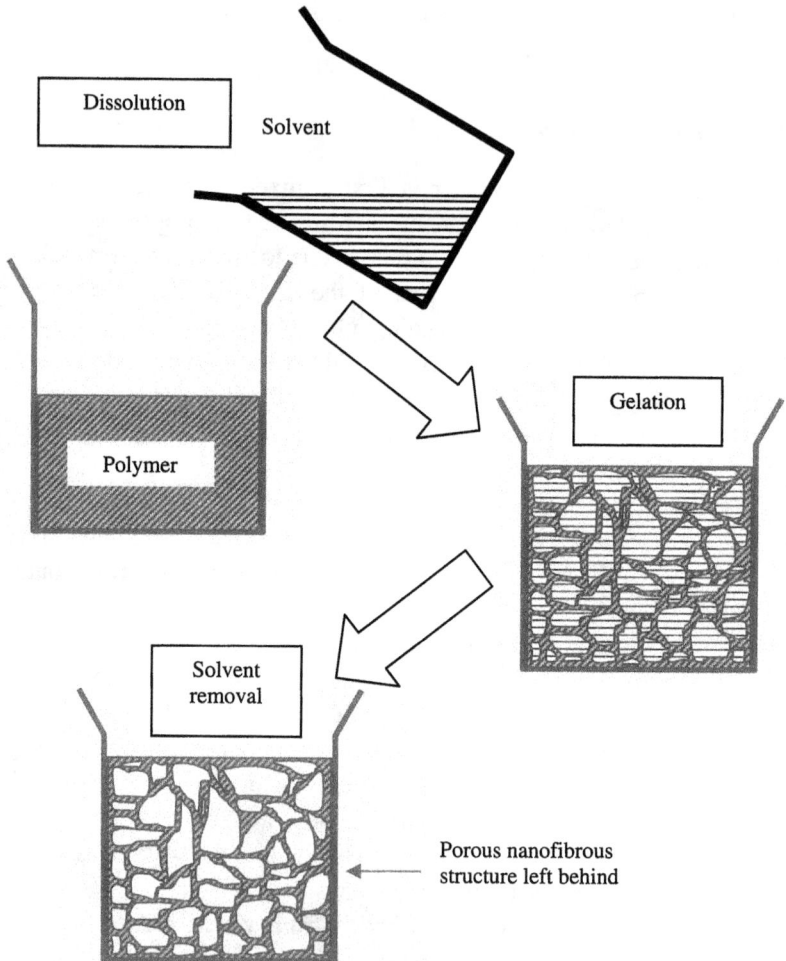

Fig. 1.4. Generic schematics of phase separation for obtaining nanofibrous structure.

1.3.4. Self-Assembly

In general, self-assembly of nanofibers refer to the build-up of nano-scale fibers using smaller molecules as basic building blocks. Various techniques have been reported by [Liu et al. (1996, 1999); Yan et al. (2001); de Moel et al. (2002) and Hartgerink et al. (2001)]. Fig. 1.5 is a simple schematic on self assembly for obtaining nanofibers, based on [Hartgerink et al. (2001)]. Here, a small molecule (Fig. 1.5 top) is arranged in a concentric manner such that bonds can form among the concentrically arranged small molecules (Fig. 1.5 middle) which, upon extension in the plane's normal gives the longitudinal axis of a nanofiber (Fig. 1.5 bottom). The main mechanism for a generic self-assembly is the intermolecular forces that bring the smaller units together and the shape of the smaller units of molecules which determine the over shape of the macromolecular nanofiber.

1.3.5. Electrospinning

A process was patented by [Formhals (1934)], wherein an experimental setup was outlined for the production of polymer filaments using electrostatic force. When used to spin fibers this way, the process is termed as electrospinning. In other words, electrospinning is a process that creates nanofibers through an electrically charged jet of polymer solution or polymer melt. Following this, investigations of the process have been carried out by a number of researchers [Baumgarten (1971); Larrondo and Manley (1981a, 1981b, 1981c) ; Reneker and Chun (1996); Fong and Reneker (1999); Reneker et al. (2000); Chen et al. (2001); Suthar and Chase (2001); Huang et al. (2003); Lim et al. (2004) and Yang et al. (2005)]. The electrospinning process, in its simplest form consisted of a pipette to hold the polymer solution, two electrodes and a DC voltage supply in the kV range, Fig. 1.6. The polymer drop from the tip of the pipette was drawn into a fiber due to the high voltage. The jet was electrically charged and the charge caused the fibers to bend in such a way that every time the polymer fiber looped, its diameter was reduced. The fiber was collected as a web of fibers on the surface of a grounded target (Fig. 1.6).

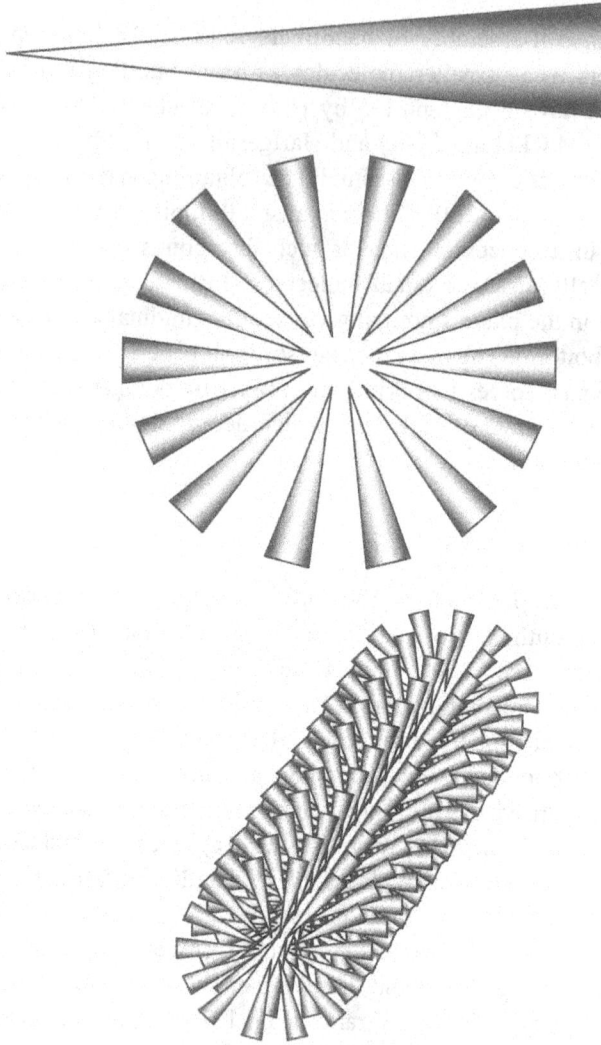

Fig. 1.5. An example of self-assembly for obtaining nanofiber [Adapted from Hartgerink et al. (2001)].

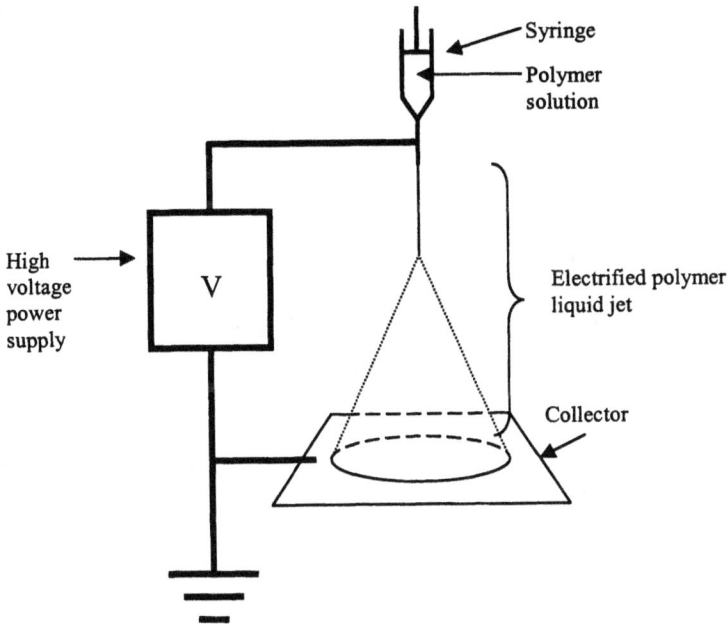

Fig. 1.6. Schematic diagram of the electrospinning process.

Important features of electrospinning are:
- Suitable solvent should be available for dissolving the polymer.
- The vapor pressure of the solvent should be suitable so that it evaporates quickly enough for the fiber to maintain its integrity when it reaches the target but not too quickly to allow the fiber to harden before it reaches the nanometer range.
- The viscosity and surface tension of the solvent must neither be too large to prevent the jet from forming nor be too small to allow the polymer solution to drain freely from the pipette.
- The power supply should be adequate to overcome the viscosity and surface tension of the polymer solution to form and sustain the jet from the pipette.

- The gap between the pipette and grounded surface should not be too small to create sparks between the electrodes but should be large enough for the solvent to evaporate in time for the fibers to form.

1.4. Scope of This Book

As the book title implies, the aims of this book are to provide an introduction to the electrospinning process for obtaining nanofibers and to interest the reader on the various potential applications of nanofibers as a result of their unique properties. This book includes both the experimental techniques and theoretical understanding on the electrospinning process.

This book can be split into two major blocks: Chapters 2 to 4 deal with electrospinning while Chapters 5 to 7 focus on the electrospun nanofibers.

Chapter 2 covers some basic aspects of the electrospinning process by firstly reviewing the various classes of materials such as polymers, composites and ceramics. Sub-topics on solution properties and electrostatics, which are part and parcel of the electrospinning process, are covered in view of their inter-relatedness with the material classes and properties. This chapter serves as a fundamental starting point in a generic manner.

The electrospinning process is covered in Chapter 3 whereby the polymer solution parameters (such as viscosity, surface tension, conductivity, etc), electrospining process parameters (such as voltage, federate, tip-to-collector distance, etc) and ambient conditions (e.g. humidity) are discussed. In addition, issues on the uniformity, productivity, patterning and the creation of various types of nanofibers (such as ribbon-like, branched, helical, porous and tubular) are covered in view of their importance in specific applications.

The purpose of this chapter is to follow up on the previous chapter by delving into greater details of the materials and processing parameters, thereby giving the reader a qualitative picture on the factors that influence the electrospinning process and the resultant nanofibers.

Chapter 4 elucidates the modeling of techniques of the electrospinning process for the purpose of introducing to the reader the more prominent modeling techniques to date. The importance of this chapter can be seen in the quantitative aspects whereby the simulated results will aid the experimentalists in obtaining the nanofiber of desired properties.

This chapter begins with the some assumptions normally adopted by eminent researchers in this field, and followed by the modeling of viscoelastic behavior with particular emphasis of free liquid jets. Detailed derivation for the conservation laws are then introduced for the physical quantities of mass, momentum and charge. This is followed by the considering the types of forces normally acted on the free liquid jet, discussions on the instability and the perturbation assumptions to induce the liquid bending instability. The result section shows some simulated results with emphasis on the effects of solidification and the instability windows. This chapter ends with a short appreciation on the future trends.

The raw materials (solution) and the electrospinning process leads to the various types and properties of the nanofibers. Chapter 5 considers the techniques for measuring the basic properties such as porosity and surface contact angle. In addition, this chapter introduces the reader to the characterization of the molecular structure (such as crystalline structure, organic group detection, etc), mechanical properties (such as for the case of single nanofiber, nanofiber yarn and nanofibrous membrane).

The purpose of this chapter is to introduce to the reader the various instrumentation techniques for measuring and characterizing the obtained electrospun nanofibers. This chapter is important as the obtained characterization results is useful for deciding the specific applications on the basis of the nanofiber properties.

In order to enable the nanofibrous' surface to capture specific molecules (such as in the case of molecular filters and sensors), there is a need for the surface to be molecularly modified to enable it to function in the desired manner. This topic is addressed in Chapter 6 with particular emphasis on the case of nanofibers.

Various techniques for surface modification (such physical coating, blending, co-polymerization, chemical vapor deposition, chemical treatment, etc) are furnished. Thereafter, the functionalized nanofibers are discussed with regard to affinity membrane, tissue engineering, sensor, protective clothing and other applications.

The last chapter expounded on the many applications of nanofibers. In view of the small fiber diameter and hence the nanofibrous structure with extremely large surface to volume ratio, this chapter discusses the current applications as well as potential for future applications in the fields of affinity membranes, filter media, tissue scaffolds, wound dressing, drug release, chemical & biological protective clothing, sensors, composite reinforcements, and energy & electrical applications.

A list of some useful websites that deal with nanofibers is given in the appendix which is correct at the point of writing. Throughout this book, the term nanofiber membrane refers a semi-transparent membrane obtained by electrospinning.

Whilst it is a membrane macroscopically, the membrane is a network of nanofibrous structure. Hence the terms nanofiber membrane, nanofiber mesh and nanofiber web are used interchangeably in order to reflect the diversity of viewpoints. These terms, as well as other technical terms, are defined and explained in the glossary of terms for the benefit of the lay readers.

Chapter 2

Basics Relevant to Electrospinning

To understand electrospinning, one can look at the mechanism behind the production of polymer fibers. Conventional fibers of large diameter involve the drawing of molten polymer out through a die. The resultant stretched polymer melt will dry to form individual strand of fiber. Similarly, electrospinning also involve the drawing of fluid, either in the form of molten polymer or polymer solution. However, unlike conventional drawing method where there is an external mechanical force that pushes the molten polymer through a die, electrospinning make use of charges that are applied to the fluid to provide a stretching force to a collector where there is a potential gradient. When a sufficient high voltage is applied, a jet of polymer solution will erupt from a polymer solution droplet. The polymer chain entanglements within the solution will prevent the electrospinning jet from breaking up. While molten polymer used in both conventional fiber production method and electrospinning method cools and solidifies to yield fiber in the atmosphere, the electrospinning of polymer solution relies on the evaporation of the solvent for the polymer to solidify to form polymer fiber.

Since electrospinning is basically the drawing of a polymer fluid, there are many different types of polymers and precursors that can be electrospun to form fibers. The materials to be electrospun will depend on the applications. Materials such as polymers and polymer nanofiber composites can be directly produced by electrospinning. Other materials such as ceramics and carbon nanotubes require post processing of the electrospun fibers.

In this chapter, material properties, solution properties and electrostatics are discussed. As each of the properties involved in electrospinning is a huge science of its own, this chapter aims to give some basic information relevant to electrospinning. However, with growing interest and research into electrospinning, new fundamental properties may be discovered.

2.1. Material Classes

The materials and application of electrospun fibers are numerous, individual material properties must be considered depending on its applications. The electrospinning process may be modified so as to yield electrospun fiber with the desired morphology and properties. When used as composite, the nanofibers can be made as a composite on its own or it can be used as reinforcement in a matrix. In the production of ceramic fibers, post processes are required after the fibers are electrospun. Thus it is important to have a basic understanding of the different group of materials before selecting the most appropriate electrospun fibers for specific applications.

2.1.1. *Polymers*

Polymers consist of long chain of molecule with repeating units called monomers that are mostly covalently bonded to one another. An example of a polymer would be polyethylene which consist of repeating units of [-CH_2CH_2-]$_n$. Such single unit is also known as monomer. The monomer must either have reactive functional groups such as amino groups (-NH_2) or the have double bonds which may react under suitable conditions to provide the covalent linkage between the repeating units. Such strong linkages form the backbone of the polymer chain. It is common to find weak secondary bonds between the molecule chains which allow the chain to slide over one another. Polymers exhibit several properties that are attractive for many applications. Most polymers are inexpensive as they contain simple elements and they are relatively easy to synthesize.

Polymers, with their low density can easily be molded into complex shape, which are strong and relatively inert. They have found applications in many areas such as clothing, food packaging, medical devices and aircraft. Natural polymers such as silks, collagen and agarose have found usage in many tissue engineering applications.

2.1.1.1. Fundamental Classification of Polymer

A widely accepted classification of polymer is their responds to heat. There are basically two types of polymer under this classification, thermoplastic and thermoset. In thermoplastics, the linear polymers melt when heat is applied but solidifies when cooled. This heating and cooling can be repeated many times without affecting the properties. Examples of thermoplastic include polyethylene, polystyrene and vinyls. However, this would impose a limiting temperature for the material in use as a structural element above which the polymer may distort over time. For thermosetting polymer, once an initial heat is applied, there is crosslinking between polymer chains. Subsequent application of heat would only degrade the polymer. Examples of thermoset include phenolics, urea and epoxies. This means that such polymer has much higher upper limiting temperature.

2.1.1.2. Polymer Crystallinity

In bulk polymer, there are usually both regions of crystalline and amorphous parts as shown in Fig. 2.1. The ratio of the two regions would determine the properties of the polymer. A polymer is said to be amorphous when the arrangement of the linear molecules is completely random. A crystalline polymer has its linear adjacent linear chains are aligned. The more commonly accepted theory for crystalline polymer is the folded chain theory. The polymer chains are first folded and stacked on top of one another held together by amorphous tie-molecules to from crystallites.

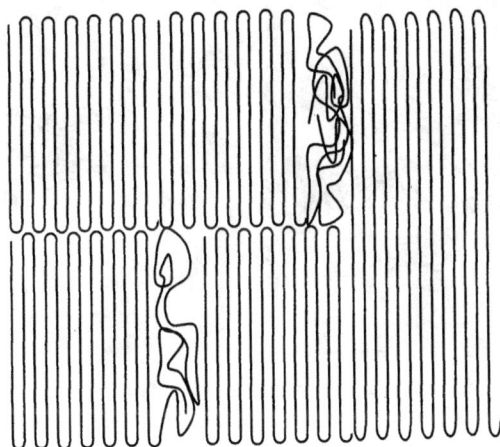

Fig. 2.1. Model of structure of partially crystalline polymer.

These are then twisted and turned to form a ribbon-like supramolecules called spherulites as shown in Fig. 2.2. Polymers that are of higher crystallinity show higher yield strength, modulus and hardness. When crystalline polymers are stretched, the polymer chains are oriented in the direction of the stress and destroy the spherulites structure. A phenomenon called necking is then observed. They also have better wear and chemical resistance. However, crystalline polymers are more brittle.

The optical properties of polymers are also affected by the crystallinity. More crystalline polymers have a higher refractive index than amorphous matrix making them either opaque or translucent. Amorphous polymer may be completely transparent [Farag (1989)].

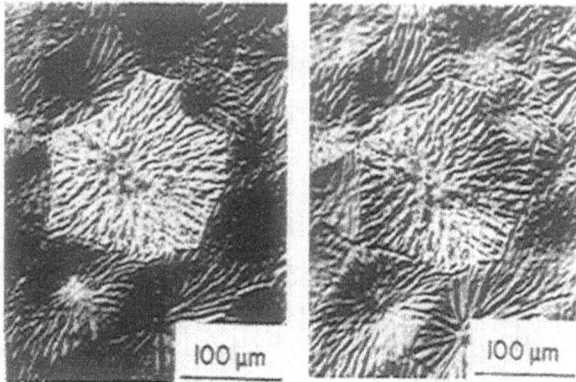

Fig. 2.2. Spherulite in polypropylene [Aboulfaraj et. al. (1993)].

2.1.1.3. Polymer Molecular Weight

As polymer chains are made of repeating units, the molecular weight of the polymer is the sum of the molecular weight of the individual monomers. Generally, a higher molecular weight increases the polymer's resistance to solvent dissolution. The molecular weight of the polymer also has a direct influence on its viscosity. There are numerous ways to obtain the molecular weight, M_n (Number average), M_v (Viscosity average), M_w (Weight average) and M_z (z average). M_n is the total weight of the individual molecular weight by the number of molecules. M_n is independent of molecular size but is highly sensitive to small molecules present in the mixture. For a heterogeneous molecular weight system,

$$M_z > M_w > M_n$$

As the heterogeneity decreases, the various molecular weights converges until for a homogenous mixture,

$$M_z = M_w = M_n$$

2.1.1.4. Glass Transition Temperature (T_g)

The glass transition temperature is a very important property of polymers. This temperature defines the mobility state of the polymer molecules. Below its T_g, the amorphous polymer is brittle as the molecules are frozen but above T_g, the polymer is ductile and the molecular chains have sufficient thermal energy to slide. This causes the elastic modulus of amorphous polymers to decrease by several orders of magnitude at temperature above T_g. It is important to note that the mechanical behavior of the polymer at temperature above T_g is affected by the loading rate. T_g affects the mobility of polymer molecules, however, the motion of the molecular chain is not instantaneous. For slow loading rate, the molecular chain have time to move at temperature near T_g but if the loading rate is fast, there may not be enough time for the molecular chain to move thus increasing the effective T_g.

At temperature below T_g, it can be said that the relaxation time of the molecule is too long for equilibrium to occur under the slowest of experimental duration. When the temperature is at T_g, the molecules within the polymer bulk move as a temperature-dependent "coorperatively rearranging" region. The size of this "coorperatively rearranging" region is dependent on the configuration restrictions due to amorphous packing [Adam and Gibbs (1965)]. Molecular dynamic simulation of a polymer melt has shown that there is an increasing clustering of mobile monomer unit as the temperature decreases to a critical temperature. When measuring the T_g, it is important to note that thermal treatment of the sample, including heating and/or cooling rate during measurement will affect the value obtained. Thus the measuring condition must be noted when making comparison between different experiments.

In the study of thin polymer films, the glass transition temperature was found to found to be lower than the bulk [Ellison et. al. (2003)] and for Polystyrene, is given by the empirical relation,

$$T_g = T_{g,bulk}\left[1-\left(\frac{a}{h}\right)^\delta\right]$$ (2.1)

where $T_{g,bulk}$ is the bulk T_g for Polystyrene
h is the thickness of the film
$a = 32\text{Å}$ and $\delta = 1.8$

Experiments have shown that in a polymer film, the mobility of the polymer molecules near its surface is higher than molecules at the bulk due to the substantially reduced T_g at the surface layer. However, the bulk dynamics beneath the surface area can slow down the mobility at the surface leading to a higher surface T_g. When the film thickness is reduced sufficiently (about 14nm), the T_g at the surface is the same as the T_g of the bulk [Ellison et. al. (2003)]. From the studies polymer film, it can be seen that T_g is affected by the surface mobility. For a nanofiber, with the surface area larger than film, it has been experimentally shown to exhibit a lower T_g than cast film [Zong et. al. (2002)].

There are many factors that affect bulk T_g. Polymers with more free-volume allows easier movement of molecular chain thus lowering the T_g. Stronger secondary bonds such as H bonds would increase the T_g. With greater the chain length, there are more entanglements within the polymer structure thus increasing the T_g. Table 2.1 shows the Tg, melting point and the molecular structure of some common synthetic polymers.

Table 2.1. Molecular Structure and Transition Temperature of Synthetic Polymers (Bulk).

NonBiodegradable Polymer	Molecular Structure
Nylon 4,6 Tm = 290 °C [Johnson et. al. (1988)]	
Nylon 6 Tg = 40°C [Gordon (1971)] Tm = 219 °C [Gordon (1971)]	
Nylon 6,6 Tg = 57°C [Callister (1997)] Tm = 265 °C [Callister (1997)]	

Structure	Polymer & Properties
	Nylon 12 $Tg = 42\,°C$ [Gordon (1971)] $Tm = 178\,°C$ [Gordon (1971)]
	Polyacrylic acid $Tg = 134\,°C$ [Al-Najjar et. al. (1996)]
	Polyacrylonitrile $Tg = 104\,°C$ [Callister (1997)] $Tm = 317\,°C$ [Callister (1997)]
	Poly(benzimidazol), PBI $Tg = 683\,°C$ [Pu (2003)]

Polycarbonate

Tg = 150 °C [Callister (1997)]
Tm = 265 °C [Callister (1997)]

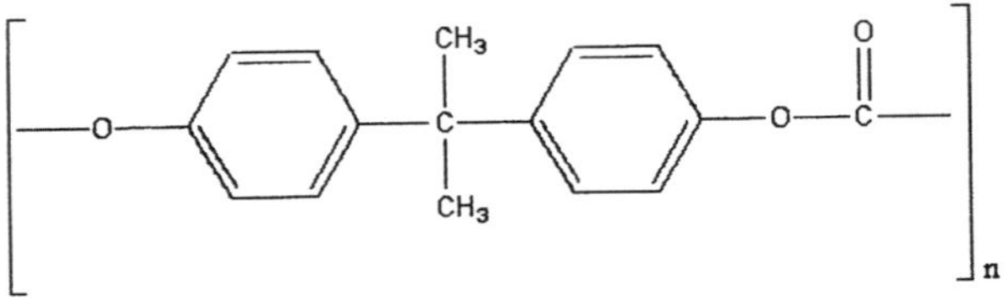

Poly(etherimide), PEI

Tg = 219 °C [Chun (1996)]

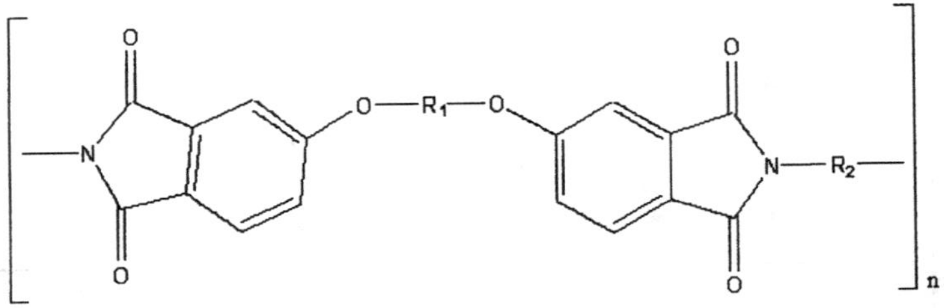

Poly(ethylene terephthalate)

Tg = 69 °C [Callister (1997)]
Tm = 265 °C [Callister (1997)]

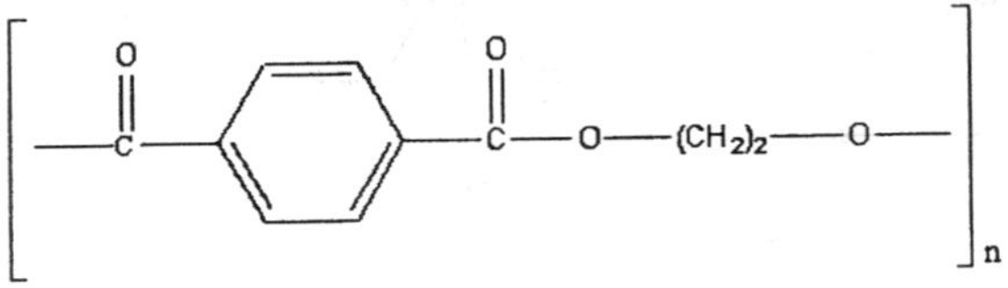

Polystyrene

Tg = 110 °C [Schaffer et. al. (1999)]
Tm = 240 °C [Schaffer et. al. (1999)]

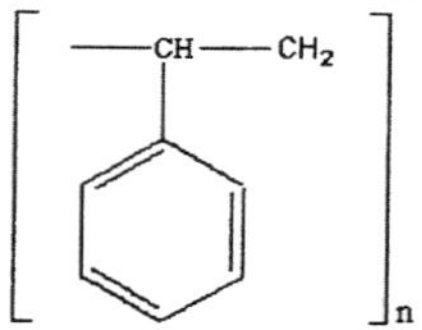

Polysulfone

Tg = 185 °C [Zhang and Wang (1994)]

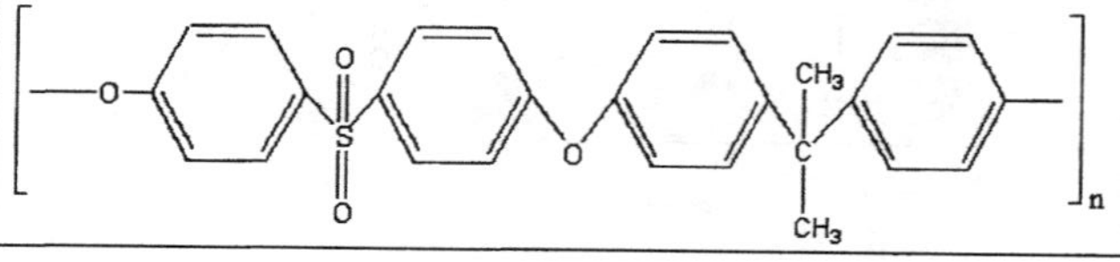

Poly(urethane)

R' = polyol segment
R = isocyanate unit

Tg (Hexamethylene
diisocyanate/Ethylene glycol) = 22 °C
Tm (Hexamethylene
diisocyanate/Ethylene glycol) = 203 °C
Tg (Hexamethylene
diisocyanate/Diethylene glycol) = -1 °C
Tm (Hexamethylene
diisocyanate/Diethylene glycol) = 123 °C
[Godovsky and Slonimsky (1974)]

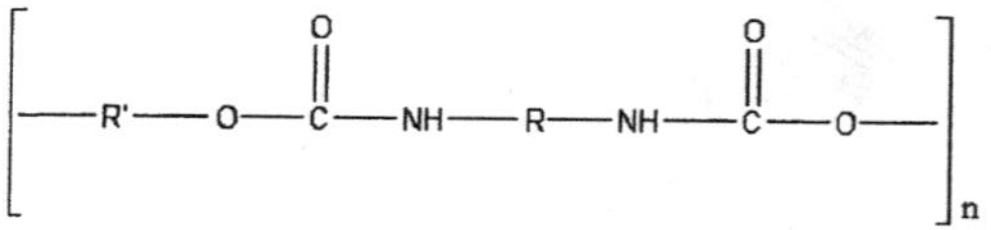

Poly(urethane urea)s

R' = soft segment
R = isocyanate unit
R" = amine unit

Tg (4,4'-Diphenylmethane
diisocyanate/polyether polyol/4,4'-
diaminodiphenyl methane)
= -57.1 °C

Tg (4,4'-Diphenylmethane
diisocyanate/polyether polyol/diethyl
toluene diamine)
= -52.1 °C

Tg (4,4'-Diphenylmethane
diisocyanate/polyether polyol/3-
chloro-3' methoxy-4,4'diamino
diphenylmethane)
= -40.2 °C
[Gao et. al. (1994)]

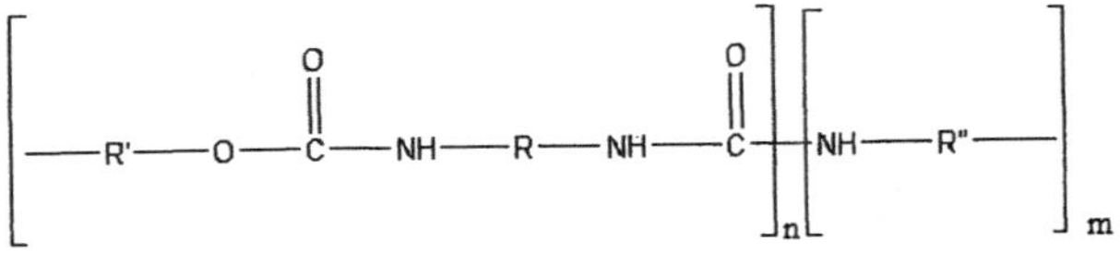

Poly(vinyl alcohol)

Tg = 40-50 °C [Mucha and Pawlak (2005)]

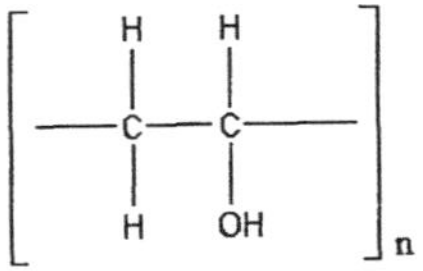

Structure	Polymer
	Poly(N-vinylcarbazole) Tg = 227 °C [Griffiths et. al. (1977)]
	Poly(vinyl chloride) Tg = 83 °C [Bureau et. al. (2005)]
	Poly(vinyl pyrrolidone) Tg = 166 °C [Nuno-Donlucas et. al. (2001)]
	Poly(vinylidene fluoride), PVDF Tg = -35 °C [Callister (1997)] Tm = 156 °C [Progelhof and Throne (1993)]

Biodegradable Polymer	Molecular Structure
Poly(ε-caprolactone) Tg = -60 °C to -63 °C [Corradini et. al. (2003); Hubbell and Cooper (1977)] Tm = 60 °C [Corradini et. al. (2003)]	
Poly(ethylene oxide) Tg = -56 °C [Progelhof and Throne (1993)] Tm = 66 °C [Progelhof and Throne (1993)]	
Polyglycolide Tg = 34 °C [Lou et. al. (2003)] Tm = 225 °C [Lou et. al. (2003)]	
Poly(L-lactic acid) Tg = 55-60 °C Lou et. al. (2003)] Tm = 170 °C Lou et. al. (2003)]	

Poly(L-lactide-co-ε-caprolactone), **90% : 10%** Tg = 41 °C [Penning et. al. (1993)] Tm = 178 °C [Penning et. al. (1993)] **Poly(L-lactide-co-ε-caprolactone),** **80% : 20%** Tg = 38 °C [Penning et. al. (1993)] Tm = 174 °C [Penning et. al. (1993)]	
Poly(lactide-co-glycolide), **80%: 20%** Tg = 54 °C [Loo et. al. (2005)] Tm = 144 °C [Loo et. al. (2005)] **Poly(L-lactide-co-glycolide),** **75%: 25%** Tg = 46.1 °C [Iwasaki et. al. (2002)] **Poly(lactide-co-glycolide),** **50%: 50%** Tg = 44 °C [Ibim et. al. (1997)]	

Table 2.2. Glass Transition Temperature of Natural Polymers (Bulk).

Natural Polymer	Tg (°C)	Reference
Cellulose Triacetate	40, 120 and 155	Russell and Kerpel (1957)
Cellulose 2.2 Acetate	55 and 115	Russell and Kerpel (1957)
Cellulose 2.5 Acetate	15, 50, 90 and 114	Daane and Barker (1964)
Chitosan (Degree of deacetylation 0.96)	203	Sakurai et. al. (2000)
Collagen	45 – 50	Campo et. al. (1963)
Gelatin	120, 180 – 190	Fraga and Williams (1985)
Wheat Gluten	167	Angell (1995)

2.1.1.5. Synthetic Polymer

For synthetic polymer, there are generally two types, the ethenic polymers and the condensation polymers.

Ethenic polymers are formed by polymerizing monomers containing the carbon to carbon double bond group. The simplest monomer that contain this structure is the olefin ethylene, $CH_2=CH_2$. Polymerization involves the breaking of the double bond of the monomer and linking up with another monomer. Thus the resultant polymer generally has a linear structure although there are some possible exceptions. One of which is unsaturated polyesters where polymer formation was initiated by condensation. The resultant structure was formed by polymerizing the carbon-carbon double bond group to give a highly cross-linked polymer. Nevertheless, the general linear structure of ethenic polymer makes it suitable for fiber and film formation. Important polymers from this class include polyethylene, vinyl chloride polymers and copolymers, and polystyrene.

Amorphous polystyrene is one of the most useful plastic due to its rigidity, low cost and excellent insulating properties. Although typical applications for polystyrenes include containers, house wares and furniture parts, electrospun polystyrene fibers have potential applications in filter membranes and protective clothing. In its fiber form polystyrene have a certain degree of flexibility. Electrospun polystyrene had shown several interesting morphologies such as beads and pores on the surface of the polymer [Casper et. al. (2004)].

For Condensation polymers, the monomers have at least two functional groups such as alcohol, amine or carboxylic acid group instead of a carbon-carbon double bond group. In condensation reaction, two units often not of the same monomer structure reacts to form a polymer at the same time releasing a small molecule such as H_2O. The reaction is slow and the growth in molecular weight is gradual.

Examples
> Polyamides = carboxylic + amine
> Polyesters = carboxylic acid + alcohol

However, not all condensation polymerization involves the liberation of small molecules. In this case, active hydrogen is transferred from one molecule to the next instead. A typical example is the polymerization of dialcohol and diisoyanate monomers to form polyurethane.

In electrospinning, the sub-micron dimension of electrospun fibers resembles that of natural extra-cellular matrix. Thus it is not surprising that there are great interests in the use of electrospun fibers in the area of bioengineering. One of the most frequently used synthetic polymers for tissue scaffolds are the biodegradable aliphatic polyesters. These degradable polyesters are derived from three monomers, namely, lactide, glycolide and caprolactone. Hydrolytic attack of the ester bond within the polymer is responsible for its degradation [Griffith (2000)]. Poly-$_L$-lactic acid for example is able to degrade to lactic acid, which is a normal intermediate of carbohydrate metabolism in man.

Polyurethanes are one of the most widely used polymers in biomedical applications especially those in contact with blood. This is due to the inherent, relative nonthrombogenicity of their surfaces and easy synthesis to different forms. Current applications include catheters, blood bags and artificial heart systems. Electrospun polyurethane fibers have shown great promises in the area of wound healing application. While previous use of polyurethane occlusive dressings on the healing of wound has the problem of significant fluid accumulation after a few days of use, nanofibrous polyurethane membrane prepared by electrospinning promoted fluid drainage. This is due to the high porosity of the nanofibrous membrane which also allows excellent oxygen permeability [Khil et. al. (2003)].

2.1.1.6. Natural Polymer

One of the greatest potential in electrospun fiber is in the area of bioengineering. For many biomedical applications, the materials used have to be biocompatible, thus natural polymers have a distinct advantage over synthetic materials. Since most natural polymer can be degraded by naturally occurring enzymes, it can be used in applications where temporary implants are desired or in drug release. It is also possible to control the degradation rate of the implanted polymer by chemical cross-linking or other chemical modifications thus allowing greater versatility in the design of the implant [Atala and Lanza (2002)]. Most polymers that have been electrospun are proteins and polysaccharides.

Proteins that have been electrospun include collagen [Matthews et. al. (2002)], gelatin [Huang et. al. (2004)], fibrinogen [Wnek et. al. (2003)] and silk [Jin et. al. (2002); Ohgo et. al. (2003)]. One of the most commonly used natural polymers used is collagen. Collagens are naturally found in connective tissues where they provide mechanical support. There are at least ten different forms of collagens and they are dominant in specific tissue.

However, all the collagens share the fundamental triple helix structure. As collagen exists naturally in fiber form, electrospun collagen fibers are able to mimic extracellular matrix in the body. In vitro studies had shown that cells respond positively to tissue scaffold made of electrospun collagen fibers. Generally, collagen is relatively strong and form stable fibers especially after cross-linking. However, till date, only Type I, II and III collagen had been successfully electrospun [Matthews et. al. (2002)] together with their blends [Fertala et. al. (2001)]. The cost of collagen makes it expensive to yield thick fiber mesh from electrospinning. A cheaper alternative to collagen would be gelatin which can also be electrospun [Huang et. al. (2004)]. Another protein that is electrospun for use in tissue engineering is fibrinogen. As this protein plays a key role in blood clotting and wound healing, electrospun fibrinogen has been explored for possible usage in wound dressings [Wnek et. al. (2003)].

Proteins such as natural silk fiber have outstanding mechanical properties. This makes it an interesting candidate for application in biomedical field where mechanical property is important. It is possible to electrospin silk to obtain fibers with average diameter less than 500nm [Ohgo et. al. (2003)]. Electrospun fibers from silk fibroin were found to promote cell adhesion and proliferation [Min et. al. (2004)]. Silk fibroin itself has several advantages biological properties such as good biocompatibility, good oxygen and water vapor permeability, biodegradability and minimal inflammatory reaction [Sakabe et. al. (1989)]. The high surface area to volume ratio of the electrospun fiber also encourages cell attachment, growth and proliferation.

A few polysaccharides and its modified form have been electrospun. Cellulose is a major constituent of nearly all form of plant matters, thus making it one of the most widely distributed and available raw materials. It is possible to modify the structure of the cellulose by reaction with the hydroxyl group or degradation of the cellulose chain. Cellulose acetate (CA) is one of the most commonly used materials for applications in where semi-permeable membranes are required such as dialysis, ultra-filtration and reverse osmosis.

CA can be electrospun and subsequent deacetylation of the fiber yield pure cellulose fibers [Son et. al. (2004c); Liu and Hsieh (2002)]. Such fine cellulose fibers have high surface area to volume ratio and the fiber structure is thought to be much less ordered than its native or regenerated cellulose. These make them highly desirable for surface-supported reactions due to the accessibility if the hydroxyl group. Cellulose fibers previously electrospun from cellulose acetate had been methacrylated to form a unique fiber mesh that has a hydrophobic surface with a hydrophilic cellulose core [Liu and Hsieh (2003)].

Hyaluronic acid (HA) is another naturally occurring polysaccharide commonly found in the specialized tissues such as synovial fluid, dermis and cartilage. This molecule plays an important role among the interstitial proteins of the extra-cellular matrix providing important biological and mechanical functions. The chemical structure of HA consists of D-glucuronic acid and N-acetylglucosamine arranged as a repeating disaccharide chain that contains as many as 30,000 or more repeating disaccharide units [Laurent (1998)].

Due to its unique rheological properties and biocompatibility, HA has been used extensively in many biomedical applications such as ophthalmology, medical implants and drug delivery. Electrospinning HA would yield HA membrane made out of sub-micron fibers [Um et. al. (2004)]. The high-surface area to volume ratio of the membrane would make it attractive in applications such as tissue scaffolds, wound dressings and artificial blood vessels.

2.1.1.7. Copolymer and Polymer Blends

Sometimes, it is beneficial to obtain a structure that shows the properties of two or more polymers. This can be achieved either through polymerization of two different homopolymers to form a copolymer or by physical mixing of two or more polymers to form a blend. In copolymers, the covalent bonding between the mers is very strong. The individual mers cannot be separated without breaking the copolymer chain.

There are generally two types of copolymers, random copolymers and block copolymers. In random copolymers, there is no sequence in the distribution between the two types of homopolymers. Thus the random copolymer exhibit properties that is intermediate to those of corresponding homopolymers. In block copolymers, the repeating homopolymers exist in long sequence within the polymer chain. The block copolymer may show property characteristics of each of the constituent homopolymer.

In blending, the polymers tend to separate into two or more distinct phases due to incompatibility. To improve compatibility and miscibility, interactive functional groups are introduced to the polymers so that the polymer chains would form stronger Hydrogen-bonding with the advantage of improving the strength of the blend. Common functional groups include carboxylic and sulfonate groups. However, as there are no chemical reactions involved in polymer blending, the links between the different polymers are not strong and leaching of one of the polymers may occur, when submerged in a solvent.

2.1.1.8. Electrospun Polymer Fiber

Till date, there are many polymers that have been electrospun including custom made polymers. Listed in Table 2.3(a) to Table 2.3(a) are the more commonly used non-biodegradable synthetic polymers with the corresponding solvent and concentration used that is able to yield fibers without beads.

Table 2.3(a). Electrospun polymer fibers.

Synthetic Polymer	Solvent	Concentration	Reference
Nylon 4,6	Formic acid	10wt%	Bergshoef and Vancso (1999)
Nylon 6 Mw 43,300	1,1,1,3,3,3 hexafluoro-2-propanol	15wt%	Stephens et. al. (2004)
Nylon 6,6	96% Formic acid	12.1 wt%	Buchko et. al. (1999)
Nylon 12 Mw, 32,000	1,1,1,3,3,3 hexafluoro-2-propanol	15wt%	Stephens et. al. (2004)
Polyacrylic acid, Mw: 250000	Ethanol	6wt%	Ding et. al. (2004b)
Polyacrylonitrile, Mv 114,000g/mol	Dimethylformamide	15wt%	Fennessey and Farris (2004)
Polyamide-6, Mw: 17000	85% v/v formic acid	34%w/v	Mit-uppatham et. al. (2004)
Polyamide-6, Mw: 20000	85% v/v formic acid	34%w/v	Mit-uppatham et. al. (2004)
Polyamide-6, Mw: 32000	85% v/v formic acid	22%w/v	Mit-uppatham et. al. (2004)
Poly(benzimidazol), PBI	Dimethyl acetamide, 185°C	20wt%	Kim et. al. (2004a)
Polycarbonate	Dichloromethane	15wt%	Bognitzki et. al. (2001)
Polycarbonate, Bisphenol-A	Chloroform	15wt%	Krishnappa et. al. (2003)
	Dimethyl formamide: Tetrahydrofuran	15wt%	Krishnappa et. al. (2003)
Poly(etherimide), PEI	1,1,2-Trichloroethane	14wt%	Choi et. al. (2004c)
Poly(ethylene oxide) Mw 400,000	Water	10wt%	Deitzel et. al. (2001a)
Poly(ethylene terephthalate), Intrinsic viscosity: 0.82+0.02	Trifluoroacetic acid	0.2g/ml	Ma et. al. (2005a)

Table 2.3(b). Electrospun polymer fibers.

Synthetic Polymer	Solvent	Concentration	Reference
Polystyrene, Mw: 299000Da	t-Butylacetate	20% (w/v)	Jarusuwannapoom et. al. (2005)
	Chlorobenzene	30% (w/v)	Jarusuwannapoom et. al. (2005)
	Chloroform	30% (w/v)	Jarusuwannapoom et. al. (2005)
	Dichloroethane	30% (w/v)	Jarusuwannapoom et. al. (2005)
	Dimethylformamide	30% (w/v)	Jarusuwannapoom et. al. (2005)
	Ethylacetate	20% (w/v)	Jarusuwannapoom et. al. (2005)
	Methylethylketone	20%-30% (w/v)	Jarusuwannapoom et. al. (2005)
	Tetrahydrofuran	20% (w/v)	Jarusuwannapoom et. al. (2005)
Poly (styrene-butadiene-styrene) triblock copolymer, Mw 151,000g/mol	75% Tetrahydrofuran : 25% dimethylformamide	14wt%	Fong and Reneker (1999)
Polysulfone, Bisphenol-A. Inherent Viscosity: 0.6dl/g	90% N,N-dimethylacetamide: 10% acetone	15% (w/v)-20%	Yuan et. al. (2005)
Poly(trimethylene terephthalate) Inherent viscosity = 0.92	50% Trifluoroacetic acid 50% Methylene Chloride	16wt%	Khil et. al. (2004)
Poly(urethane)	60% Tetrahydrofuran: 40% N,N-Dimethylformamide	13wt%	Lee et. al. (2003a)
Poly(urethane urea)s Mw 42,000	20%THF: 80% Isopropyl alcohol	35wt%	McKee et. al. (2005)
Poly(vinyl alcohol) Mw 13,000-50,000	Water at 80°C	25wt%	Koski et. al. (2004)
Poly(vinyl carbazole)	Dichloromethane	7.5wt%	Bognitzki et. al. (2001)
Poly(vinyl chloride), PVC	60% Tetrahydrofuran: 40% N,N-Dimethylformamide	13wt%	Lee et. al. (2003a)
75% Poly(vinyl chloride) 25% Poly(urethane) blend	60% Tetrahydrofuran: 40% N,N-Dimethylformamide	13wt%	Lee et. al. (2003a)

Table 2.3(c). Electrospun polymer fibers.

Synthetic Polymer	Solvent	Concentration	Reference
50% Poly(vinyl chloride): 50% Poly(urethane) blend	60% Tetrahydrofuran: 40% N,N-Dimethylformamide	13wt%	Lee et. al. (2003a)
25% Poly(vinyl chloride): 75% Poly(urethane) blend	60% Tetrahydrofuran: 40% N,N-Dimethylformamide	13wt%	Lee et. al. (2003a)
Poly(vinyl pyrrolidone)	Ethanol	4wt%	Yang et. al. (2004c)
	65%ethanol: 35% DMF	4wt%	Yang et. al. (2004c)
Poly(vinylidene fluoride), PVDF	N,N-Dimethylacetamide	25wt%	Choi et. al. (2004b)
Poly(vinylidene fluoride-co-hexafluoropropylene, P(VDF-HFP)	70% Acetone: 30% DMAC	12-18wt%	Kim et. al. (2005)

Electrospun fibers are commonly used in the field of tissue engineering due to their small diameters which are able to mimic natural extracellular matrix.   Thus there are two groups of polymers that are commonly electrospun. These are the biodegradable polymers and natural polymers. Many different types of polymers from these two classes have been successfully electrospun which showed the versatility of electrospinning. Table 2.3(d) shows the list of commonly available biodegradable polymers that have been successfully electrospun. Table 2.3(e) and 2.3(f) are the list of natural polymers such as collagen, chitosan and their blends that have been electrospun.

Table 2.3(d). Electrospun polymer fibers.

Biodegradable Polymer	Solvent	Concentration	Reference
Degradable Polyesterurethane, DegraPol®	Chloroform	30 wt%	Riboldi et. al. (2005)
Poly(ε-caprolactone) Mw 80,000	Chloroform	10 wt%	Yoshimoto et. al. (2003)
Poly(ε-caprolactone) Mn 80,000	85% N,N-dimethylformamide : 15% Methylene Chloride	7-9 wt%	Ohkawa et. al. (2004b)
Polydioxanone	1,1,1,3,3,3 hexafluoro-2-propanol	42-167g/ml	Boland et. al. (2005)
Polyglycolide Mw 14,000-20,000	1,1,1,3,3,3 hexafluoro-2-propanol	8wt%	You et. al. (2005)
Poly(L-lactic acid) Mw 300,000	70% Dichloromethane : 30% n,n-dimethyl-formamide	2-5 wt%	Fang et. al. (2004A)
Poly(L-lactic acid) Mw 450,000	1,1,1,3,3,3 hexafluoro-2-propanol	5 wt%	You et. al. (2005)
Poly(L-lactide-co-ε-caprolactone) [75 : 25] Block copolymer	Acetone	3-9wt%	Mo XM et al., 2004
Poly(D,L-lactide-co-glycolide) [85 : 15, PLGA]	50% Tetrahydrofuran : 50% Dimethylformamide	0.05g/ml	Li et. al. (2002b)
Poly(L-lactide-co-glycolide) [10 : 90, PLGA] Mw 100,000	Hexafluoroisopropanol	5-7wt%	Bini et. al. (2004)
Poly(lactic-co-glycolic acid) [50 : 50 PLGA] Mw 25,000	1,1,1,3,3,3-Hexafluoro-2-propanol	15%	Min et. al. (2004)
Poly(L-lactic-co-glycolic acid) [50 : 50 PLGA] Mw 108,000	Chloroform	15%	You et. al. (2005)

Table 2.3(e). Electrospun polymer fibers.

Natural Polymer	Solvent	Concentration	Reference
Bombyx mori silk fibroin	Formic acid	9-12wt%	Ayutsede et. al. (2005)
80% Bombyx mori silk fibroin: 20% Poly(ethylene oxide), Mw, 900000g/mol	Water	7.1wt%	Wang et. al. (2004b)
20% Casein : 80% Poly(ethylene oxide) Mv 600,000	5% aqueous triethanolamine	5 wt%	Xie and Hsieh. (2003)
80% Casein : 20% Poly(ethylene oxide) Mv 600,000	5% aqueous triethanolamine	10 wt%	Xie and Hsieh (2003)
30% Casein : 70% Poly(vinyl alcohol) Mv 124,000-186,000	5% aqueous triethanolamine	10 wt%	Xie and Hsieh (2003)
50% Casein : 50% Poly(vinyl alcohol) Mv 124,000-186,000	5% aqueous triethanolamine	10 wt%	Xie and Hsieh (2003)
Cellulose Acetate	N,N-dimethylacetamide : Acetone [1 : 2]	15wt%	Wang and Hsieh (2004)
	85% acetone : 15% water	17wt%	Son et. al. (2004b)
Chitosan Mv 210,000, degree of deacetylation 0.78	70% Trifluoroacetic acid 30% Methylene Chloride	8 wt%	Ohkawa et. al. (2004a)
50% Chitosan with 90% degree of deacetylation : 50% Poly(ethylene Oxide), 1500 kDa to 4000 kDa	2wt % acetic acid	6 wt%	Duan et. al. (2004)
Collagen Type I	1,1,1,3,3,3 hexafluoro-2-propanol	0.083g/ml	Matthews et. al. (2002)
Collagen Type II	hexafluoropropanol	Not indicated	Shield et. al. (2004)
Collagen Type III	1,1,1,3,3,3 hexafluoro-2-propanol	0.04g/ml	Matthews et. al. (2002)
50% Collagen Type I : 50% Collagen Type III	1,1,1,3,3,3 hexafluoro-2-propanol	0.06g/ml	Matthews et. al. (2002)

Table 2.3(f). Electrospun polymer fibers.

Natural Polymer	Solvent	Concentration	Reference
Fibrinogen Fraction I	90% 1,1,1,3,3,3 hexafluoro-2-propanol : 10% 10x minimal essential medium, Earle's without L-glutamine and sodium bicarbonate	0.083g/ml	Wnek et. al. (2003)
Gelatin Type A	2,2,2-trifluorethanol	10-12.5wt%	Zhang et. al. (2005)
50% Gelatin Type A: 50% Poly(caprolactone)	2,2,2-trifluorethanol	10wt%	Zhang et. al. (2005)
Wheat Gluten	1,1,1,3,3,3-hexafluoro-2-propanol	10%(w/v)	Woerdeman et. al. (2005)

2.1.2. *Composites*

Composites are combination of two distinct material phases, a bulk phase, also known as a matrix and a reinforcement phase. It is the combination of the strength of the reinforcement and the toughness of the matrix that gives composite its superior properties that are not available in any single conventional material. Both matrix and reinforcement phases can be metal, ceramic or polymer. Generally, the matrix binds the reinforcement together to give the composite its shape, surface appearance and resistance to environmental damage. While the matrix is usually ductile or tough, the reinforcements are strong with low densities. It is the reinforcement that carries most of the load thus giving the composite its stiffness and strength. When fiber reinforcements of less than 100nm are used, it is possible to produce transparent composites [Bergshoef and Vancso (1999)] although they are generally opaque due to light scattering. In most cases, composites are designed for load-bearing applications although there are other classes of composites that are used for their interesting electrical, thermal or magnetic properties.

2.1.2.1. Composite Reinforcement

There are generally two types of composite reinforcements, fibrous reinforcement and particulate reinforcement. In fibrous reinforcement, the fibers arrangement can be of many different forms. The simplest arrangement of the fibers in the matrix is to have the fibers aligned in a certain direction to form a laminate composite. Thin sheets of unidirectional composites can be stacked in an arrangement such that the fibers are oriented at $0°$, $90°$ and $\pm45°$ directions. Such composite laminates are strong in all directions within the plane containing the fibers but are weak in the direction normal to the plane of the laminates. Other types of fiber arrangements include weaving to produce fabrics of different shapes and weave configurations, knitting as well as braiding as seen in Fig. 2.3. Depending on the application of the composite, different fiber arrangements are used as reinforcement in the composite. For example, braided fiber can be used as reinforcement for composite where high torsional stiffness is desired. Randomly distributed fibers in the form of non-woven mat can also be used as reinforcement in composite.

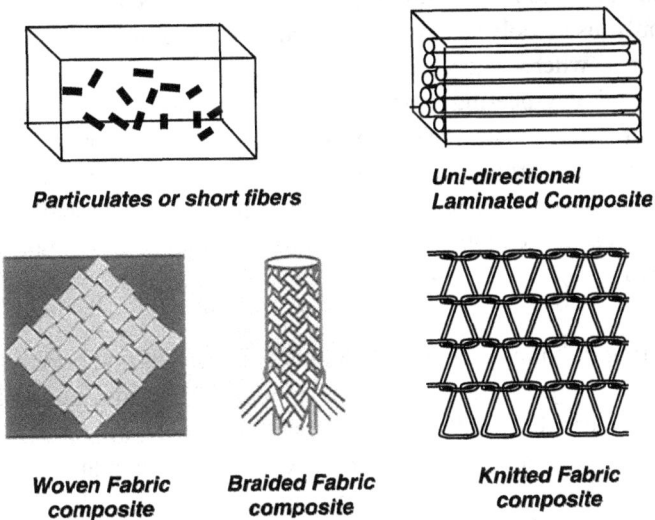

Particulates or short fibers

Uni-directional Laminated Composite

Woven Fabric composite

Braided Fabric composite

Knitted Fabric composite

Fig. 2.3.   Composite reinforcements.

In particulate reinforcement, the reinforcing phase has roughly equal dimensions in all directions. The material is known as aggregate composite. The reinforcing particles need not be equal sizes as in the case of concrete and the volume fraction of the particles can be higher than that of the matrix. The irregularly shaped particles provide strength through mechanical interlocking. Another form of aggregate composite is known as particulate composite. In this case, the volume fraction of the particles is much lower than that of the matrix. In conventional composites, the thermal expansion mismatch of the particle matrix material will result in high dislocation density in the matrix during cooling from the fabrication temperature. The resultant strain hardening contributes significantly to the strength of the composite. Nanofiber composites of nylon 6 with Closite 30B [Fong et. al. (2002)] and polyimide with single wall nanotubes [Park et. al. (2002)] were fabricated by electrospinning.

2.1.2.2. Polymer-Matrix composite

Polymer-matrix composite is the most common form of composite to be used in the industry. Although thermosetting resins is the most commonly used polymer matrix, thermoplastic is able to dominate in composites where short fibers are incorporated. Despite the huge selection of engineering plastics, only a few are selected as matrix. Important issues when selecting polymer matrix depends mainly on reinforcement-matrix compatibility in terms of bonding, mechanical properties, thermal properties, cost etc. Although traditionally used in load-bearing applications, polymer-matrix composite has found uses in other areas.

Conducting polymer such as polyaniline is blend with other polymers to form conducting composites [Chipara et. al. (2003)]. Enzymes embedded in polymer can be used as biosensors [Sawicka et. al. (2005)]. Metal particles are also added polymer matrix to produce conductive polymer composite which can be used to provide shielding for electromagnetic interference and electrostatic discharge [Bigg (1979)].

Another way is to use metals as fillers on top of the polymer matrix and a reinforcement material such as carbon fibers [Chen et. al. (2002)]. Carbon fiber is preferred in this case as it has good conductivity and at the same time provides excellent mechanical strength to the composite. This has lead to the development of carbon nanofibers for use as reinforcement in composites [Dzenis and Wen (2002)]. Presence of carbon nanotubes in electrospun fibers has shown significant improvement in the strength of the fibers [Ye et. al. (2004); Ko et. al. (2003)]. Composite consisting of polymer matrix with ceramic particles can also be used to form two-phase piezoelectric composites [Tressler et. al. (1999)].

2.1.3. *Ceramics*

Unlike polymer, where there is usually no need for post-electrospinning process, ceramics nanofibers can be made from electrospinning of the ceramic precursors and followed by sintering of the electrospun fibers to derive ceramic fibers [Dai et. al. (2002); Madhugiri et. al. (2004); Shao et. al. (2004c)]. Ceramics are materials that composed of both metallic and non-metallic elements and commonly exist as compounds of oxides, nitrides and carbides. While most ceramics are crystalline, there are common non-crystalline ceramics like window glass which is made of silicon dioxide. As the atomic bonding is either ionic and/or covalent, there are no "free" electrons in ceramics making them, excellent insulators. The strong ionic and covalent bonding gives the complex structure of ceramics many advantages such as high temperature stability, resistance to chemical attacks and adsorption of foreign. Such rigid configuration also gives ceramics their brittleness.

With advances in technology, ceramics have move away from traditional applications which depends largely on its insulating properties and mechanical hardness. Ceramics have found uses as biomaterials [Niklason (2000)] such as calcium carbonate-based ceramics and hydroxyapatite ceramics. Nanoscale ceramics are being manufactured and found huge potential in many areas due to its high surface area to volume ratio.

2.1.3.1. Crystalline Structure

In most ceramic materials where the atomic bonding is predominantly ionic, the crystals are made of electrically charged ions. The metallic ions are positively charged, also known as cation as they have given up their valence electrons to the non-metallic ions, which are negatively charged, also known as anion. The overall charges of the cations and anions must be electrically balanced. As there is a tendency of particles to maintain a spatial arrangement due to forces of interaction, a crystalline state is preferred. Depending on the anions and cations, there are equilibrium positions in crystals which give it a specific lattice structure [Pampuch (1991)].

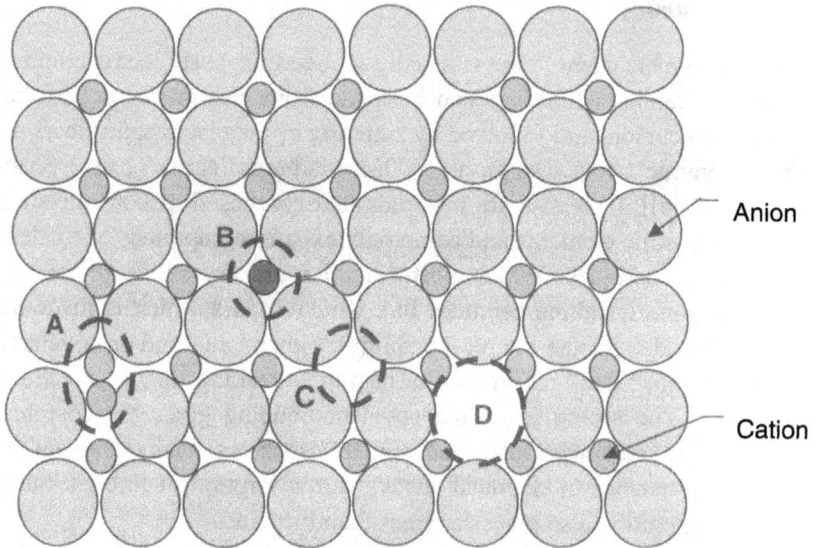

Fig. 2.4.   Schematic representations of cation and anion. (A) Cation interstitial. (B) Cation substitutional defect. (C) Cation vacancy. (D) Anion vacancy.

In real crystals, there are a number of defects in their structure. Atomic point defects in the form of vacancies and interstitials are possible in ceramic crystals as shown in Fig. 2.4. However, as there are both anions and cations in a ceramic crystal, defects for each ion type may occur. While defects in the form of anion or cation vacancy and cation interstitial are possible, it is highly improbable for anion interstitial to occur due to the large anion size. Impurities may also be found in the ceramic crystal. The impurity may cause either interstitial defects or substitutional defects. For an interstitial defect, the impurity must be relatively small compared to the anion. As for substitutional defect, the impurity will displace the host ion which it is most similar in electrical sense. For example, if the impurity usually forms a cation, it will displace the host cation.

2.1.3.2. Amorphous Structure

At the other extreme of crystalline structure is the highly disordered structure, which gives rise to amorphous materials. Typical amorphous ceramics include one component glasses, multi-component alkali-silica glasses, chalcogenide-based glasses, amorphous semi-conducting materials like silicon and amorphous metals. Amorphous glasses are generally made from solidification of liquids under conditions which does not favor the formation of crystals. For window glasses, cooling rate of a few $10^{-2}Ks^{-1}$ will avoid crystallization while metallic glasses require 10^5 to 10^6Ks^{-1}.

Recently, metal-organic polymers are used as precursors to form ceramics after calcinations at high temperature [Viswanathamurthi et. al. (2004c)]. The resulting ceramic after the decomposition process is usually amorphous. However, this phase is only stable up to a temperature, T_c, when a microcrystalline structure will start to appear [Soraru et. al. (1988)].

2.1.3.3. Ceramic Biomaterials

Ceramics including glasses have been widely used in many biomedical applications such as diagnostic instruments, porous glasses as carriers for antibodies and restorative dental materials. Due to its excellent wear resistance, good biocompatibility and corrosion resistance, ceramics are also frequently used in load bearing prosthesis. Alumina has been used in orthopedic surgery for more than two decades. Recently, Zirconia oxide is used as the articulating ball in total hip prostheses.

The use of porous ceramic implant is used to encourage bone growth within the interconnecting pore channels with pore size exceeding $100\mu m$. Such ceramics are used in nonload-bearing applications. Porous ceramics are commonly manufactured from the microstructures of certain corals [Niklason (2000)]. Other methods have also been explored to create porous ceramic implants. For example bacterial thread has been infiltrated with silica nanoparticles to form ordered porous ceramic architecture [Davis et. al. (1997)].

Bioresorbable ceramics have also been widely tested as artificial bone substitute. Hydroxyapatite ceramics is one of the most commonly used bioresorbable ceramics as it can be found naturally in bones. In fact, natural hydroxyapatite ceramics has been studied in great details for use as implants [Joschek et. al. (2000)]. Although man-made hydroxyapatite ceramics are slightly different from its natural counterpart [McConnell (1965)], investigations have shown promising results as bone substitute. Man-made silicon-stabilized calcium phosphate ceramics with interconnected morphology can be resorbed when osteoclasts are cultured on it. Bonding of the implant to new bone was found to be excellent with evidence of remodeling of implant interface by the osteoclasts [Langstaff et. al. (2001)].

2.1.3.4. Nanostructured Ceramics

Traditionally, ceramics are often used as reinforcement materials in composites. Ceramic nanoparticles have already been widely used as particle reinforcements. Other ceramic nanostructures such as fibers [Choi and Lee (2003); Dai et. al. (2002)] and tubes [Wang et. al. (2004a); Li et. al. (2004a)] have also been developed which can be used as fiber reinforcement or ceramic filters. Nanoscale Alumina-borate oxide fibers have successfully been electrospun to form a random mesh [Dai et. al. (2002)] which may have the potential for use as ceramic filters.

Nanostructured ceramics are also desirable in other applications such as microelectronic devices, chemical and biological sensing and diagnosis, energy conversion, catalysis and drug delivery. The huge range of ceramics with their unique properties has found new applications as well as enhanced performance when reduced to nanoscale. For example, the sensitivity of ZnO gas sensor is affected by the specific surface area of ZnO [Xu et. al. (2000)]. Silica being hydrophilic, easy formation of colloidal suspension and surface functionalizing has been fabricated into nanotubes [Wang et. al. (2004a)] and nanofibers [Choi and Lee (2003)].

Various ceramic materials have been expected to have applications as catalyst, sensors, solar cells and electro-optical devices, to name a few. As such, the researcher's emphasis has been placed on the fabrication of ceramics nanofibers which have high total surface area as compared to bulk ceramics, as shown in Table 2.4. To date, most reports describe the fabrication method of ceramics nanofibers and investigated the average fiber diameter through Scanning Electron Microscope (SEM), fiber surface morphology through Atomic Force Microscope (AFM) and the existence of ceramics phase in the nanofibers through X-ray Diffraction (XRD) and Fourier Transform Infrared Spectroscopy (FT-IR).

Table 2.4 (a) Precursor materials of ceramics nanofibers with sintering process which were developed by the group of Northeast Normal University, China.

Sol-Gel Precursor Material	Diameter of Ceramics Nanofibers after Sintering	Targeted Applications	Reference
PVA solution / Cu(CH₃COO)₂·H₂O sol	CuO Fibers 100 ~ 200 nm	Gas Sensors, Magnetic Storage Media, Solar Energy Transformation, Semiconductors, Catalysis	Guan et. al (2003a)
PVA solution / Co(CH₃COO)₂·4H₂O sol	Co₃O₄ Fibers 50 ~ 200 nm	Solid-State Sensors, Heterogeneous Catalysts, Electrochemical Devices, Solar Energy Absorbers	Guan et. al (2003b)
PVA solution / Co(CH₃COO)₂·4H₂O & Ni(CH₃COO)₂·4H₂O sol	NiCo₂O₄ Fibers 50 ~ 150 nm	Electrode for Oxygen Reduction and Water Electrolysis	Guan et. al (2004)
PVA solution / ZrOCl₂ sol	ZrO₂ Fibers 50 ~ 200 nm	Transparent Optical Devices, Fuel Cell, Catalyst, Electrode, Oxygen Sensor, Electrochemical Capacitor	Shao et. al (2004a)
PVA solution / Nickel acetate solution & Zinc acetate sol	NiO – ZnO Fibers 50 ~ 150 nm	High Active Catalyst, High Sensitive Sensors, Electrooptic Devices	Shao et. al (2004b)
PVA solution / Mn(CH₃COO)₂·4H₂O sol	Mn₂O₃ Fibers Mn₃O₄ Fibers 50 ~ 200 nm	Electrical Catalyst for the Decomposition of Nitrogenoxide and Selective Reduction of Nitrobenzene	Shao et. al (2004c)
PVA solution / Zinc acetate sol	ZnO Fibers 50 ~ 100 nm	Solar Cells, Nanocluster	Yang et. al (2004d)
PVA solution / H₄SiW₁₂O₄₀	H₄SiW₁₂O₄₀ Fibers ------------	Catalyst	Gong et. al (2003)

Polyvinyl alcohol) (PVA) solution was prepared with distilled water solvent.

Table 2.4 (b). Precursor materials of ceramics nanofibers with sintering process which were developed by the group of Chonbuk National University, South Korea.

Precursor Gel Material	Diameter of Ceramics Nanofibers after Sintering	Targeted Applications	Reference
Poly(vinyl acetate) (PVac) solution / Silica gel & Titanium isopropoxide sol	TiO₂-SiO₂ Fibers 50 ~ 400 nm	Anti-Reflective Coating, Optical Chemical Sensors	Ding et. al. (2003)
PVac solution / Nickel acetate tetrahydrate sol	NiTiO₃ Fibers 150 ~ 200 nm	Electrode of Solid Oxide Fuel Cell, Metal Air Batteries	Dharmaraj et. al. (2004a)
PVac solution / Magnesium Ethoxide and Titanium Isopropoxide	MgTiO₃ Fibers 200 ~ 400 nm	Dew Sensors, Pigments, On-Chip Capacitors, High Frequency Capacitors, Temperate Compensating Capacitors	Dharmaraj et. al. (2004b)
PVac solution / Niobium oxide sol	Nb₂Oₓ Fibers -------------	Electrochromic Device, Optical Waveguide, Modulator	Viswanathamurthi et. al. (2003a)
PVac solution / (VO(OC₃H₇)₃) sol	V₂O₅ Fibers -------------	Electrical Transport Devices	Viswanathamurthi et. al. (2003b)
Polycarbonate (PC) solution / Palladium Acetate	Palladium Oxide Fibers	Catalyst, Sensors, Photoelectrolysis	Viswanathamurthi et. al. (2004a)
PVac solution / Titanium Isopropoxide sol	Rithenium doped TiO₂ Fibers -------------	Solar Cells, Sensors	Viswanathamurthi et. al. (2004b)

Poly(vinyl acetate) (PVac) solution was prepared with dimethylformamide (DMF).

Table 2.4 (c).　Precursor materials of ceramics nanofibers with sintering process which were developed by other research groups.

Precursor Material	Diameter of Ceramics Nanofibers	Targeted Applications	Reference
Poly(vinyl pyrrolidone) (PVP) solution / $Ti(OiPr)_4$ sol	TiO_2 Fibers $20 \sim 200$ nm	Environmental Cleaning and Protection, Photocatalysis, Gas Sensors, Solar Cells, Batteries	Li and Xia (2003)
PVP solution / Nickel ethylhexaisopropoxide + iron (III) ethylhexano-isopropoxide sol	$NiFe_2O_4$ Fibers 46 nm	Ultrahigh-Density Data Storage, Sensors, Spintronic Devices	Li et. al. (2003a)
PVP solution / $Ti(OiPr)_4$ sol	TiO_2-Au Fibers (after dip-coating of TiO_2 Fibers into $HAuCl_4$ solution) 30 nm	Photo Catalytic and Photoelectrochemical Devices	Li and Xia (2004)
Zirconium n-propoxide solution / Titanium isopropoxide /	$Pb(Zr0.52T0.48)O_3$ 500 nm ~ several microns	High Performance Hydrophone Ultrasonic Transducer	Wang and Santiago-Aviles (2004)
Lead 2-ethylhexanoate / Xylene	$Pb(Zr0.52T0.48)O_3$ several hundreds nm ~ 10 mm	High Voltage & High Power Capacitor	Wang et. al. (2004d)
Poly(ethylene oxide) (PEO) solution / $C_2H_44O_4Sn$ sol	SnO_2 Fibers 100 nm ~ 10 mm	Gas Sensors	Wang et. al. (2004e)

Other than fabrication study of ceramics nanofibers, no detailed discussion about the property of targeted applications has been done except the work of [Li et al., 2003a] which investigated the magnetic property of nickel ferrite nanofibers. Therefore, property investigation of potential applications should be the next challenge of fabricated ceramics nanofibers. Preparation of ceramics nanofibers includes the preparation of sol-gel precursor of ceramics and sintering process of the electrospun nanofibers.

For example, Shao et al. (2004c) prepared aqueous manganese acetate ($Mn(CH_3COO)_2 \cdot 4H_2O$) solution (manganese acetate : distilled water = 33:66 wt%) and aqueous poly(vinyl alcohol) (PVA) solution with 10wt% polymer concentration. 20 g PVA solution was slowly lowered into manganese acetate solution and chemical reaction proceeded in a water bath for a period of 5 hours at a temperature of $50^{\circ}C$.

The obtained PVA/manganese acetate gel was electrospun onto the aluminum foil with an applied voltage of 20kV. The resultant nanofiber webs were dried at a temperature of $70^{\circ}C$ for a period of 12 hours under vacuum condition and then sintering process was performed at $300^{\circ}C$, $700^{\circ}C$ and $1000^{\circ}C$ for 10 hours. During the sintering process, the organic belonged to PVA and CH_3COO group of manganese acetate and other volatiles (such as H_2O, CO_x, etc) were removed. From the XRD and FT-IR results, sintering process at $700^{\circ}C$ resulted in the production of Mn_2O_3 nanofibers whilst sintering at $1000^{\circ}C$ resulted in the production of Mn_3O_4 ceramics nanofibers. It was shown that sintering condition significantly influences the reaction of ceramics gel and the structure of ceramics nanofibers.

2.1.3.5. *Carbon*

Carbon is the basic building blocks of naturally occurring organic compounds. As such, pure carbon can be manufactured from a variety of carbonaceous precursors. In its pure carbon can exist in three allotropic forms, carbon, graphite and diamond. While carbon as an allotropic form has an "amorphous" or non-ordered atomic structure, graphite consists of parallel stacking of carbon atoms in fused hexagonal rings. Because of the weak attractive forces between the parallel layers of hexagonal rings, graphite is softer than carbon. In the form of diamond, its tetrahedrally bonded carbon atoms makes it the hardest naturally occurring material known to man. Recently, there has been great interest in carbon nanotubes due to its huge potential for many different applications. Carbon nanotube has also been successfully embedded in polymer nanofibers by means of electrospinning [Dror et. al. (2003)] or grown from the surface of electrospun fibers [Hou et. al. (2004)].

Carbon in its graphite form is a well known stable electrode material as it is free of functional groups which interact with ions in the electrolyte. The large surface area and low resistivity of carbon nanotubes make it an interesting candidate in electrochemistry. When an electrode is immersed in an electrolyte, there is a potential induced adsorption of solutes onto the surface of the electrode called electroadsorption. In its simplest case, opposing-charged ions are adsorbed onto the electrode while like-charged ions are repelled. Two layers of charges with a potential drop are formed as a result of the separation of charge with the attracted ions forming a layer that balanced the charge on the electrode. Thus, an electrochemical double layer is said to be formed.

Generally, the specific electroadsorption capacity of ions is very small. However, by increasing the electrode surface, such as porous carbon electrodes, the capacities can be significantly increased [Soffer and Folman (1972)]. While it was also been demonstrated that porous carbon electrodes can be used for purification of solutions containing suspended bacteria [Oren et. al. (1983)], carbon nanotube may be used to perform the same function.

Electrochemical double layer can act as a capacitor [Kotz and Carlen. (2000)] if the potential difference across it is less than that needed to dissociate the electrolyte. Since capacitance is given by,

$$C = \frac{\varepsilon A}{d} \qquad (2.2)$$

where ε is the dielectric constant
   A is the electrode surface
   d is the electrochemical double layer distance

The capacitance can be huge for carbon electrode that has a high area and low electrochemical double layer distance. Thus this has led to the investigation of using carbon nanotube [Emmenegger et. al. (2003)] and electrospun activated carbon nanofibers [Kim (2004a)] as super capacitor.

While carbon in the form of graphite is known to be a good conductor of electricity, recent interest has been on using carbon nanotubes in electronics applications [McEuen (1998)]. This is especially so with the miniaturization of electronic devices. Single walled carbon nanotubes have been experimentally determined to exhibit either metallic behavior or semiconductor behavior depending on the hexagonal-ring structure of the wall [Odom et. al. (1998) Dresselhaus (1998)]. Carbon nanotubes have also been made into field-effect transistors by using a semi-conducting silicon substrate covered with a layer of silicon dioxide as a back-gate and making contacts using lead electrodes [Tan et. al. (1998)].

However, this design implies that all the devices are switched on simultaneously. Subsequent improvement involves having a separate 'top' gate over the individual nanotube so that the device can be switched individually. The transconductance of the carbon nanotube field-effect transistor was found to be 20 times better than that of conventional silicon metal-oxide-semiconductor field-effect transistors [Wind et. al. (2002)].

Carbon nanofibers possess numerous possible applications such as filters, high temperature catalysts, heat management materials in aircraft, and semiconductor device.   [Dzenis and Wen (2002)] made carbon nanofibers with 100 ~ 500 nm fiber diameter and the utilization of carbon nanofibers. Carbon nanofibers have been expected as reinforcement of traditional fiber reinforced polymer composites. The high electrical conductivity of carbon nanofibers also opens up the potential for application such as sensor devices [Wang et. al. (2002a)]. The fabricated carbon nanofibers demonstrated a high electrical conductivity (490 S/m).   Fabrication process of carbon nanofibers contains electrospinning of polyacrylonitrile (PAN) nanofibers and carbonization of deposited nanofibers.   A research group at the University of Pennsylvania [Wang et. al. (2002a); Wang et. al. (2003); Santiago-Aviles and Wang (2003)] prepared Dimethylformamide (DMF) solution mixed with PAN powder (PAN : DMF = 600 mg : 10ml). This mixture was well stirred at room temperature.   After PAN nanofiber deposition through electrospinning process (15kV applied voltage, 15cm distance), the fibers were carbonized at 560°C, 800°C, 1000°C and 1200°C respectively for 30 minutes with 10^{-6} Torr pressure under vacuuming [Wang et. al. (2003)].   Raman spectra characterization showed that the graphite domain size in carbon nanofibers was enlarged from 1.5 to 2.6nm with higher carbonization temperature.   The conductivity of carbon nanofibers was also influenced by the carbonization time. It was concluded that the transformation of disordered carbon into graphitic carbon is a highly kinetically controlled process.   Hence carbonization condition (i.e. temperature and time) is very important to control the structure and property of carbon nanofibers.

2.2. Solution Property

In order to carry out electrospinning, the polymer must first be in a liquid form, either as molten polymer or as polymer solution. The property of the solution plays a significant part in the electrospinning process and the resultant fiber morphology. During the electrospinning process, the polymer solution will be drawn from the tip of the needle. The electrical property of the solution, surface tension and viscosity will determine the amount of stretching of the solution. The rate of evaporation will also have an influence on the viscosity of the solution as it is being stretched. The solubility of the polymer in the solvent not only determines the viscosity of the solution but also the types of polymer that can be mixed together.

2.2.1. *Surface Tension*

In electrospinning, the charges on the polymer solution must be high enough to overcome the surface tension of the solution. As the solution jet accelerates from the tip of the source to the collector, the solution is stretched while surface tension of the solution may cause the solution to breakup into droplets [Shummer and Tebel (1983); Christanti and Walker (2001)]. When droplets are collected, a different process called electrospraying [Morozov et. al. (1998)] is taking place rather than electrospinning, where fibers are collected instead. Surface tension has also been attributed to the formation of beads on the electrospun fibers. Thus it is important to understand the role of surface tension in a fluid.

When a very small drop of water falls through the air, the droplet generally takes up a spherical shape. The liquid surface property that causes this phenomenon is known as surface tension. For a liquid molecule submerged within the solution, there is uniform attractive forces exerted on it be other liquid molecules surrounding it. However, for a liquid molecule at the surface of the solution, there is a net downward force as the liquid molecules below exert a greater attractive force than the gas molecules above as shown in Fig. 2.5.

Thus the surface is in tension and this causes a contraction at the surface of the solution, which is balanced by repulsive forces that arise from the collisions of molecules from the interior of the solution. The net effect of the pulling of all the surface liquid molecules causes the liquid surface to contract thereby reducing the surface area. Therefore, for a droplet of water, a spherical shape is the lowest surface area to volume ratio.

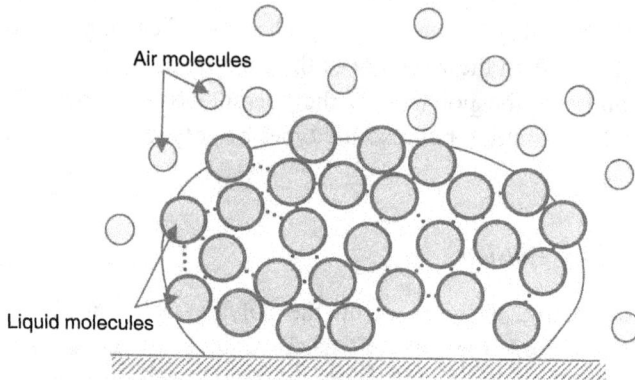

Fig. 2.5.   Attractive forces between the liquid molecules are stronger than the attractive forces of the air molecules.

The most common quantitative index of surface tension ξ is defined by the force exerted in the plane of the surface per unit length. We can consider a reversible isothermal system where at a constant pressure and temperature, the surface area A_s, is increased when the surface is pulled apart to allow the liquid molecule to enter. Taking σ_g as surface Gibbs energy per unit area, the differential reversible work is $\xi\, dA_s$. For an equilibrium system, the Gibbs energy is at its minimum, thus ξA_s will also be at its minimum. For a fixed ξ, the surface area, A_s of the system is at its lowest [Reid et. al. (1988)]. A drop of water on a horizontal surface is flattened rather then spherical. This is due to the force of gravity. Other factors that cause the droplet to flatten on the surface include higher liquid density, decrease in surface tension and increasing the volume of the droplet.

2.2.1.1. Effect of Temperature on Surface Tension

For a pure liquid system, the surface tension of the liquid would decrease with increasing temperature. When the temperature is raised, the equilibrium between the surface tension and the vapor pressure would decrease. At a critical point, the interface between the liquid and the gas disappears [Clark (1938)]. From the molecular point of view, at a higher temperature, the liquid molecules gain more energy and start to move more rapidly in the space. As a result, the fast moving molecules do not bound together as strongly as the molecules in a cooler liquid. With the reduction in the bonding between the molecules, the surface tension drops. The effect of temperature on surface tension for pure liquid may be different from mixtures. For a mixture of methane and nonane, the surface tension actually increases with increasing temperature except at the lowest pressure. At a lower temperature, methane is more soluble than nonane and the effect of liquid mixture composition is more pronounce than temperature in the determination of the surface tension [Deam and Maddox (1970)]

2.2.1.2. Surface Tension of Solvent Mixtures

For a mixture, the surface tension is not a simple function of the surface tension of the pure components. In a mixture, the composition at the surface is often different from that of the bulk. The surface tension of the mixture of non-aqueous solution can be approximated by a linear dependence on the mole fraction average of the surface tension of the pure components. For aqueous solution, the surface tension shows pronounced nonlinear characteristics. In an organic-aqueous system, a small amount of organic liquid may significantly affect the mixture surface tension. This is due to hydrophobic behavior of a small amount of organic molecules, which tend to be rejected from the water phase and concentrating on the surface. As a result, the surface composition of the mixture is very different from that of bulk.

66 An Introduction to Electrospinning and Nanofibers

2.2.2. *Polymer Solubility*

Although molten polymers can be electrospun [Lyons and Ko. (2004); Larrondo and Manley (1981a); Larrondo and Manley (1981b); Larrondo and Manley (1981c)], it is more common to electrospin using polymer solution. Since different solvents have different level of electrospinnability [Jarusuwannapoom et. al. (2005)] it is important to use an appropriate solvent that can dissolve the polymer and at the same time electrospinnable. The solubility of the polymer in a particular solvent may also affect the resultant fiber morphology [Wannatong et. al. (2004)].

Polymer solubility is more complex than those of low-molecular weight compound due to size difference between polymer and solvent molecules, viscosity of the system, effects of the structure and molecular weight of the polymer. There are two stages when a polymer dissolves in the solvent. Firstly, solvent molecules diffuse slowly into the polymer bulk to produce a swollen gel. If the polymer-polymer intermolecular forces are high as a result of cross-linking, crystallinity or strong hydrogen bonding, the polymer-solvent interactions may not be strong enough to break the polymer-polymer bond. The second stage of solution will only take place when the polymer-polymer bond is broken to give true solution.

2.2.2.1. *Effect of Polymer Structure on Solubility*

The structure of the polymer has an impact on its solubility in the solvent. Generally, a polymer with higher molecular weight is less soluble and takes a much longer time to dissolve than one with a lower molecular weight using the same solvent. The intermolecular forces between longer chain molecules are stronger and the solvent molecules take a longer time to diffuse into the polymer bulk. Crosslinked polymers do not dissolve, as covalent bonding between the molecules is much stronger than the secondary forces exerted from polymer-solvent interactions.

Crystallinity of the polymer measures the degree of orderliness of the polymer chain packed within the bulk. Polymer of higher crystallinity has lower solubility as the solvent molecules have difficulty in penetrating the interior of the polymer bulk.

2.2.2.2. *Gibbs Free Energy*

In discussing the solubility of the polymer in a system, it is useful to make use of Gibbs free energy.

$$\Delta G = \Delta H - T \Delta S \qquad\qquad (2.3)$$

where, ΔG is the Gibbs free energy,
ΔH is the heat of mixing or enthalpy
T is Temperature
ΔS is Entropy of mixing

A good solvent causes the polymer to expand to reduce the overall Gibbs free energy of the system. In a poor solvent, the polymer molecules will curl up or collapse to reduce the overall Gibbs free energy. Solubility occurs when the Gibbs free energy give a negative value. As seen from equation (2.3), the temperature has an influence on whether the solvent is a "good" or "poor" solvent. The lowest temperature at which a polymer of infinite molecular weight is completely soluble in a specific solvent is known as Flory-Huggins temperature, θ. At a higher temperature, the polymer will expand while at a lower temperature, the polymer will collapse.

The entropy of mixing ΔS is long thought to be positive thus the sign of ΔG is also determined by the sign and magnitude of the heat of mixing ΔH. For non-polar molecules, ΔH is assumed to be positive and the same as those derived from the mixing of small molecules. The enthalpy from mixing a solute and a solvent is proportional to the square of the difference in solubility parameters [Hildebrand and Robert (1950)].

$$\Delta H = v_s v_p \left(\delta_s - \delta_p \right)^2 \qquad (2.4)$$

where ΔH is the heat of mixing or enthalpy
v_s and v_p is the partial volume of solvent and polymer respectively
δ_s and δ_p is the solubility parameter of solvent and polymer respectively

Seen from equation of Gibbs free energy (2.3), ΔG will have a negative value if ΔH approaches zero and thus, solution will occur. For non-polar solvents, the solubility parameter δ_s, is a function of enthalpy of evaporation, temperature and molar volume.

$$\delta_s = \sqrt{\left(\Delta H_v - RT \right)/V_M} \qquad (2.5)$$

where ΔH_v is the Enthalpy of evaporation
R is the ideal gas constant
T is temperature in Kelvin
V_M is the molar volume

To determine the solubility, δ_p, for a non-polar polymer of known structure, molar-attraction constants E_M, are summed over the structural configuration of the repeating unit in the polymer chain,

$$\delta_p = \frac{\rho \sum E_M}{M} \qquad (2.6)$$

where δ_p is the solubility of polymer
ρ is the density
E_M is the molar-attraction constants
M is the repeat molecular weight.

2.2.3 *Viscosity*

The viscosity of the solution has a profound effect on electrospinning and the resultant fiber morphology. Generally, the viscosity of the solution is related to the extent of polymer molecule chains entanglement within the solution. When the viscosity of the solution is too low, electrospraying may occur and polymer particles are formed instead of fibers. At lower viscosity where generally the polymer chain entanglements are lower, there is a higher likelihood that beaded fibers are obtained instead of smooth fibers. Therefore, factors that affect the viscosity of the solution will also affect the electrospinning process and the resultant fibers.

A liquid under normal experimental considerations will respond to shear stress by flowing. The shearing stress f is the force required to move a plane relative to another which is dependent on a proportionality factor, called coefficient of shear viscosity or viscosity is introduced. Therefore,

$$f = \eta \left(\frac{ds}{dt} \right) \tag{2.7}$$

where η is viscosity
        ds/dt is the rate of shear

Viscosity is a measure of the resistance of a material to flow. There are several forms of viscosities, relative viscosity (η_r) is the ratio of the viscosity of a polymer solution to that of its solvent. Relative viscosity minus the viscosity of water (1) is the specific viscosity (η_{sp}). Reduced viscosity or viscosity number (η_{red}) is by dividing η_{sp} by the concentration of the solution, c. Finally, the intrinsic viscosity or limiting viscosity number ($[\eta]$) is obtained by extrapolating reduced η_{red} to zero concentration.

The intrinsic viscosity is related to the average molecular weight of the polymer (M) of the polymer by a proportionality constant, K, which is the characteristic of the polymer and solvent and the exponential a, which is a function of the shape of polymer coil in a solution.

$$[\eta] = KM^a \qquad (2.8)$$

$[\eta]$ is also found to be directly proportional to the effective hydrodynamic volume of the polymer in solution and inversely proportional to the molecular weight (M). The effective hydrodynamic volume is the cube of the root-mean-square of the end-to-end distance of the molecule, $\left(\sqrt{\overline{r^2}}\right)^3$.

$$[\eta] = \frac{K\left(\overline{r^2}\right)^{3/2}}{M} \qquad (2.9)$$

where K is the proportionality constant in the Flory equation for hydrodynamic volume.

Table 2.5 shows a summary of the various viscometry terms and their corresponding formula.

In a polymer solution, the configuration that is adopted by the polymer chain will affect the intrinsic viscosity of the solution. An extended or uncurled configuration of the polymer chain molecules is associated with an increase in the intrinsic viscosity of the solution. While polymer chains that adopt a curled configuration in the solvent will result in a drop in the intrinsic viscosity.

Table 2.5. Viscometry term and its formulas.

Viscometry Term	Formula
Relative viscosity or viscosity ratio, η_{rel} or η_r	$\dfrac{\eta_1}{\eta_2}$
Specific viscosity, η_{sp}	$\dfrac{\eta_1}{\eta_2}-1 \;\; or \;\; \eta_r -1$
Reduced viscosity or viscosity number, η_{red}	$\dfrac{\eta_{sp}}{c}$
Intrinsic viscosity, $[\eta]$	$KM^a \;\; or \;\; \lim\limits_{c\to 0}\left(\eta_{sp}\Big/ C\right) \;\; or$ $[\eta]=\dfrac{K\left(\overline{r^{-2}}\right)^{3/2}}{M}$

1 is the polymer
2 is the solvent
c is the concentration of the solution

2.2.3.1. *Solvent Effect on Intrinsic Viscosity*

For a polymer whose molecules are long chain hydrocarbons such as polystyrene, the effect of the solvent on the intrinsic viscosity can be described in relatively simple terms. Generally, the intrinsic viscosity is high in "good" solvents but low in "poor" solvents [Alfrey (1946)]. For a good solvent, the long chain molecules are surrounded by continuous, energetically indifferent solvent molecules. This reduces polymer-polymer contacts therefore favoring uncurled configurations.

However, in a poor solvent, which is energetically unfavorable, the dissolving of the polymer is an endothermic process and the polymer segments will be attracted to one another in the solution and squeeze out the solvent between them. The polymer chains will adopt a curled configuration. Depending on the solubility of the polymer in the solvent, a good solvent allows a higher concentration of the polymer while maintaining a fluid, stable solution whereas a poorer solvent will have more polymer-polymer contacts between different chains and lead to gelation. The effect of solvent on the intrinsic viscosity also depends on the flexibility of the polymer chain. If the polymer chain is more rigid, the solvent effect on the intrinsic viscosity is less [Alfrey et. al. (1942)].

2.2.3.2. Temperature on Intrinsic Viscosity

Generally, when the temperature of the solution is increased, the solubility of the polymer in the solvent will increase. In pure solvents, specific viscosity decreases with increase in temperature. This is due to higher polymer chain mobility. The dependency of intrinsic viscosity on temperature can be shown by,

$$[\eta] = Be^{E_a / RT} \qquad (2.10)$$

where E_a is an activation energy for viscous flow
B is a constant
R is the ideal gas constant
T is the temperature in Kelvin

However, if the initial temperature is below the Flory-Huggins temperature θ, the viscosity will increase when the mixture of polymer and solvent is heated to slightly above θ. In a mixed system where the mixture of good and poor solvents is used to dissolve the polymer, specific viscosity may increase with temperature. This is the result of uncurling of the molecule due to the increase of entropy factors over energetic factors when the temperature is increased [Alfrey et. al. (1942)] to θ of the solvent mixtures.

2.2.3.3. *Viscometry*

Viscometry is concerned with the determination of the relationship between shear stress and rate of shear in a fluid. There are several methods that are used to determine the viscosity of the fluid. In viscometric flow, there are two cases, the Poiseuille flows and Coutte flows.

For Poiseuille flows, the motion of the fluid is caused by a pressure gradient acting parallel to a fixed boundary, while in Coutte flows, there is no pressure gradient and the flow of the fluid is caused by the movement of the walls of the system.

Poiseuille flow in Capillary Design [Ferguson and Zemblowski (1991)]

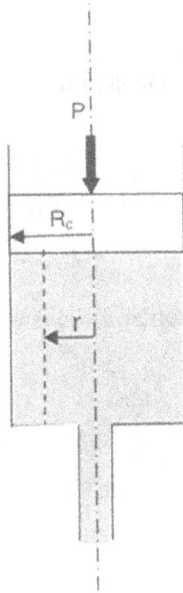

Fig. 2.6.   Diagram of Capillary.

In a capillary design, the fluid is forced through a cylindrical tube with a smooth inner surface. The flow of the fluid is assumed to be in steady-state, isothermal and laminar. The shear stress and the shear rate are determined from the interaction between the fluid and the wall of the capillary. The flow velocity at the axis of the capillary is the fastest while the flow velocity is slowest at the walls of the capillary. The shear stress of the fluid is given by,

$$\tau = \frac{r\Delta p}{2L} \qquad (2.11)$$

where τ is the shear stress at radius r

r is the radius

$\frac{\Delta p}{L}$ is the axial pressure due to friction

The boundary condition, at the capillary axis, $\tau = 0$, at the capillary wall $(r = R)$, $\tau = \tau_w$,

$$\tau_w = \frac{R_c \Delta p}{2L} \qquad (2.12)$$

where R_c is the radius of the capillary.

The shear rate of non-Newtonian fluid is given by the following equation [Rabinowitsch (1929)],

$$\dot{\gamma}_w = \left(-\frac{du}{dr}\right)_w = 3\left(\frac{8Q}{\pi D^3}\right) + \frac{D\Delta p}{4L}\frac{d(8Q/\pi D^3)}{d(D\Delta p/4L)} \qquad (2.13)$$

where Q is the volumetric flow rate

D is the diameter of the capillary

$\frac{\Delta p}{L}$ is the axial pressure due to friction

Coutte flow from Coaxial Cylinder

In a coaxial cylinder, there is one cylinder with a smaller diameter enclosed by a bigger cylinder. The gap between the smaller cylinder and outer cylinder is filled with the test fluid. Assuming that the inner cylinder is stationary while the outer cylinder is rotating at a constant velocity of Ω. The angular velocity is such that there is a laminar flow of the fluid developed in the gap. Assuming that there is no slip at the walls of the cylinder in contact with the fluid, the fluid layer adjacent to the inner cylinder will be stationary while the fluid layer adjacent to the outer cylinder will move at an angular velocity of Ω [Harris (1977)].

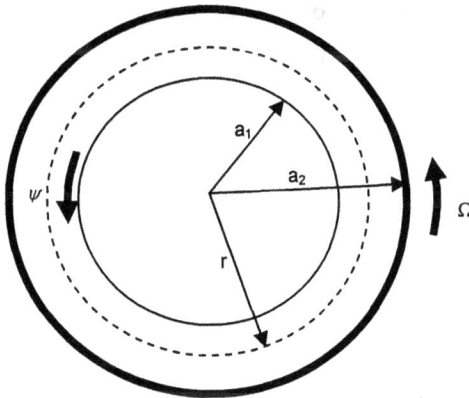

Fig. 2.7 Diagram of coaxial cylinder

The shear stress of the fluid is given by,

$$\tau = \frac{\psi}{2\pi h_E r^2} \qquad (2.14)$$

where ψ is the torque resulting from the tangential force due to internal friction

h_E is the effective depth of the liquid

r is the radial distance

The boundary condition, at the wall of the inner cylinder, $\tau(a_1) = \tau_1$, at the wall of the outer cylinder, $\tau(a_2) = \tau_2$, therefore,

$$\tau_1 = \frac{\psi}{2\pi h_E a_1^2} \tag{2.15}$$

$$\tau_2 = \frac{\psi}{2\pi h_E a_2^2} \tag{2.16}$$

For the shear rate of non-Newtonian fluids if the ratio of the outer to inner cylinder radius, $S < 1.2$, then [Krieger and Maron (1954); Ferguson and Zemblowski (1991)],

$$\dot{\gamma} = \frac{4\pi N}{1 - \frac{1}{S^2}} \left\{ 1 + K_1 \left(\frac{1}{n''} - 1 \right) + K_2 \left[\left(\frac{1}{n''} - 1 \right)^2 + \frac{d\left(\frac{1}{n''} - 1 \right)}{d(\log \Psi)} \right] \right\} \tag{2.17}$$

where the constants,

$$K_1 = \frac{S^2 - 1}{2S^2} \left(1 + \tfrac{2}{3} \ln S \right)$$

$$K_2 = \frac{S^2 - 1}{6S^2} \ln S$$

$$S = \frac{a_2}{a_1}$$

n'' is the slope of a logarithmic plot of torque vs rotational speed N

Elongational Viscoelasticity

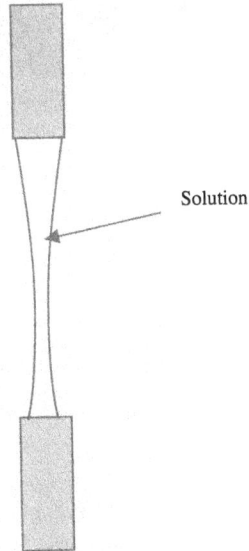

Solution

Fig 2.8.   Elongation of viscous solution.

The usual measurement of shear flow viscosity is not sufficient to characterize the deformation behavior of fluid. In processes such as fiber spinning and blow molding, it is the elongational rather then shear deformation that is the dominant mode of deformation as shown in Fig. 2.8. There are several methods to study the elongation of viscous solutions [Ferguson and Zemblowski (1991)]. However, in process such as electrospinning, the viscosity of the solution is much lower. In a method proposed by [Yentov et. al. (1988)], the method of stretching a highly viscous cylindrical sample is modified to measure less viscous solution. The stresses in the viscoelastic material from stretching at constant rate is,

$$\sigma = G\left(\frac{t}{t_0}\right)^2 \exp\left(-\frac{t}{\tau_r}\right), \quad t_0 = \frac{l_0}{v_0}, \quad \frac{t}{t_0} = \lambda \qquad (2.18)$$

and for relaxation is,

$$\sigma = \sigma_* \exp\left(-\frac{t}{\tau_r}\right) \qquad (2.19)$$

where G is the elasticity of modulus
         l_0 is the initial value of the length
         λ is the degree of elongation
         v_0 is the speed of movement of the active clamp
         σ_* is the initial stress level
         τ_r is the relaxation time

Using polyethylene oxide and polyacrylamide solution, the elasticity modulus G, was found to be monotonically raising function of concentration. The relaxation time for all the solutions is about 0.1s and is at least one to two orders of magnitude smaller than the relaxation time obtained from stationary shear experiments for slow strains. It was interesting to note that the relaxation time of the solutions were independent from the concentration of the polymer.

2.2.4. *Volatility (Evaporation) of Solution*

During the electrospinning process, the solvent will evaporate as the electrospinning jet accelerates towards the collector. When most of the solvents have evaporated when the jet reaches the collector, individual fibers are formed. However, if the rate of evaporation of the solvent is too low such that the solution has not evaporated sufficiently when the electrospinning jet reaches the collector, fibers may not be formed at all and a thin film of polymer solution are deposited on the collector.

The evaporation rate of a solvent is dependent on many factors:
- Vapor pressure
- Boiling point
- Specific heat
- Enthalpy and heat of vaporization of the solvent
- Rate of heat supply
- Interaction between solvent molecules and between solvent and solute molecules
- Surface tension of liquid
- Air movement above the liquid surface

The vapor pressure of a solvent is dependent on its molecular weight, heat of vaporization and temperature as given by Clausius-Clapeyron equation,

$$\frac{d(\ln p)}{dT} = \frac{M\Lambda}{RT^2} \tag{2.20}$$

where p is the vapor pressure

T is temperature

M is the molecular mass of solvent

Λ is the heat of vaporization

R is the ideal gas constant

A more reliable vapor pressure estimate can be given by the Antoine equation at a specific temperature,

$$\log P = A_{an} - \frac{B_{an}}{T + C_{an}} \tag{2.21}$$

where A_{an}, B_{an} and C_{an} are Antoine coefficients

T is temperature

Boiling point of solvent, heat of vaporization and enthalpy are all affected by the molecular weight of the solvent molecules. Due to the complex function of the evaporation mechanism, experimental methods are used to determine the volatility of the solution.

2.2.5. *Conductivity of Solution*

For electrospinning process to be initiated, the solution must gain sufficient charges such that the repulsive forces within the solution are able to overcome the surface tension of the solution. Subsequent stretching or drawing of the electrospinning jet is also dependent on the ability of the solution to carry charges.

Generally, the electric conductivity of solvents is very low (typically between 10^{-3} to 10^{-9} ohm^{-1}m^{-1}) as they contain very few free ions, if any, which are responsible for the electric conductivity of solution. The presence of acids, bases, salts and dissolved carbon dioxide may increase the conductivity of the solvent. The electrical conductivity of the solvent can be increased significantly through mixing chemically non-interacting components. Substances that can be added to the solvent to increase its conductivity includes mineral salts, mineral acids, carboxylic acids, some complexes of acids with amines, stannous chloride and some tetraalkylammonium salts. For organic acid solvents, the addition of a small amount of water will also greatly increase its conductivity due to ionization of the solvent molecules.

There is another solution behavior known as the leaky dielectric which deals with poorly conducting liquids [Allan et. al. (1962)]. For a conducting liquid such as strong electrolytes, all the neutral species will dissociate fully to give free ions. However, for a leaky dielectric, there are both a forward reaction where positive and negative ions are produced from the dissociation of neutral species as well as a recombination reaction where the ions recombines to form back the neutral species.

Unlike perfect conductors or dielectrics where the electric stress is perpendicular to the interface and the alterations of the interface shape together with the interface tension balance the electric stress, free charge accumulated at the interface of a leaky dielectric and modifies the field. A subsequent viscous flow is induced to provide the stress to balance the tangential components of the field acting on the interface charge. The electric stresses produced by the conduction of the charges to the interface are different from those of the perfect dielectrics or perfect conductors. Leaky dielectric model is generally derived from Stokes equation to describe the fluid motion and the Ohmic conductivity to describe the conservation of current [Saville (1997)].

2.3. Electrostatics

Electrospinning is possible only if there is a potential difference between the solution and the collector. Very often, an external electric field is used to control the charged electrospinning jet. Factors that affect the ability of the solution to carry charges, the electric field that surrounds the electrospinning jet and the dissipation of charges on the polymer fibers that are deposited on the collector will have an impact on the electrospinning process.

In order to understand electric field and electrostatic, one has to start from the most fundamental unit, charge. In an atom, there exist protons and electrons with the protons located inside the nucleus and electrons outside it. The charge of a proton is positive while that of the electron is negative. This is the smallest existing amount of charge and is measured in the unit of coulomb, C where 1 coulomb is defined as 1 ampere.second, and the electron charge, e, is given by,

$$e = -1.608.10^{-19} C$$

The proton charge has the same numerical value but have a positive charge instead. Therefore, if a body has excess electrons, it is negatively charged while a deficit of electrons will give it a positive charge.

2.3.1. *Electric Field*

The region where there is an electric force caused by the presence of electric charges is known as an electric field. The force can either be attraction between opposite charges or repulsion between charges of the same polarity and is given by the Coulomb's Law,

$$F = \frac{q_1 q_2}{4\pi\varepsilon_p d^2}$$

(2.22)

where q is the charge
        ε_p is the absolute permittivity of the space between the charges
        d is the distance between the charges

However, this law is only valid if the charges involved are point charges. In most practical cases, the electric field is more widely used and it is defined as a region where a charge feels a force created by other charges. The magnitude of the field is given by its field strength.

$$F = qE$$

(2.23)

where F is the force
        q is the charge
        E is the electric field strength

For a positive charge, the force has the same direction as the field strength while a negative charge will have the force at the opposite direction.

Given a charge q, the field strength at a distance d is written as,

$$E = \frac{q}{4\pi\varepsilon d^2}$$

(2.24)

where ε is the permittivity of the material

2.3.2. *Potential Difference and Electric Field Representations*

The potential of a point in space is given by the work required for a charge to move a test charge infinitely slowly from a reference point to this point. Thus the potential can be given by,

$$\Phi = \frac{dW}{dQ} \tag{2.25}$$

The voltage between two points in space equal to the potential difference between them and is given by,

$$U_{1,2} = \Phi_2 - \Phi_1 \tag{2.26}$$

When there are just two point charges, it is easy to describe the interaction between them using the Coulomb's law. However, when more charges are involved, it will be advantages to use electric field and potential to describe their interactions. While it is possible to use vectorial representation of the electric field, it is more common to use field lines instead. The direction of the electric field along any point on the field line is given by the tangent to the field line at that point although the field direction is indicated by arrows on the field lines. The field lines always begin at the positive charges and end at the negative charges as shown in Fig. 2.9.

Another way to represent the electric field is by a descriptive geometric quantity called an equipotential surface. Equipotential surfaces are used to link all the points in space of the same potential with one another. The field lines at all the points must be perpendicular to the equipotential surfaces.

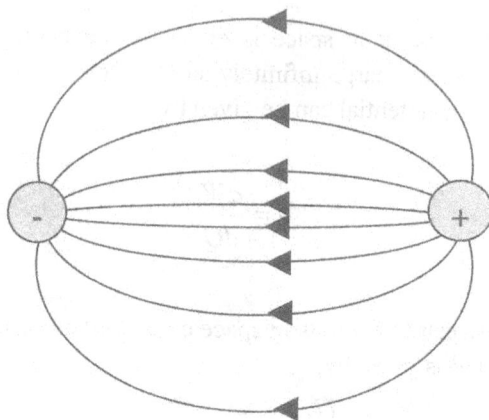

Fig. 2.9. Field line plot of a negative point charge and positive point charge.

2.3.3. *Surface Charge of Insulator*

As electrospun fibers are deposited on the collector, there will be residual charges left on the mesh of fibers. The build-up of residual charges on the fiber mesh [Tsai et. al. (2002)] may have an influence on the electrospinning process especially when large volumes of fibers are to be collected. Most polymers are good insulating materials thus charges on the fibers does not dissipate easily on their on.

The surface charge density of a material is the charge per unit area,

$$\sigma_e = \frac{dQ_c}{dA} \qquad (2.27)$$

where σ_e is the surface charge density
Q_c is the total charge on the surface
A is the total surface area

The total charge on any surface is therefore given by,

$$Q_c = \int_A \sigma_e dA \qquad (2.28)$$

For an insulating material, the surface potentials will usually vary from point to point. Since the point potential at the interior of the insulating material is unknown, it is not possible to determine the voltage on the insulating materials as a whole.

When a charge is introduced to an insulator, the charge will create a field both on the surface of the insulator as well as its interior. The resultant field will cause the individual charges to move due to mutual repulsion such that equilibrium is reached where the resulting force on any charge is zero. The charges in the interior of the insulator will move to the surface such that the field in any point inside the insulator is zero. The charge on the surface will orientate itself on the surface until the field is perpendicular to the surface. Therefore, charge will accumulate and the field will be strongest around sharp point or protrusions of the surface.

2.3.4. Field Ionization

In electrospinning, the force that causes the stretching of the solution is due to the repulsive forces between the charges on the electrospinning jet. Thus the presence of ions that neutralizes the charges during the electrospinning process would have an impact on it [Fong et. al. (1999)]. Under atmospheric condition using plane electrodes, an electric field of about 3×10^6 V/m is required for electrons freed by natural radioactive ionization to ionize molecules of the air. The freed ions will then ionize other air molecules in the region where the field strength is high enough. It must be emphasized that ionization depends on high field strength and not high voltage. Also, the discharge condition for positive and negative voltage is different. Generally, the positive voltage required for ionization is 30% higher than negative voltage.

The geometry of the conducting electrodes will have an effect on the breakdown field strength. The breakdown field strength, E_b has the values,

$$E_b \approx \left(300 + \frac{18}{\sqrt{r}} \right) x 10^4 \text{ V/m for points}$$

$$E_b \approx \left(300 + \frac{9}{\sqrt{r}} \right) x 10^4 \text{ V/m for wires}$$

where r is the radius of the points and wires respectively in meters

It seems that the breakdown field strength of points and wires electrode is higher than parallel electrodes, but for sharp electrode, the field is strongly inhomogeneous and the voltage necessary is much lower. The voltage required to generate the breakdown field for point electrodes is 5kV to 6kV if the nearest grounded object is some meters away [Jonassen (2002)].

2.4. Conclusions

Although electrospinning is a very simple process, requiring just simple laboratory equipment to yield fibers down to the nanoscale, the science behind it is not simple at all. Electrospinning process involves the understanding of electrostatics, fluid rheology and polymer solution properties such as rate of solvent evaporation, surface tension and solution conductivity. These fundamental properties are constantly interacting and influencing each other during the electrospinning process.

The versatility of electrospinning also meant that fibers of different morphology and made of different materials can be made directly or indirectly from electrospinning. Therefore, different polymers, blends, mixtures or precursors can be used to make into fibers to suit specific applications. Understanding the basics behind the materials and the fundamentals that affect electrospinning will open new avenues and applications for electrospun fibers.

Notation

A	Electrode surface
B	Constant
A_{an}, B_{an}, C_{an}	Antoine constant
c	Concentration
C	Capacitance
C_0	Capacitance vacuum
D	Diameter of the capillary
d	Distance
E	Electric field strength
E_a	Activation energy
E_b	Breakdown field strength
E_M	Molar-attraction constants
F	Force
G	Elasticity of modulus
h	Height or thickness
h_E	Effective depth
K	Proportionality constant
l	Length
M	Molecular weight
M_n	Number average
M_v	Viscosity average
M_w	Weight average
M_z	z average
N	Rotational Speed
p	Vapor pressure

q	Charge
r, a_1, a_2	Radius
Q	Volumetric flow rate
Q_c	Total charge on the surface
R	Ideal gas constant
R_c	Radius of the capillary.
t	Time
t_o	Initial Time
T	Temperature in Kelvin
T_g	Glass Transition temperature
$U_{1,2}$	Potential difference
V	Surface potential
V_m	Molar volume
W	Work
α	Ratio of the volume of the component to the volume of mixture
δ_s	Solubility parameter of solvent
δ_p	Solubility parameter of polymer
ε	Permittivity of material
ε_p	Absolute permittivity of the space between the charges
ε_r	Dielectric constant or relative permittivity
$\dot{\gamma}$, ds/dt	Rate of shear
η	Viscosity
$[\eta]$	Intrinsic viscosity
λ	Elongation
θ	Flory-Huggins temperature

ρ	Density
σ	Stress
σ_e	Surface charge density
τ	Shear stress
τ_r	Relaxation time
v	Velocity
v_s	Partial volume of solvent
v_p	Partial volume of polymer
ψ	Torque
ΔH	Heat of mixing or enthalpy
ΔH_v	Enthalpy of evaporation
$\dfrac{\Delta p}{L}$	Axial pressure due to friction
Φ, Φ_1, Φ_2	Potential
Λ	Heat of vaporization

Chapter 3

Electrospinning Process

In electrospinning, a high voltage is applied to a polymer fluid such that charges are induced within the fluid. When charges within the fluid reached a critical amount, a fluid jet will erupt from the droplet at the tip of the needle resulting in the formation of a Taylor cone. The electrospinning jet will travel towards the region of lower potential, which in most cases, is a grounded collector. There are many parameters that will influence the morphology of the resultant electrospun fibers, from beaded fibers to fibers with pores on its surface. Although electrospinning has been carried out on molten polymer [Lyons et. al. (2004); Larrondo and Manley (1981a); Larrondo and Manley (1981b); Larrondo and Manley (1981c)], most electrospinning are carried out using polymer solution. Thus, the parameters affecting electrospinning of polymer solution is of greater interest and will be discussed in greater details in this chapter. The parameters affecting electrospinning and the fibers may be broadly classified into polymer solution parameters, processing conditions which include the applied voltage, temperature and effect of collector, and ambient conditions. With the understanding of these parameters, it is possible to come out with setups to yield fibrous structures of various forms and arrangements. It is also possible to create nanofiber with different morphology by varying the parameters.

3.1. Polymer Solution Parameters

The properties of the polymer solution have the most significant influence in the electrospinning process and the resultant fiber morphology. The surface tension has a part to play in the formation of beads along the fiber length. The viscosity of the solution and its electrical properties will determine the extent of elongation of the solution. This will in turn have an effect on the diameter of the resultant electrospun fibers.

3.1.1. Molecular Weight and Solution Viscosity

As mentioned in Chapter 2, one of the factors that affect the viscosity of the solution is the molecular weight of the polymer. Generally, when a polymer of higher molecular weight is dissolved in a solvent, its viscosity will be higher than solution of the same polymer but of a lower molecular weight. One of the conditions necessary for electrospinning to occur where fibers are formed is that the solution must consists of polymer of sufficient molecular weight and the solution must be of sufficient viscosity. As the jet leaves the needle tip during electrospinning, the polymer solution is stretched as it travels towards the collection plate. During the stretching of the polymer solution, it is the entanglement of the molecule chains that prevents the electrically driven jet from breaking up thus maintaining a continuous solution jet. As a result, monomeric polymer solution does not form fibers when electrospun [Buchko et. al. (1999)].

The molecular weight of the polymer represents the length of the polymer chain, which in turn have an effect on the viscosity of the solution since the polymer length will determine the amount of entanglement of the polymer chains in the solvent. Another way to increase the viscosity of the solution is to increase the polymer concentration. Similar to increasing the molecular weight, an increased in the concentration will result in greater polymer chain entanglements within the solution which is necessary to maintain the continuity of the jet during electrospinning.

The polymer chain entanglements were found to have a significant impact on whether the electrospinning jet breaks up into small droplets or whether resultant electrospun fibers contain beads [Shenoy et. al. (2005)]. Although a minimum amount of polymer chain entanglements and thus, viscosity is necessary for electrospinning, a viscosity that is too high will make it very difficult to pump the solution through the syringe needle [Kameoka et. al. (2003)]. Moreover, when the viscosity is too high, the solution may dries at the tip of the needle before electrospinning can be initiated [Zhong et. al. (2002)].

Many experiments have shown that a minimum viscosity for each polymer solution is required to yield fibers without beads [Megelski et. al. (2002); Fong et. al. (1999)]. At a low viscosity, it is common to find beads along the fibers deposited on the collection plate. When the viscosity increases, there is a gradual change in the shape of the beads from spherical to spindle-like until a smooth fiber is obtained [Fong et. al. (1999); Mit-uppatham et. al. (2004)] as shown in Fig. 3.1. At a lower viscosity, the higher amount of solvent molecules and fewer chain entanglements will mean that surface tension has a dominant influence along the electrospinning jet causing beads to form along the fiber. When the viscosity is increased which means that there is a higher amount of polymer chains entanglement in the solution, the charges on the electrospinning jet will be able to fully stretch the solution with the solvent molecules distributed among the polymer chains. With increased viscosity, the diameter of the fiber also increases [Baumgarten (1971); Jarusuwannapoom et. al. (2005); Demir et. al. (2002); Deitzel et. al. (2001b); Megelski et. al. (2002)]. This is probably due to the greater resistance of the solution to be stretched by the charges on the jet [Jarusuwannapoom et. al. (2005)]. Table 3.1 shows the effect of concentration on beads formation.

Table 3.1 (a). The boundary condition of polymer concentration between uniform and beaded nanofibers.

Polymer	Molecular Weight	Solvent	Boundary Condition of Polymer Concentration	Other Spinning Parameters
Poly(ethylene oxide), **PEO** [Fong et. al. (1999)]	Mw = 900,000	Water	Uniform Fibers: 3.8~4.3 wt% Beaded Fibers: 1~3 wt%	Voltage: 15 kV Distance: 21.5 cm
Poly(ε-caprolactone), **PCL** [Lee et. al. (2003b)]	Mn = 80,000	Dichloromethane (MC)/ Dimethylformamide (DMF)	Uniform Fibers: 13 wt% Beaded Fibers: 10 wt%	----------
Poly(D,L-lactic acid), **PDLA** [Zong et. al. (2002)]	----------	Dimethylformamide	Uniform Fibers: 35 wt% Beaded Fibers: 20~30 wt%	Voltage: 20 kV Distance: 15 cm Feed Rate: 1.2 ml/h
Poly (p-dioxanone-co-L-lactide)-block-poly (ethylene glycolide) [Bhattarai et. al. (2003)]	Mw = 42,000	Dichloromethane	Uniform Fibers: 20 wt% Beaded Fibers: 15 wt%	Voltage: 15 kV Distance: 11 cm
Bombyx mori silk [Sukigara et. al. (2003)]	----------	50% aqueous Calcium Chloride	Uniform Fibers: 15~20 wt% Beaded Fibers: 8~12 wt%	Voltage: 28 kV Distance: 7 cm
Dextran [Jiang et. al. (2004)]	Mw = 64,000 ~ 76,000	Water	Uniform Fibers: 0.75~1.0 g/ml Beaded Fibers: 0.5 ~0.65 g/ml	Voltage: 25 kV Distance: 15 cm Feed Rate: 1.2 ml/h

Mw: Weight-average molecular weight, Mn: Number-average molecular weight

Table 3.1 (b). The boundary condition of polymer concentration between uniform and beaded nanofibers.

Polymer	Molecular Weight	Solvent	Boundary Condition of Polymer Concentration	Other Spinning Parameters
Poly(trimethylene terephthalate), PTT, [Khil et. al. (2004)]	--------	Trifluoroacetic Acid (TFA) / MC (50:50)	Uniform Fibers: 13~16 wt% Beaded Fibers: 5~10 wt%	Voltage: 13 kV Distance: 13 cm
Poly(ethlene terephthalate-co-ethylene isophtalate), PET-co-PEI [McKee et. al. (2003)]	Mw = 76,000	Chloroform / DMF (70:30)	Uniform Fibers: 20 wt% Beaded Fibers: 16~18 wt%	Voltage: 18 kV Distance: 24 cm Feed Rate: 3 ml/h
Poly(3-hydroxybutyrate-co-hydroxyvalerate), PHBV, [Choi et. al. (2004a)]	Mw = 680,000	Chloroform	Uniform Fibers: 20 wt% Beaded Fibers: 13~17 wt%	Voltage: 15 kV Distance: 15 cm Feed Rate: 5 ml/h
Polyamide-6, PA-6 [Mit-uppatham et. al. (2004)]	17,000 Da	Formic Acid	Uniform Fibers: 32~46 w/v% Beaded Fibers: 20~24 w/v%	Voltage: 21 kV Distance: 10 cm
Poly(vinyl chloride), PVC [Lee et. al. (2002)]	--------	THF / DMF (50:50)	Uniform Fibers: 15 wt% Beaded Fibers: 10 & 13 wt%	Voltage: 15 kV Distance: 10 cm
Polystylene, PS [Lee et. al. (2003c)]	Mw = 140,000	THF / DMF (50:50)	Uniform Fibers: 15 wt% Beaded Fibers: 7~13 wt%	Voltage: 15 kV Distance: 12 cm
Poly(methyl methacrylate, PMMA [McKee et. al. (2004)]	Mw = 210,000	Chloroform / DMF (40:60)	Uniform Fibers: 8~10 wt% Beaded Fibers: 5~7 wt%	Voltage: 22 kV Distance: 24 cm Feed Rate: 6 ml/h
Poly(methyl methacrylate-co-SCMHB methacrylic acid), PMMA-co-SCMHB [McKee et. al. (2004)]	Mw = 183,000	Chloroform / DMF (40:60)	Uniform Fibers: 8.5~9 wt% Beaded Fibers: 4~7 wt%	

Da (Dalton): A unit of mass that equals the weight of a hydrogen atom, or 1.657 x 10-24 grams

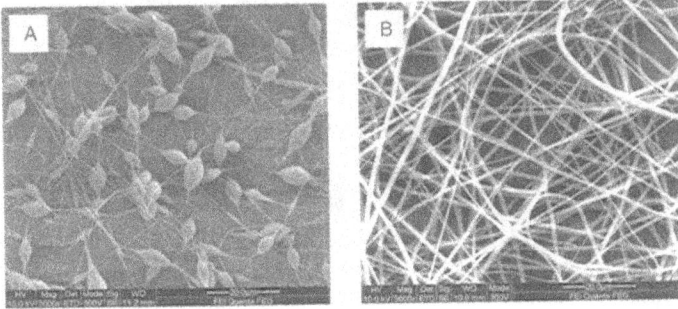

Fig. 3.1. Polycaprolactone electrospun fibers with [A] beads at concentration at 0.1g/ml and [B] beadless fibers at 0.12g/ml. [Courtesy of Teo and Ramakrishna, National University of Singapore].

The interaction between the solution and the charges on the jet will determine the distribution of the fiber diameters obtained. During electrospinning, there may be secondary jet erupting from the main electrospinning jet [Reneker et. al. (2000)] which is stable enough to yield fibers of smaller diameter at certain viscosity. This may explain the differential fiber diameter distribution observed in some cases [Kim et. al. (2005); Demir et. al. (2002); Deitzel et. al. (2001b)]. However, when the viscosity is high enough, it may discourage secondary jets from breaking off from the main jet which may contribute to the increased fiber diameter [Zhao et. al. (2004)].

Another effect of higher concentration is seen by a smaller deposition area. Increased concentration means that the viscosity of the solution is strong enough to discourage the bending instability to set in for a longer distance as it emerges from the tip of the needle. As a result, the jet path is reduced and the bending instability spreads over a smaller area [Mituppatham et. al. (2004)]. This reduced jet path also means that there is less stretching of the solution resulting in a larger fiber diameter.

Although viscosity has an important role in the formation of smooth fibers, it may not determine the concentration at which fibers are formed during electrospinning.

For Polyethylene Oxide, it was found that the minimum concentration for smooth fiber formation was the same despite a 3.5-fold increased in viscosity when the molecular weight of the polymer was increased from 8K to 8M. Thus in this case, it seems that concentration had a greater effect on the formation of smooth fibers [Morozov et. al. (1998)].

3.1.2. Surface Tension

The initiation of electrospinning requires the charged solution to overcome its surface tension. However, as the jet travels towards the collection plate, the surface tension may cause the formation of beads along the jet (See Table 3.2). Surface tension has the effect of decreasing the surface area per unit mass of a fluid. In this case, when there is a high concentration of free solvent molecules, there is a greater tendency for the solvent molecules to congregate and adopt a spherical shape due to surface tension. A higher viscosity will means that there is greater interaction between the solvent and polymer molecules thus when the solution is stretched under the influence of the charges, the solvent molecules will tend to spread over the entangled polymer molecules thus reducing the tendency for the solvent molecules to come together under the influence of surface tension as shown in Fig. 3.2.

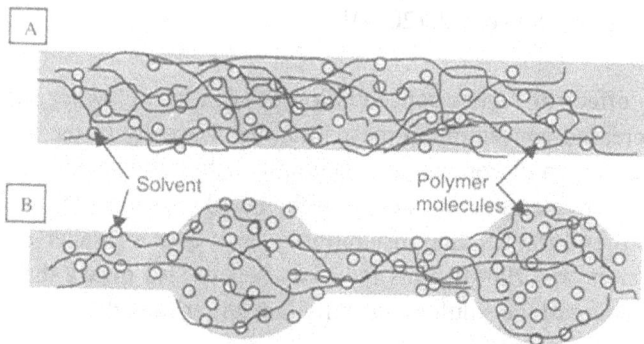

Fig. 3.2. [A] At high viscosity, the solvent molecules are distributed over the entangled polymer molecules. [B] With a lower viscosity, the solvent molecules tend to congregate under the action of surface tension.

Table 3.2.   The boundary condition of viscosity and surface tension with solvent composition between uniform and beaded nanofibers.

Polymer	Molecular Weight	Solvent	Boundary Condition of Viscosity & Surface Tension	Other Spinning Parameters
Poly(ethylene oxide), **PEO** [Fong et al. (1999)]	Mw = 900,000	Water/Ethanol (100:0 ~ 59:41)	Uniform Fibers: Water:Ethanol (69:31~59:41) Viscosity (1129~1179 cPs) Surface Tension (55~51 mN/m) Beaded Fibers: Water:Ethanol (100:0~79:21) Viscosity (402~889 cPs) Surface Tension (76~60 mN/m)	Polymer Concentration: 3 wt% Voltage: 11 kV Distance: 21.5 cm
Poly (p-dioxanone-co-L-lactide)-block-poly (ethylene glycolide), [Bhattarai et. al. (2003)]	Mw = 42,000	Dichloromethane (DM)/DMF (100:0 ~ 40:60)	Uniform Fibers: DM:DMF (100:0, 75:25) Viscosity (39, 35 mN/m) Surface Tension (125, 175 cPs) Beaded Fibers: DM:DMF (40:60) Viscosity (37 mN/m) Surface Tension (170 cPs)	Polymer Concentration: 20 wt% Voltage: 15 kV Distance: 11 cm
Polystylene, PS [Lee et. al. (2003c)]	Mw = 140,000	Tetrahydrofuran (THF)/DMF (100:0 ~ 0:100)	Uniform Fibers: THF:DMF (50:50) Surface Tension (45 cPs) Beaded Fibers: THF:DMF(100:0, 0:100) Surface Tension (42, 48cPs)	Polymer Concentration: 13 wt% Voltage: 15 kV Distance: 12 cm

Solvent such as ethanol has a low surface tension thus it can be added to encourage the formation of smooth fibers [Fong et. al. (1999)]. Another way to reduce the surface tension is to add surfactant to the solution. The addition of surfactant was found to yield more uniform fibers. Even when insoluble surfactant is dispersed in a solution as fine powders, the fiber morphology is also improved [Zeng et. al. (2003)].

3.1.3. *Solution Conductivity*

Electrospinning involves stretching of the solution caused by repulsion of the charges at its surface. Thus if the conductivity of the solution is increased, more charges can be carried by the electrospinning jet. The conductivity of the solution can be increased by the addition of ions. Moreover, most drugs and proteins form ions when dissolved in water. As previously mentioned, beads formation will occur if the solution is not fully stretched. Therefore, when a small amount of salt or polyelectrolyte is added to the solution, the increased charges carried by the solution will increase the stretching of the solution. As a result, smooth fibers are formed which may otherwise yield beaded fibers. The increased in the stretching of the solution also will tend to yield fibers of smaller diameter [Zhong et. al. (2002)]. However, there is a limit to the reduction in the fiber diameter. As the solution is being stretched, there will be a greater viscoelastic force acting against the columbic forces of the charges [Choi et. al. (2004a)]. Table 3.3 shows the effect of the addition of salt to the solution.

Since the presence of ions increases the conductivity of the solution, the critical voltage for electrospinning to occur is also reduced [Son et. al. (2004c)]. Another effect of the increased charges is that it results in a greater bending instability. As a result, the deposition area of the fibers is increased [Choi et. al. (2004a)]. This will also favor the formation of finer fibers since the jet path is now increased.

Table 3.3. The boundary condition of electrical conductivity between uniform and beaded nanofibers.

Polymer	Molecular Weight	Solvent	Boundary Condition of Electrical Conductivity	Other Spinning Parameters
Poly(ethylene oxide), PEO [Fong et. al. (1999)]	$Mw = 900,000$	Water (NaCl addition)	Uniform Fibers: NaCl addition (0.3~1.5 wt%) Solution Resistivity (1.9~0.462 Wm) Beaded Fibers: NaCl addition (0.0015~0.15 wt%) Solution Resistivity (83.4~3.61 Wm)	Polymer Concentration: 3 wt% Voltage: 15 kV Distance: 21.5 cm
Collagen-Polyethyleneoxide, Collagen-PEO [Huang et. al. (2001a)]	900 kDa	Hydrogen Chloride (NaCl addition)	Uniform Fibers: NaCl addition (68 mmol/l) Solution Conductivity(--- ms) Beaded Fibers: NaCl addition (15~34 mmol/l) Solution Conductivity(1.5 ~2.3 ms)	Polymer Concentration: 2 wt% Voltage: 18 kV Distance: 15 cm Feed Rate: 6 ml/h
Poly(D,L-lactic acid), PDLA [Zong et. al. (2002)]	-------	DMF (1wt% KH_2PO_4, NaH_2PO_4 and NaCl addition)	Uniform Fibers: KH_2PO_4 addition (AFD = 1000 nm) NaH_2PO_4 addition (AFD = 330 nm) NaCl addition (AFD = 210nm)	Polymer Concentration: 30 wt% Voltage: 20 kV Distance: 15 cm Feed Rate: 1.2 ml/h

Although organic solvents are known to be non-conductive, many of them do have a certain level of conductivity as mentioned in Chapter 2. Solution prepared using solvents of higher conductivity generally yield fibers without beads while no fibers are formed if the solution has zero conductivity [Jarusuwannapoom et. al. (2005)]. Table 3.4 shows the conductivity of some solvents commonly used in electrospinning.

Table 3.4. Electrical conductivity of solvents.

Solvent	Conductivity (mS/m)	Reference
1,2-Dichloroethane	0.034	Jarusuwannapoom et. al. (2005)
Acetone	0.0202	Theron et. al. (2004)
Butanol	0.0036	Prego et. al. (2000)
Dichloromethane/ Dimethylformamide (40/60)	0.505	Theron et. al. (2004)
Dichloromethane/ Dimethylformamide (75/25)	0.273	Theron et. al. (2004)
Dimethylformamide	1.090	Jarusuwannapoom et. al. (2005)
Distilled Water	0.447	Theron et. al. (2004)
Ethanol	0.0554	Prego et. al. (2000)
Ethanol (95%)	0.0624	Theron et. al. (2004)
Ethanol/Water (40/60)	0.150	Theron et. al. (2004)
Methanol	0.1207	Prego et. al. (2000)
Propanol	0.0385	Prego et. al. (2000)
Tetrahydrofuran/Ethanol (50/50)	0.037	Theron et. al. (2004)

The size of the ions may have an influence in the fiber morphology. Electrospun fibers from a solution with dissolved NaCl was found to have the smallest diameter while fibers from a solution with dissolved KH_2PO_4 had the largest diameter and fibers electrospun from solution with NaH_2PO_4 dissolved had intermediate diameter. As sodium and chloride ions have a smaller atomic radius than potassium and phosphate ions, they may have a greater mobility under an external electrostatic field. As a result, the greater elongational force on the electrospinning jet caused by the more mobile smaller ions could yield fibers with smaller diameter [Zhong et. al. (2002)].

To increase the conductivity of the solution at the same time reducing the surface tension, ionic surfactant [Lin et. al. (2004)] such as triethyl benzyl ammonium chloride [Zeng et. al. (2003)] can be added. This was found to cause a reduction in fiber diameter. Another way to increase the conductivity of the solution is by changing the pH of the solution. Under a basic condition, electrospinning cellulose acetate (CA) solution results in a significant reduction in fiber diameter compared to those obtained in a neutral condition [Son et. al. (2004c)]. Since CA will undergo deacetylation under basic condition, OH$^-$ ions may be able to exert a greater influence in the conduction and the stretching of the solution.

However, the interaction between the reagent that is added to improve the conductivity of the solution and the original solution may also have a significant impact on the resultant fiber. When CA was dissolved in acidic condition, the electrospun fiber had a slight increased in fiber diameter although the conductivity of the solution was improved [Son et. al. (2004c)]. In some cases, the addition of ionic salt may cause an increase in the viscosity of the solution. Thus although the conductivity of the solution is improved, the viscoelastic force is stronger than the columbic force resulting in an increased in the fiber diameter instead [Mit-uppatham et. al. (2004)].

3.1.4. Dielectric Effect of Solvent

The dielectric constant of a solvent has a significant influence on electrospinning. Generally, a solution with a greater dielectric property reduces the beads formation and the diameter of the resultant electrospun fiber [Son et. al. (2004a)]. Solvents such as N,N-Dimethylformamide (DMF) may added to a solution to increase its dielectric property to improve the fiber morphology [Lee et. al. (2003b)]. The bending instability of the electrospinning jet also increases with higher dielectric constant. This is shown by increased deposition area of the fibers. This may also facilitate the reduction of the fiber diameter due to the increased jet path [Hsu et. al. (2004)]. The dielectric constant of some common solvents used in electrospinning is shown in Table 3.5.

However, if a solvent of a higher dielectric constant is added to a solution to improve the electrospinnability of the solution, the interaction between the mixtures such as the solubility of the polymer will also have an impact on the morphology of the resultant fibers. When DMF is added to polystyrene (PS) solution, beads are formed even though electrospinnability should improve due to the higher dielectric constant of DMF. This could be the result of the retraction of PS molecule due to poor interaction between PS and the solvent molecules [Wannatong et. al. (2004)].

Table 3.5. Dielectric constant of solvents.

Solvent	Dielectric constant	Reference
2-Propanol	18.3	MERCK technical data sheet
Acetic acid	6.15	Wannatong et. al. (2004)
Acetone	20.7	Berkland et. al. (2004)
Acetonitrile	35.92-37.06	Wannatong et. al. (2004)
Chloroform	4.8	Berkland et. al. (2004)
Dichloromethane	8.93	Yang et. al. (2004c)
Dimethylformamide	36.71	Yang et. al. (2004c)
Ethyl acetate	6.0	Berkland et. al. (2004)
Ethanol	24.55	Yang et. al. (2004c)
m-Cresol	11.8	Wannatong et. al. (2004)
Methanol	32.6	MERCK technical data sheet
Pyridine	12.3	MERCK technical data sheet
Tetrahydrofuran	7.47	Wannatong et. al. (2004)
Toluene	2.438	Wannatong et. al. (2004)
Trifluoroethanol	27.0	Berkland et. al. (2004)
Water	80.2	MERCK technical data sheet

3.2. Processing Conditions

Another important parameter that affects the electrospinning process is the various external factors exerting on the electrospinning jet. This includes the voltage supplied, the feedrate, temperature of the solution, type of collector, diameter of needle and distance between the needle tip and collector. These parameters have a certain influence in the fiber morphology although they are less significant than the solution parameters.

3.2.1. *Voltage*

A crucial element in electrospinning is the application of a high voltage to the solution.. The high voltage will induce the necessary charges on the solution and together with the external electric field, will initiate the electrospinning process when the electrostatic force in the solution overcomes the surface tension of the solution. Generally, both high negative or positive voltage of more than 6kV is able to cause the solution drop at the tip of the needle to distort into the shape of a Taylor Cone during jet initiation [Taylor (1964)]. Depending on the feedrate of the solution, a higher voltage may be required so that the Taylor Cone is stable. The columbic repulsive force in the jet will then stretch the viscoelastic solution. If the applied voltage is higher, the greater amount of charges will cause the jet to accelerate faster and more volume of solution will be drawn from the tip of the needle. This may result in a smaller and less stable Taylor Cone [Zhong et. al. (2002)]. When the drawing of the solution to the collection plate is faster than the supply from the source, the Taylor Cone may recede into the needle [Deitzel et. al. (2001b)].

As both the voltage supplied and the resultant electric field have an influence in the stretching and the acceleration of the jet, they will have an influence on the morphology of the fibers obtained. In most cases, a higher voltage will lead to greater stretching of the solution due to the greater columbic forces in the jet as well as the stronger electric field. These have the effect of reducing the diameter of the fibers [Lee et. al. (2004); Buchko et. al. (1999); Megelski et. al. (2002)] and also encourage faster solvent evaporation to yield drier fibers [Pawlowski et. al. (2005)]. When a solution of lower viscosity is used, a higher voltage may favor the formation of secondary jets during electrospinning. This has the effect of reducing the fiber diameter [Demir et. al. (2002)]. Another factor that may influence the diameter of the fiber is the flight time of the electrospinning jet. A longer flight time will allow more time for the fibers to stretch and elongates before it is deposited on the collection plate. Thus, at a lower voltage, the reduced acceleration of the jet and the weaker electric field may increase the flight time of the electrospinning jet which may favor the formation of finer fibers. In this case, a voltage close to the critical voltage for electrospinning may be favorable to obtain finer fibers [Zhao et. al. (2004)].

At a higher voltage, it was found that there is a greater tendency for beads formation [Deitzel et. al. (2001b); Demir et. al. (2002); Zhong et. al. (2002)]. It was also reported that the shape of the beads changes from spindle-like to spherical-like with increasing voltage [Zhong et. al. (2002)]. Given the increased stretching of the jet due to higher voltage, there should be less beads formation as reported in some cases [Jarusuwannapoom et. al. (2005)] as shown in Fig. 3.3. The increased in beads density due to increased voltage may be the result of increased instability of the jet as the Taylor Cone recedes into the syringe needle [Deitzel et. al. (2001b); Zhong et. al. (2002)]. In an interesting observation, [Krishnappa et. al. (2003)] reported that increasing voltage will increased the beads density, which at an even higher voltage, the beads will join to form a thicker diameter fiber.

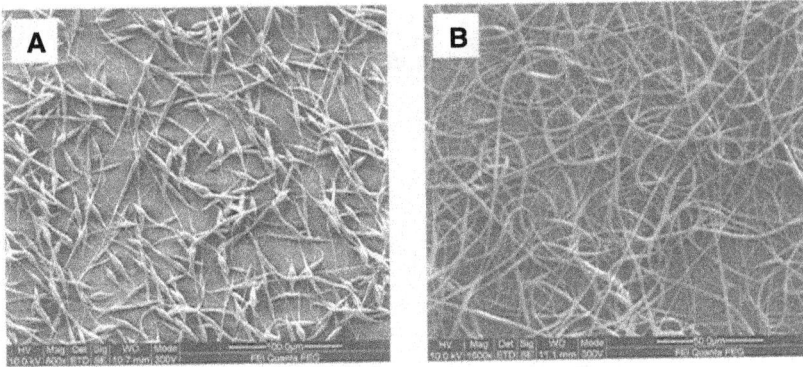

Fig. 3.3. Polycaprolactone fibers with [A] beads for electrospinning at a voltage of 6kV and [B] beadless fibers at 22kV. [Courtesy of W.E. Teo and S. Ramakrishna, National University of Singapore].

The effect of high voltage is not only on the physical appearance of the fiber, it also affects the crystallinity of the polymer fiber. The electrostatic field may cause the polymer molecules to be more ordered during electrospinning thus induces a greater crystallinity in the fiber. However, above a certain voltage, the crystallinity of the fiber is reduced. With increased voltage, the acceleration of the fibers also increases. This reduces the flight time of the electrospinning jet. Since the orientation of the polymer molecules will take some time, the reduced flight time means that the fibers will be deposited before the polymer molecules have sufficient time to align itself. Thus, given sufficient flight time, the crystallinity of the fiber will improve with higher voltage [Zhao et. al. (2004)].

Since electrospinning is caused by charges on the jet, these charges can be influenced by the external electric field which will in turn affect the jet path. It is thus not surprising that there are several attempts to control the electrospinning jet through changing the electric field profile between the source of the electrospinning jet and the collector.

This can be achieved by using auxiliary electrodes or by changing the orientation or shape of the collector. Aligned and even patterned nanofibers can be obtained by clever manipulation of the electric field. More detailed discussions to obtain various fiber meshes are discussed in Section 3.8

While DC voltage supply is most commonly used in electrospinning, it is also possible to use AC potential for electrospinning. The jet initiation, stretching and bending instability is caused by charges present in the solution. The charging of the solution is very rapid and jet initiation will occur before the voltage alternates in an AC supply. As the jet travels towards the collection plate, regular segments of the jet will contain positive or negative charges on it. Since bending instability is the result of columbic repulsive forces in the jet, the regular segments in the jet of either positive or negative voltage will reduces the repulsive forces thus reducing the bending instability in the jet. Since there is less bending instability and less stretching of the jet, the resultant fibers have a higher diameter than fibers that are formed by DC supply of the same voltage. Another advantage of an AC supply is that there is fewer tendencies for accumulation of like-charges on the fiber after it has been deposited. Thus, a thicker layer of electrospun fiber can be collected especially when an insulating collection plate is used [Kessick et. al. (2004)].

3.2.2. Feedrate

The feedrate will determine the amount of solution available for electrospinning. For a given voltage, there is a corresponding feedrate if a stable Taylor cone is to be maintained. When the feedrate is increased, there is a corresponding increase in the fiber diameter or beads size as shown in Fig. 3.4. This is apparent as there is a greater volume of solution that is drawn away from the needle tip [Zhong et. al. (2002); Rutledge et. al. (2001)].

However, there is a limit to the increase in the diameter of the fiber due to higher feedrate [Rutledge et. al. (2001)]. If the feedrate is at the same rate which the solution is carried away by the electrospinning jet, there must be a corresponding increased in charges when the feedrate is increased. Thus there is a corresponding increased in the stretching of the solution which counters the increased diameter due to increased volume.

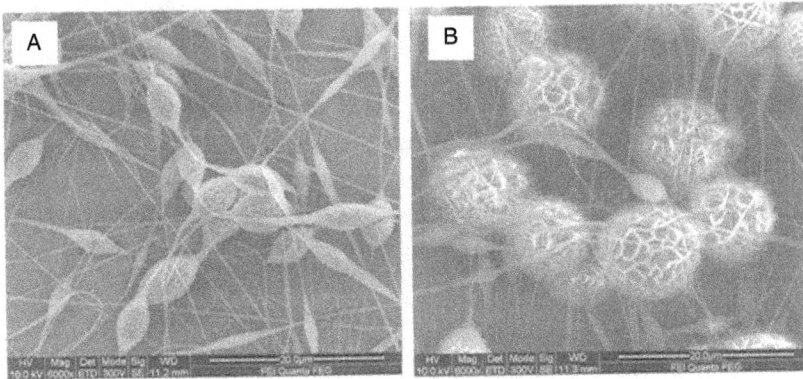

Fig. 3.4.   Polycaprolactone fibers with increasing beads size with increasing feedrate at [A] 0.5ml/hr and [B] 2ml/hr. [Courtesy of Teo and Ramakrishna, National University of Singapore].

Due to the greater volume of solution drawn from the needle tip, the jet will takes a longer time to dry. As a result, the solvents in the deposited fibers may not have enough time to evaporate given the same flight time. The residual solvents may cause the fibers to fuse together where they make contact forming webs. A lower feedrate is more desirable as the solvent will have more time for evaporation [Yuan et. al. (2004)].

3.2.3. *Temperature*

The temperature of the solution has both the effect of increasing its evaporation rate and reducing the viscosity of the polymer solution. When polyurethane is electrospun at a higher temperature, the fibers produced have a more uniform diameter [Demir et. al. (2002)]. This may be due to the lower viscosity of the solution and greater solubility of the polymer in the solvent which allows more even stretching of the solution. With a lower viscosity, the Columbic forces are able to exert a greater stretching force on the solution thus resulting in fibers of smaller diameter [Mit-uppatham et. al. (2004)]. Increased polymer molecules mobility due to increased temperature also allows the Columbic force to stretch the solution further. However, in cases where biological substances such as enzymes and proteins are added to the solution for electrospinning, the use of high temperature may cause the substance to lose its functionality.

3.2.4. *Effect of Collector*

There must be an electric field between the source and the collector for electrospinning to initiate. Thus in most electrospinning setup, the collector plate is made out of conductive material such as aluminum foil which is electrically grounded so that there is a stable potential difference between the source and the collector. In the case when a non-conducting material is used as a collector, charges on the electrospinning jet will quickly accumulates on the collector which will result in fewer fibers deposited [Kessick et. al. (2004); Liu and Hsieh (2002)]. Fibers that are collected on the non-conducting material usually have a lower packing density compared to those collected on a conducting surface. This is caused by the repulsive forces of the accumulated charges on the collector as more fibers are deposited. For a conducting collector, charges on the fibers are dissipated thus allowing more fibers to be attracted to the collector. The fibers are able to pack closely together as a result [Liu and Hsieh (2002)].

For a non-conducting collector, the accumulation of the charges may result in the formation of 3D fiber structures due to the repulsive forces of the like-charges. When there is a sufficient density of charges on the fiber mesh that is formed initially, repulsion on the subsequent fibers may result in the formation of honey comb structures [Deitzel et. al. (2001b)]. However, even for conductive collector, when the deposition rate is high and the fiber mesh is thick enough, there will also be a high accumulation of residual charges on the fiber mesh since polymer nanofibers are generally non-conductive. This may result in the formation of dimples on the fiber mesh as seen in Fig. 3.5.

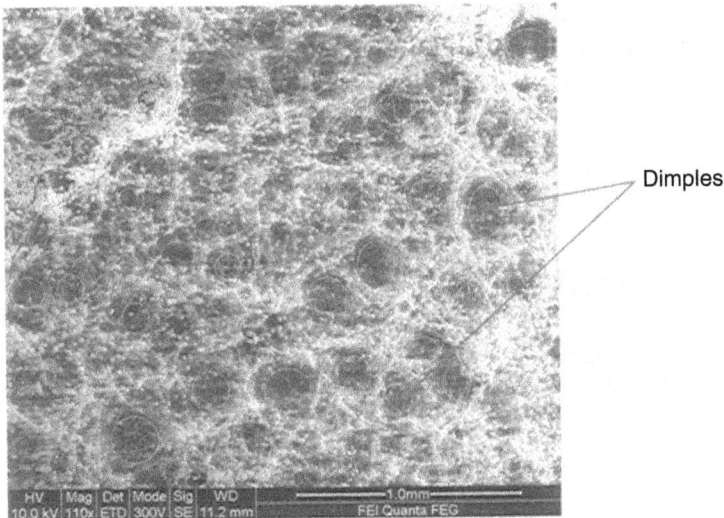

Fig. 3.5 Dimples formed by electrospun Polycaprolactone fibers at a high feedrate of 8mL/hr. [Courtesy of Teo and Ramakrishna, National University of Singapore].

The porosity of the collector seems to have an effect on the deposited fibers. Experiments with porous collector such as paper and metal mesh had shown that the fiber mesh collected had a lower packing density than smooth surfaces such as metal foils. This can be attributed to the diffusion and rate of evaporation of the residual solvents on the fibers collected. In a porous target, there is faster evaporation of residual fibers due to higher surface area while smooth surfaces may cause an accumulation of solvents around the fibers due to slow evaporation rate. Due to the wicking and diffusion of the residual solvents on the fibers, the fibers may be pulled together to give a more densely packed structure [Liu and Hsieh (2002)]. As the fibers dry faster on a porous collector, it is more likely that the residual charges remain on the fiber which will repel subsequent fibers. However, on a smooth surface, the residual solvents will encourage the residual charges to be conducted away to the collector.

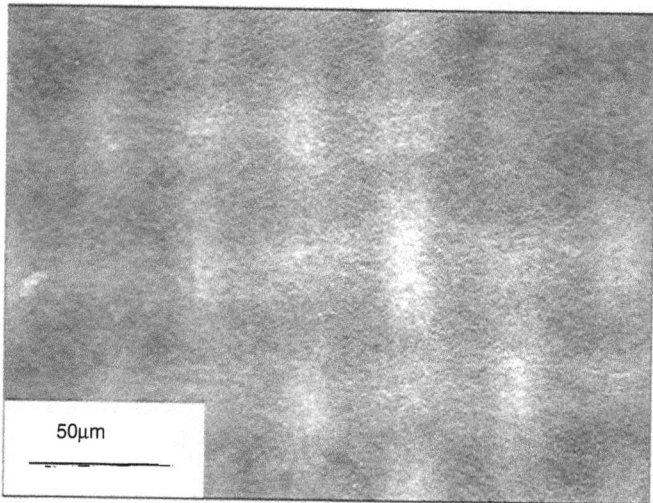

Fig. 3.6 Collagen collected on braided Teflon sheet. [Courtesy of Teo and Ramakrishna, National University of Singapore].

The texture of the fiber mesh may also be varied by using patterned collector as shown in Fig. 3.6. The fibers are deposited on the braided Teflon sheet and the resultant mesh has a topography that takes the pattern of the surface of the Teflon sheet.

Whether or not the collector is static or moving also have an effect on the electrospinning process. While rotating collector has been used to collect aligned fibers, which is covered in more details in section 3.8, it was found to assist in yielding fibers that are dry. This is useful because certain solvents such as Dimethylformamide (DMF) which is good for electrospinning but have a high boiling point that may result in the fibers being wet when they are collected. A rotating collector will give the solvent more time to evaporate [Wannatong et. al. (2004)] and also increase the rate of evaporation of the solvents on the fibers. This will improve the morphology of the fiber where distinct fibers are required.

3.2.5. *Diameter of Pipette Orifice / Needle*

The internal diameter of the needle or the pipette orifice has a certain effect on the electrospinning process. A smaller internal diameter was found to reduce the clogging as well as the amount of beads on the electrospun fibers [Mo et. al. (2004)]. The reduction in the clogging could be due to less exposure of the solution to the atmosphere during electrospinning. Decrease in the internal diameter of the orifice was also found to cause a reduction in the diameter of the electrospun fibers. When the size of the droplet at the tip of the orifice is decreased, such as in the case of a smaller internal diameter of the orifice, the surface tension of the droplet increases. For the same voltage supplied, a greater columbic force is required to cause jet initiation. As a result, the acceleration of the jet decreases and this allows more time for the solution to be stretched and elongated before it is collected. However, if the diameter of the orifice is too small, it may not be possible to extrude a droplet of solution at the tip of the orifice [Zhao et. al. (2004)].

3.2.6. *Distance between Tip and Collector*

In several cases, the flight time as well as the electric field strength will affect the electrospinning process and the resultant fibers. Varying the distance between the tip and the collector will have a direct influence in both the flight time and the electric field strength. For independent fibers to form, the electrospinning jet must be allowed time for most of the solvents to be evaporated. When the distance between the tip and the collector is reduced, the jet will have a shorter distance to travel before it reaches the collector plate. Moreover, the electric field strength will also increase at the same time and this will increase the acceleration of the jet to the collector. As a result, there may not have enough time for the solvents to evaporate when it hits the collector. When the distance is too low, excess solvent may cause the fibers to merge where they contact to form junctions resulting in inter and intra layer bonding as shown in Fig. 3.7 [Buchko et. al. (1999)]. This interconnected fiber mesh may provide additional strength to the resultant scaffold.

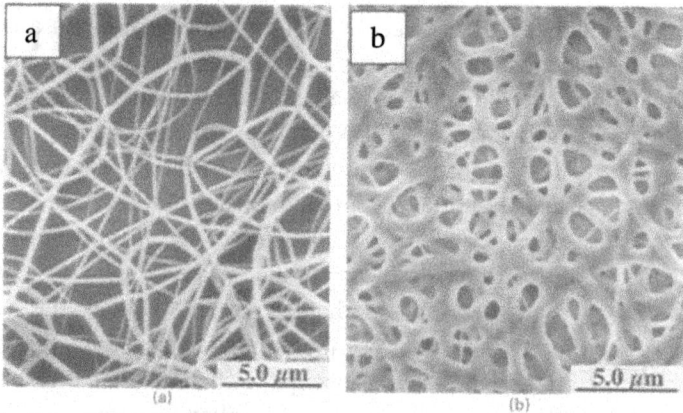

Fig. 3.7.   Nylon 6,6 at (a) 2 cm deposition distance and (b) 0.5cm deposition distance [Buchko et. al. (1999)].

Depending on the solution property, the effect of varying the distance may or may not have a significant effect on the fiber morphology. In some cases, changing the distance has no significant effect on the fiber diameter. However, beads were observed to form when distance was too low [Megelski et. al. (2002)]. The formation of beads may be the result of increased field strength between the needle tip and the collector. Decreasing the distance has the same effect as increasing the voltage supplied and this will cause an increased in the field strength. As mentioned earlier, if the field strength is too high, the increased instability of the jet may encourage beads formation [Deitzel et. al. (2001b); Zhong et. al. (2002)]. However, if the distance is such that the field strength is at an optimal value, there is less beads formed as the electrostatic field provides sufficient stretching force to the jet [Jarusuwannapoom et. al. (2005)]. In other circumstances, increasing the distance results in a decrease in the average fiber diameter [Ayutsede et. al. (2005)]. The longer distance means that there is a longer flight time for the solution to be stretched before it is deposited on the collector [Zhao et. al. (2004); Reneker et. al. (2000)]. However, there are cases where at a longer distance, the fiber diameter increases. This is due to the decrease in the electrostatic field strength resulting in less stretching of the fibers [Lee et. al. (2004)]. When the distance is too large, no fibers are deposited on the collector [Zhao et. al. (2004)]. Therefore, it seems that there is an optimal electrostatic field strength below which the stretching of the solution will decrease resulting in increased fiber diameters.

3.3. Ambient Parameters

The effect of the electrospinning jet surrounding is one area which is still poorly investigated. Any interaction between the surrounding and the polymer solution may have an effect on the electrospun fiber morphology. High humidity for example was found to cause the formation of pores on the surface of the fibers. Since electrospinning is influenced by external electric field, any changes in the electrospinning environment will also affect the electrospinning process.

3.3.1. *Humidity*

The humidity of the electrospinning environment may have an influence in the polymer solution during electrospinning. At high humidity, it is likely that water condenses on the surface of the fiber when electrospinning is carried out under normal atmosphere. As a result, this may have an influence on the fiber morphology especially polymer dissolved in volatile solvents [Megelski et. al. (2002); Bognitzki et. al. (2001)]. Experiments using Polysulfone (PS) dissolved in Tetrahydrofuran (THF) shows that at humidity of less than 50%, the fiber surfaces are smooth. However, an increased in the humidity during electrospinning will cause circular pores to form on the fiber surfaces. The sizes of the circular pores increases with increasing humidity until they coalescence to form large, non-uniform shaped structures as shown in Fig 3.8. The depth of the pore also increases with increasing humidity as determined by atomic force microscopy. However, above certain humidity, the depth of pores, its diameter and numbers start to saturate [Casper et. al. (2004)].

In the study of thin films, formation of pores on the surface is attributed to breath figures [Srnivasarao et. al. (2001)]. This may also happen during electrospinning as water vapor may condense on the surface of the jet due to cooling of the surface of the jet as a result of rapid evaporation of the volatile solvent. Pores are created when both water and solvent eventually evaporates. Although pores formed on thin film due to breath figures are uniform, pores seen on electrospun fibers are not and this may be due to the dynamic condition of the electrospinning jet as compared to static condition where the effects of breath figures are seen on thin films [Megelski et. al. (2002); Casper et. al. (2004)].

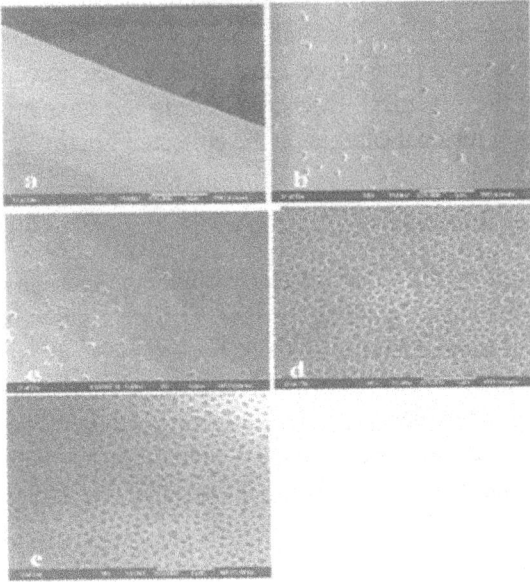

Fig. 3.8. FESEM micrographs of 190 000 g/mol Polysulfone/Tetrahydrofuran fibers electrospun under varying humidity: (a) <25%, (b) 31-38%, (c) 40-45%, (d) 50-29%, (e) 60-72% [Casper et. al. (2004)].

The humidity of the environment will also determine the rate of evaporation of the solvent in the solution. At a very low humidity, a volatile solvent may dries very rapidly. The evaporation of the solvent may be faster than the removal of the solvent from the tip of the needle. As a result, the electrospinning process may only be carried out for a few minutes before the needle tip is clogged [Baumgarten (1971)].

The effect of humidity on electrostatic charges on non-conducting surfaces has been widely studied. Studies on glass particles transported in a grounded copper pipe had found that at higher relative humidity (>76%), there were no charges on the particles. With decreasing humidity, there was an increased in the amount of charge on the particle [Nieh and Nguyen (1988)].

It has also been suggested that the high humidity can help the discharge of the electrospun fiber [Li and Xia (2004a); Li et. al. (2005a)]. However, more tests have to be carried out to determine the effect of humidity on the electrical discharge during electrospinning and the accumulation of residual charges on the collected fibers.

3.3.2. *Type of Atmosphere*

The composition of the air in the electrospinning environment will have an effect on the electrospinning process. Different gases have different behavior under high electrostatic field. For example, helium will breaks down under high electrostatic field and thus electrospinning is not possible. However, when a gas with higher breakdown voltage is used such as Freon®-12, the fibers obtained have twice the diameter of those electrospun in air given all other conditions equal [Baumgarten (1971)].

3.3.3. *Pressure*

Under enclosed condition, it is possible to investigate the effect of pressure on the electrospinning jet. Generally, reduction in the pressure surrounding the electrospinning jet does not improve the electrospinning process. When the pressure is below atmospheric pressure, the polymer solution in the syringe will have a greater tendency to flow out of the needle and there causes unstable jet initiation. As the pressure decreases, rapid bubbling of the solution will occur at the needle tip. At very low pressure, electrospinning is not possible due to direct discharge of the electrical charges.

3.4. Melt-Electrospinning

Although polymer solutions are more often used in electrospinning, there are also several investigations on electrospinning using polymer melts. Generally, most conditions that affect electrospinning of polymer solution are also applicable to electrospinning of molten polymer. Similar to electrospinning of polymer solution, polymers with higher molecular weight formed the largest diameter fibers. However, there are some differences between electrospinning of polymer melts and polymer solution.

In polymer melt, there must be a constant heat supplied to the reservoir containing the polymer solution for electrospinning so that the polymer will remains in its molten state. The distance between the tip of the needle and the collector for electrospinning of molten polymer is generally much closer (2cm) [Larrondo and Manley (1981); Lyons et. al. (2004)] than conventional electrospinning using polymer solution (>10cm). Depending on the temperature that maintains the polymer in its molten form, polymer melt is more viscous than polymer solution thus a greater charge is required for electrospinning jet initiation. With increasing field strength, it was found that the diameters of the resultant fibers were reduced [Lyons et. al. (2004)].

3.5. Creation of Different Nanofibers

Just by varying the parameters of electrospinning, it is possible to get some variation in the morphology of the nanofibers. Several parameters have been linked to the formation of beaded fibers as covered in the earlier part of the chapter. However, other than beaded and non-beaded fibers, electrospinning is able to produce other types of fibers with interesting morphology as well as different types of nanofibers.

3.5.1. Porous Nanofibers

As mentioned in the section 3.3.1, the humidity of the electrospinning environment plays a part in the formation of porous nanofibers. However, there are other factors that may contribute to the formation of pores on nanofibers during electrospinning.

Fig. 3.9 shows a unique nanometer scale porous structures that have been created on the surface of polymer fiber during the electrospinning process. It was found that the size of the surface structures ranges from a few tens of a nanometer to 1 micron, and is influenced by the type of polymers, solvent used, and electrospinning conditions, as listed in Table 3.6 (a) ~ (c). [Bognitzki et. al. (2001)] and [Megelski et. al. (2002)] pointed out that the mechanism that forms porous surface on polymer casting film is applicable to the phenomenon on electrospun nanofibers. When a polymer solution is cast on a support, convective evaporation takes place. During the event of solvent evaporation, the solution becomes thermodynamically unstable. This leads to the occurrence of phase separation into two phases: a polymer-rich phase and a polymer-deficient phase.

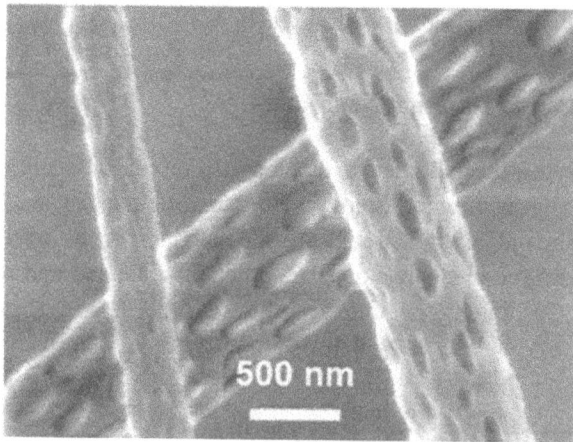

Fig. 3.9 PLLA porous nanofibers electrospun from a solution of PLLA in dichloromethane [Bognitzki et. al. (2001)].

Table 3.6 (a). Electrospun polymer fibers with porous surface structure.

Polymer	Molecular Weight	Solvent	Electrospinning Conditions	Pore Size (Fiber Diameter)	Reference
Polystyrene, PS	Mw = 190,000	Tetrahydrofuran (THF)	Polymer Concen.:18~35 wt% Voltage: 10kV Distance: 35cm Feed Rate: 4 ml/h	50 ~ 200 nm (5 ~ 15 µm)	Megelski et. al. (2002)
		Acetone / Cyclohexane	Polymer Concen.: 30 wt% Voltage: 10 kV Distance: 35 cm Feed Rate: 4 ml/h	300 ~ 1000 nm (15 ~ 30 µm)	
		THF / Dimethylformamide (DMF)	Polymer Concen.: 30 wt% Voltage: 10 kV Distance: 35 cm Feed Rate: 4 ml/h	> 50 nm (< 10 µm)	
		Carbon Disulfide	Polymer Concen.: 30 wt% Voltage: 20 kV Distance: 12 cm Feed Rate: 6 ml/h	100 ~ 750 nm (4 ~ 25 µm)	
	190,000	THF	Polymer Concen.: 35 wt% Voltage: 10 kV Distance: 35 cm Feed Rate: 4 ml/h	[Humidity] 60-190 nm [31-38%] 90-230 nm [40-45%] 50-270 nm [50-59%] 50-280 nm [66-72%]	Casper et. al. (2004)
	560,000			150-650 nm [31-38%] 150-600 nm [40-45%] 100-850 nm [50-59%] 200-1800 nm [66-72%]	

Table 3.6 (b). Electrospun polymer fibers with porous surface structure.

Polymer	Molecular Weight	Solvent	Electrospinning Conditions	Pore Size (Fiber Diameter)	Reference
Poly(methyl methacrylate), PMMA	Mw = 540,000	Tetrahydrofuran (THF)	Polymer Concen.: 10 wt% Voltage: 10kV Distance: 35cm Feed Rate: 4 ml/h	<1000 nm (10 ~ 35 μm)	Megelski et. al. (2002)
		Acetone	Polymer Concen.: 10 wt% Voltage: 10 kV Distance: 35 cm Feed Rate: 4 ml/h	<250 nm (<7.5 μm)	
		Chloroform	Polymer Concen.: 10 wt% Voltage: 10 kV Distance: 35 cm Feed Rate: 5 ml/h	75 ~ 250 nm (0.75 ~ 13 μm)	
Polycarbonate, PC	Mw = 60,000	Chloroform	Polymer Concen.: 20 wt% Voltage: 10 kV Distance: 35 cm Feed Rate: 5 ml/h	100 ~ 250 nm (10 ~ 12 μm)	
		Tetrahydrofuran (THF)	Polymer Concen.:20 wt% Voltage: 10kV Distance: 35cm Feed Rate: 4 ml/h	micropores (<20μm)	
	Mw = 1,100,000	Dichloromethane	Polymer Concen.: 15 wt%	200 nm (>3 μm)	Bognitzki et. al. (2001)

Table 3.6 (c). Electrospun polymer fibers with porous surface structure.

Polymer	Molecular Weight	Solvent	Electrospinning Conditions	Pore Size (Fiber Diameter)	Reference
Poly(L-lactic acid), PLLA	Mw = 150,000	Dichloromethane	Polymer Concen.: 5 wt%	100 ~ 250 nm (< 1µm)	Bognitzki et. al. (2001)
	---------	Dichloromethane	---------	< 250 nm (0.3 ~ 3.5 µm)	Caruso et. al. (2001)
Polyvinyl carbozole	Mw = 1,100,000	Dichloromethane	Polymer Concen.: 7.5 wt%	Regular pits with ellipsoidal shape	Bognitzki et. al. (2001)
Ethyl – Cyanoethyl Cellulose, (E-CE)C	Mn = 97,000	Tetrahydrofuran (THF)	Polymer Concen.: 17 wt% Voltage: 30kV Distance: 15cm	(200nm)	Zhao et. al. (2004)

Whilst the concentrated polymer-rich phase solidifies shortly after phase separation and forms the matrix, the polymer-deficient phase forms the pores. As such, the solvent vapor pressure has a critical influence on the process of pore formation. For the case of poly(L-lactic acid) (PLLA) porous nanofibers [Bognitzki et. al. (2001)], it was reported that the replacement of dichloromethane by chloroform with lower vapor pressure effectively reduce the formation of pores.

3.5.2. *Flattened or Ribbon-like Fibers*

As shown in Table 3.7, there have been reports of flattened or ribbon-like fibers made from several polymers under certain processing conditions. SEM photos of flattened poly(vinyl alcohol) (PVA) nanofibers and ribbon-like elastic-mimetic peptide polymer fibers are furnished in Fig. 3.10. The mechanism by which the flattened or ribbon-like fibers were obtained can be related to solvent evaporation during electrospinning process.

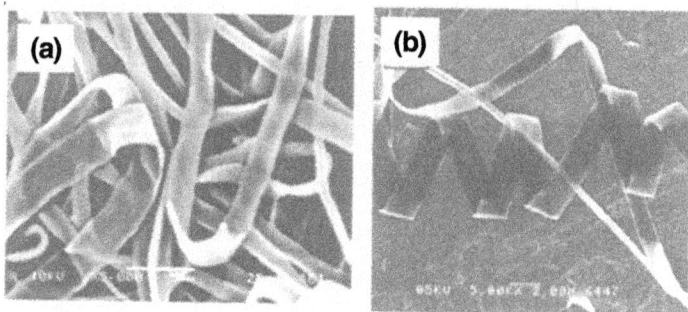

Fig. 3.10(a). Flattened Poly(vinyl alcohol) [Koski et. al. (2005)] and (b) Ribbon-like Elastin Peptide polymer [Huang et. al. (2000)] electrospun nanofibers.

For poly(vinyl alcohol) (PVA) nanofiber [Koski et. al. (2004)], flattened nanofibers were created with higher molecular weight and higher polymer concentration.

It was elucidated that water solvent evaporation reduces with higher solution viscosity and wet fibers that reach fiber collector are flattened by the impact. Fig. 3.11 illustrates the mechanism of forming ribbon-like fibers [Koombhongse et. al. (2001)].

Thin polymer skin was formed on the liquid jet as a result of solvent evaporation at the jet surface, thereby resulting in a thin layer of solid skin with a liquid core (Fig. 3.11 (a)). As a result of atmospheric pressure, the tube collapses in tandem with solvent evaporation from the core. During the course of tube collapse, the circular cross section initially became elliptical shape (Fig. 3.11 (b)), and thereafter into ribbon-like shape (Fig. 3.11 (c)).In some cases, small tubes formed at each edge of the ribbon and a web made from the skin connected the two tubes (Fig. 3.11 (d)).

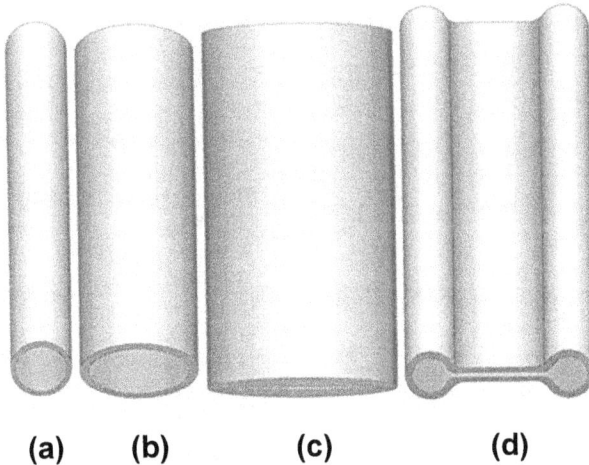

(a)   (b)   (c)   (d)

Fig. 3.11.   Mechanism of forming ribbon-like fibers
[Koombhongse et. al. (2001)].

Table 3.7.   Electrospun polymer fibers with flattened or ribbon-like morphology.

Polymer	Molecular Weight	Solvent	Electrospinning Conditions	References
Elastin-mimetic peptide polymer	81kDa	Water	Polymer concentration: 15 wt% Voltage: 18 kV Distance: 15 cm Feedrate : 3 ml/h	Nagapudi et. al. (2002)
Elastin-mimetic peptide polymer	81kDa	Water	Polymer concentration : 15~20 wt% Voltage : 18 kV Distance : 15 cm Feedrate : 3~12 ml/hr	Huang et. al. (2000)
Poly(vinyl alcohol), **PVA**	Mw = 13,000-23,000 Mw = 31,000-50,000 Mw = 50,000-89,000	Water	Polymer concentration : 31 wt% Polymer concentration : 25 wt% Polymer concentration : 13 & 17 wt% Voltage : 30 kV Distance : 10 cm	Koski et. al. (2004)
Poly (ether imide), **PEI**	-----------	Hexafluoro-2-propanol	Polymer concentration : 10 wt% Voltage : approx. 20kV Distance : 20 cm	Koombhongse et. al. (2001)
Polystyrene, **PS**	-----------	Dimethyl-formamide	Polymer concentration : 30 wt% Voltage : approx. 20kV Distance: 20 cm	Koombhongse et. al. (2001)
Polystyrene, **PS**	Mw = 190,000	Tetrahydrofuran	Polymer Concentration: 35 wt% Voltage: 10 kV Distance: 35 cm Feedrate : 4 ml/h	Megelski et. al. (2002)
Nylon 6	-----------	Hexafluoro-2-propanol	Polymer concentration : 10 wt% Voltage : 20 kV Distance : 25 cm	Fong et. al. (2002)

3.5.3. Branched Fibers

As shown in Fig. 3.12, branched fibers can be made by ejecting smaller jets from the surface of the primary jets, which is comparable to the ejection of an initial jet from the surface of a charged droplet. Split fibers are also obtained by the separation of a primary jet into two smaller jets. The elongation of the jet and evaporation of the solvent alter the shape and the charge per unit area carried by the jet. The balance between the electrical forces and surface tension can shift, thereby causing the shape of a jet to be unstable. Such instability can decrease its local charge per unit surface area by ejecting a smaller jet from the surface of the primary jet or by splitting apart into two smaller jets.

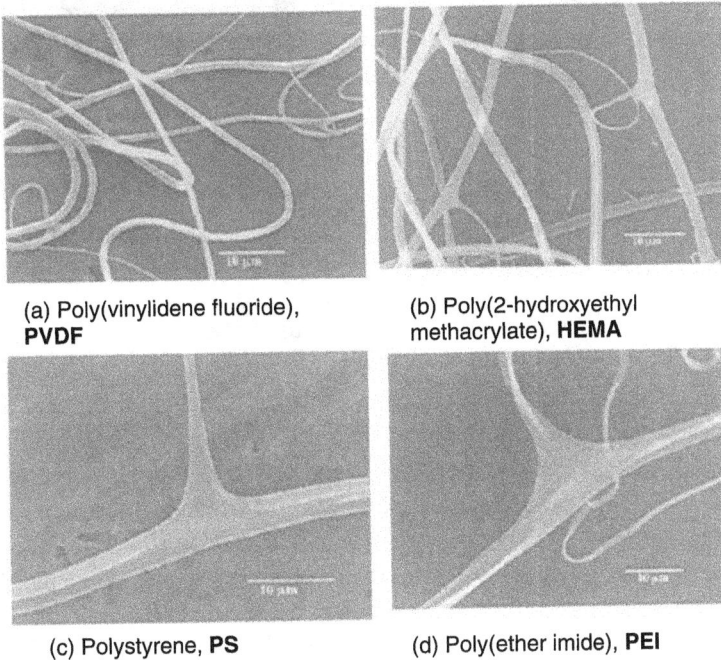

(a) Poly(vinylidene fluoride), **PVDF**

(b) Poly(2-hydroxyethyl methacrylate), **HEMA**

(c) Polystyrene, **PS**

(d) Poly(ether imide), **PEI**

Fig. 3.12. Branched fibers of (a) 20wt% PVDF in a 50:50 mixture of dimethylformamide (DMF)/demethylacetamide (DMA), (b) 16wt% HEMA in ethanol, (c) 30wt%PS in DMF and (d) 10wt% PEI in hexafluoro-2-propanol (HFIP) [Koombhongse et. al. (2001)].

3.5.4. Helical Fibers

Recently, [Kessick and Tepper (2004)] reported their discovery of helical fibers which can be of use in microelectromechanical system devices, advanced optical components and drug delivery systems. Helical fibers made of a mixture of poly(ethylene oxide) (PEO) and poly(aniline sulfonic acid) (PASA) were displayed as shown in Fig. 3.13. By changing the PEO concentration, the diameter of the loop could be controlled.

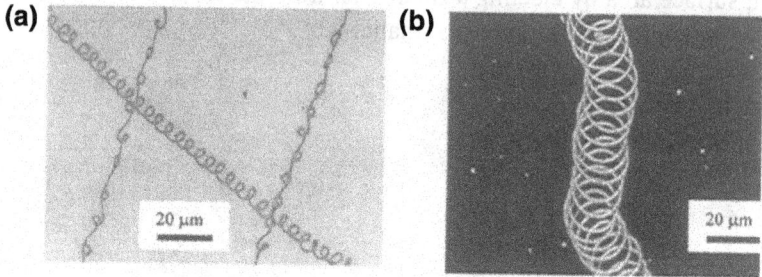

Fig. 3.13. Helical fibers electrospun from (a) 6wt% PEO solution and (b) 8.5wt% PEO solution with 1.5wt% PASA in each solution [Kessick and Tepper (2004)].

The creation of such helical loops is intriguing enough. It was considered that the electrochemical property of conductive PASA polymer played an essential role as helical fiber structure was not achieved from neat PVA solution. One possible mechanism was that there was equilibrium between the electrostatic repulsive force and the viscoelastic restoring force in electrospun polymer jet. The charges from PASA regions were transferred to the fiber surface when a fiber was collected at a conductive collector. As a result, a force imbalance occurred, which promoted fiber structural rearrangement in order to retain the force equilibrium.

3.5.5. Hollow Nanofibers

Nanotubes processed from various materials such as carbon, ceramics, metals, and polymers are essential in industrial applications, such as separation, gas storage and energy conversion. By employing the electrospinning process, chemical vapor deposition (CVD) method [Bognitzki et. al. (2000); Hou et. al. (2002)] and direct co-axial spinning method [Li et. al. (2005a)], hollow nanofibers can be made. The method by CVD is shown in Fig. 3.14. First, the template polymer nanofibers are electrospun before they are coated by sheath material by CVD. Hollow nanofibers are finally formed when the template is removed via annealing. To attain this, the template nanofiber should be stable during the coating but degradable or extractable without destroying the coating layer.

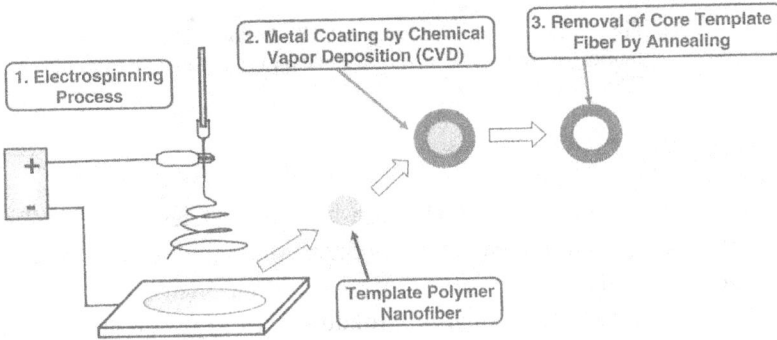

Fig. 3.14. Hollow fibers made by coating of template polymer nanofibers using chemical vapor deposition method [Bognitzki et. al. (2000); Hou et. al. (2002)].

[Hou et. al. (2002)] used poly(L-lactic acid) (PLLA) nanofibers as template and poly(p-xylylene) (PPX) as coating material by CVD (see Fig. 3.15). The created PPX/PLLA (sheath/core) fibers subsequently underwent annealing in order to remove the core PLLA template at a temperature of 280°C under vacuum. This led to the formation of PPX hollow nanofibers. The outer and inner diameters of PPX hollow nanofibers were found to be about 55 nm and 7 nm, respectively.

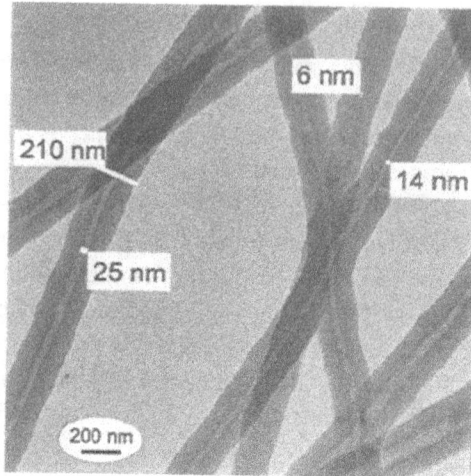

Fig. 3.15. Poly(p-xylylene) (PPX) hollow fibers fabricated by chemical vapor deposition CVD method [Hou et. al. (2002)].

Direct co-axial spinning method offers one-step processing of hollow nanofibers and the other feature is to utilize mineral oil as core material [Li and Xia (2004a); Li et. al. (2005a)] (see Fig. 3.16). [Li and Xia, (2004a)] prepared poly(vinyl pyrrolidone) (PVP) solution which contained precursor of titania [Ti(O$_i$ Pr)$_4$] as sheath material and electrospinning was simultaneously performed using oil eject. Hollow PVP/ Ti(O$_i$ Pr)$_4$ nanofibers were obtained after the core oil was removed by octane. Fig. 3.17 shows that if composite hollow fibers further undergo calcinations, TiO$_2$ ceramic hollow nanofibers could be created.

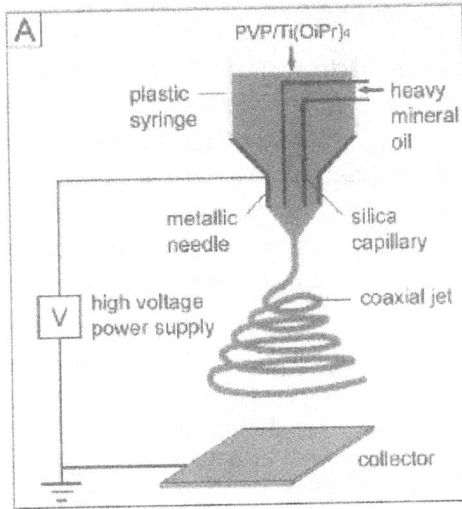

Fig. 3.16. Direct co-axial electrospinning method to create a core/sheath structure. Core and sheath materials are immiscible each other. Core material is removed by chemical solvent or annealing [Li and Xia (2004a); Li et. al. (2005)].

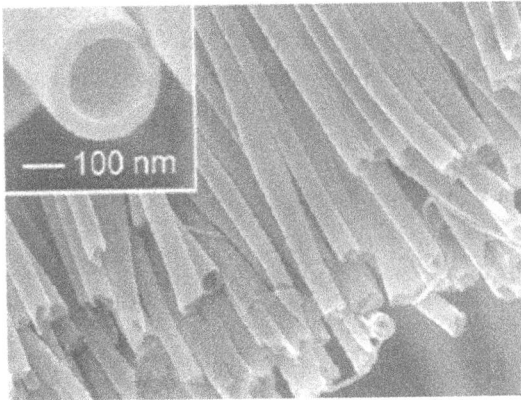

Fig. 3.17. TiO_2 hollow fibers fabricated by direct co-axial electrospinning [Li and Xia (2004)].

3.5.6. Fiber With Different Compositions

By using the co-axial electrospinning method shown in Fig. 3.16, it is possible to achieve electrospun fibers with different polymers [Zhang et. al. (2004); Sun et. al. (2003)]. One advantage of using co-axial electrospinning other then to produce hollow fibers is to produce core materials that will not form fibers via electrospinning. The outer shell polymeric material in this case, will serve as the template. Potential applications for this include drug delivery and photocatalysis.

Using a similar concept as co-axial electrospinning, another method of producing fiber with different compositions is to use side-by-side bicomponent electrospinning method [Gupta and Wilkes (2003)]. In this form of electrospinning, instead of having a capillary within a capillary, the two capillaries are placed side by side such that the resultant fiber is composed of two materials, side by side. The two polymer solutions do not come into physical contact until they reach the end of the spinneret where the electrospinning process is carried out. By using two different polymers, the fiber obtained will exhibit the properties of the two polymers. However, the potential difference between the tip of the needle and the collector must be optimized to ensure the formation of a combined electrospinning jet. If the potential difference is too high, two separate electrospinning jets will form. The resultant electrospun fibers showed that the amount of each given component can vary significantly along the fiber.

3.6. Uniformity and Productivity of Nanofiber Webs

Although electrospinning is a very simple process with huge potential for applications in different areas, one of its main disadvantages is its low productivity. The usual feedrate for electrospinning is about 1.5ml/hr. Given a solution concentration of 0.2g/ml, the mass of nanofiber collected from a single needle after an hour is only 0.3g. In order for electrospinning to be commercially viable, it is necessary to increase the production rate of the nanofibers.

To do so, multiple-spinning setup is necessary to increase the yield while at the same time maintaining the uniformity of the nanofiber mesh.

3.6.1. *Jet Stability*

To yield individual fibers, most, if not all of the solvents must be evaporated by the time the electrospinning jet reaches the collection plate. As a result, volatile solvents are often used to dissolve the polymer. However, clogging of the polymer may occur when the solvent evaporates before the formation of Taylor cone during the extrusion of the solution from several needles. In order to maintain a stable jet while still using a volatile solvent, an effective method is to use a gas jacket around the Taylor cone through two coaxial capillary tubes [Larsen et. al. (2004)]. The outer tube which surrounds the inner tube will provide a controlled flow of inert gas which is saturated with the solvent used to dissolve the polymer. The inner tube is then used to deliver the polymer solution. For 10 wt% poly(L-lactic acid) (PLLA) solution in dichloromethane, electrospinning was not possible due to clogging of the needle. However, when N_2 gas was used to create a flowing gas jacket, a stable Taylor cone was formed and electrospinning was carried out smoothly.

3.6.2. *Multiple-spinning Setup*

An obvious method of increasing the productivity of the nanofibers is to use multiple needles for electrospinning. As shown in Fig. 3.18, current authors make use of a multiple-spinning setup with a rotating mandrel. Polycaprolactone (PCL) solution with a feedrate of 24ml/hr was delivered to 8 needles using a single syringe pump. A nanofiber mesh of 280μm thickness was fabricated within 45 minutes. In a separate experiment [Ding et. al. (2004c)], a similar setup where four syringes were clamped to a movable speed that moves from side to side at a speed of 20m/min was used to create a uniform fiber mesh. When multiple syringes are used, the stability of the jet must be examined as the charges carried by the individual jet will distort the electrostatic field and subsequently influence other jets close to it [Theron et. al. (2005)]. The distance between the needles was kept at 3cm as it was considered that there would be less interference between the adjacent jets at this distance. When multiple jets are employed, it is possible to spin different type of polymers simultaneously. The ratio of the polymer fibers is determined by the ratio of the syringes containing the different polymers. Should there be any clogging of the needles during electrospinning, gas jacket can be used to improve the stability of the electrospinning process [Larsen et. al. (2004)]. Other similar mass production of electrospun fiber has been described in US Patent 6616435 B2 [Lee et. al. (2003b)] and US Patent 6713011 B2 [Chu et. al. (2004)].

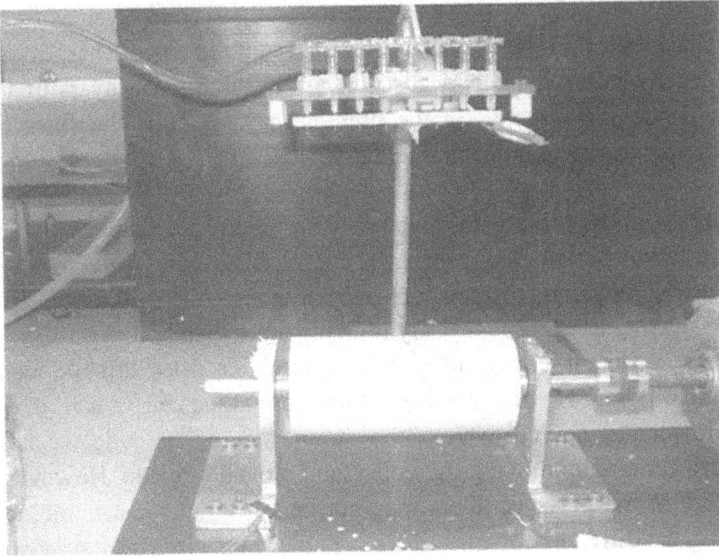

Fig. 3.18. Multiple-spinning setup using 8 syringes with a rotating cylinder. Polycaprolactone (PCL) nanofibers were electrospun with 15 kV applied voltage and 24 ml/h feeding rate. A PCL nanofiber membrane with 280 μm thickness was fabricated within 45 minutes [Courtesy of Fujihara and Ramakrishna, National University of Singapore].

3.7. Mixed Electrospun Fiber Mesh

Under normal electrospinning setup, only electrospun fibers of the same material are produced. However, to further improve on the versatility of electrospinning, some researches had produced fiber mesh composed of different types of polymer fibers. The fiber mesh may consist of layers formed by different types of polymer or they can be randomly mixed together. The production method is straight forward.

To produce a fiber mesh that consists of layers of different fibers, the fibers are sequentially electrospun [Kidoaki et. al. (2005)]. This resultant fiber mesh may be used in artificial blood vessel scaffold

However, care must be taken to ensure that the second layer of fibers is deposited on top of the first fiber mesh. Accumulations of charges on the first layer of fiber mesh may discourage the subsequent layers of fibers from depositing directly on the previous layers. Thus the fiber mesh layers may not be of uniform thickness.

To obtain a fiber mesh with mixed polymer fibers, the simplest method is to use two different electrospinning source filled with the two different polymer solution and electrospun onto a moving collector [Kidoaki et. al. (2005)]; Ding et. al. (2004c)]. By utilizing more than two electrospinning sources, different composition of the polymer fiber mixture can be obtained [Ding et. al. (2004c)]. Current methods are such that the electrospinning sources were placed at one side of the collector and the mixing is done by moving the collector from side to side. However, the mixture of the polymer fibers may not be uniform. A better method of mixing the different polymer fibers is to place the electrospinning source around a rotating drum as shown in Fig. 3.19. This method will ensure that the fibers from the two different electrospinning sources are evenly distributed throughout the fiber mesh.

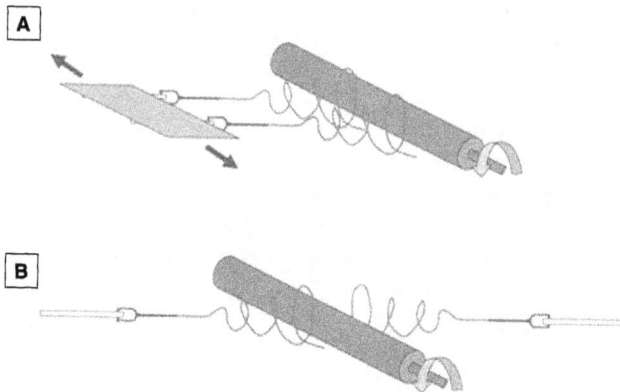

Fig. 3.19. (A) Production of polymer fiber mixture by lateral movement of the electrospinning sources. (B) A simpler method of fiber mixture production.

3.8. Patterning

Although the morphology of the nanofibers can be controlled to a certain extent, basic electrospinning can only yield randomly aligned nanofibers. This will limit the areas which electrospun fiber mesh can be used in. Thus to fully explore the potential of electrospinning, it is necessary to have some control on the deposition of the fibers. The ability to obtain one-dimensional (1D), two-dimensional (2D) as well as three-dimensioanl (3D) fiber architecture with controlled alignment and pitch will greatly increase the potential application of electrospun fibers. In this section, several methods of obtaining fiber patterns are introduced.

3.8.1. *Cylinder Collector*

The most basic form of getting aligned fibers rather than random mesh is through the use of a rotating mandrel as shown in Fig. 3.20. This is a simple, mechanical method of aligning the fibers along the circumference of the mandrel. As the fibers are formed from the electrospinning jet, the mandrel that is used to collect the fibers is rotated at a very high speed up to thousands of rpm (revolution per minute). This method has been successful in obtaining aligned ligned electrospun poly(glycolic acid) (PGA) (1000rpm rotating speed) and type I collagen (4500rpm rotating speed) fibers [Boland et. al. (2001); Matthews et. al. (2002)]. Although there is obvious alignment of the fibers, the degree of alignment is not very good as there are still a substantial number of misaligned fibers collected.

In electrospinning, the jet is traveling at a very high speed. As a result, to align the fibers around the mandrel, it is necessary that the mandrel is rotating at a very high speed so that the fibers can be taken up on the surface of the mandrel and wounded around it. Such a speed can be called as alignment speed. If the rotation of the mandrel is slower than the alignment speed, the fibers deposited will be randomly oriented. As more fibers are collected on the mandrel, accumulation of charges may caused incoming fibers to be repelled resulting in less oriented fibers.

Syringe

Electrospinning jet

**Rotating Cylinder
Collector**

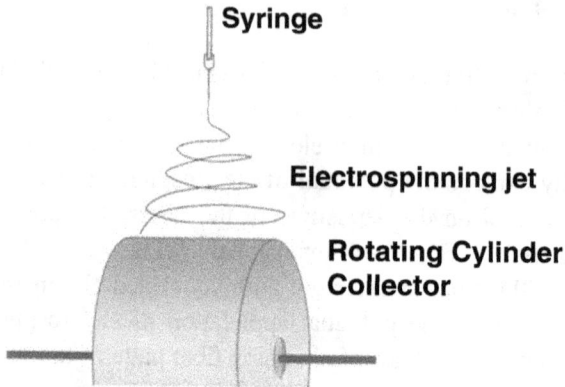

Fig. 3.20. A rotating cylinder collector to obtain unidirectional nanofiber alignment along the rotating direction.

To improve on the alignment of the fibers, the introduction of a sharp pin with a negative potential applied in the rotating mandrel can be used to create an electric field that leads from the tip of the needle and converges at the sharp pin in the rotating mandrel [Sundaray et. al. (2004)]. The pin in the mandrel was mounted vertically and it lies directly below the positively charged syringe needle as shown in Fig. 3.21. The mandrel was rotating at a speed of 2000rpm. Poly(methyl methacrylate) (PMMA) collected was better aligned than fibers collected on just a rotating mandrel.

However, the distance between the two electrodes was less than 2.5cm. The short distance is required to maintain the electric field from the needle to the sharp pin. If the distance is too great, the sharp pin will not be able to alter the electric field such that the field lines converge towards it. Since the sharp pin was inside the mandrel, the distance from the needle tip to the collector would be even shorter. Thus this method of collection may not be suitable when solvent which is less volatile is used to dissolve the polymer.

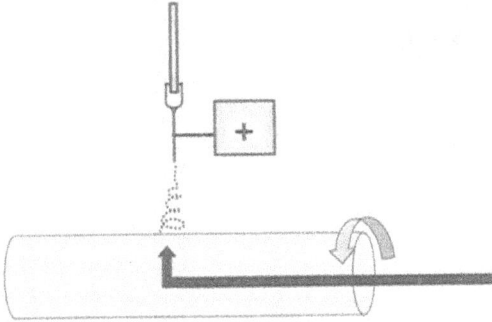

Fig. 3.21. Rotating mandrel with sharp pin inside to control the direction of fiber deposition.

Katta et. al. (2004) used a rotating wire drum (See Fig. 3.22) to collect aligned electrospun fibers. The rotation speed of the drum was reported to be 1rpm. This is a much slower rotating speed compared to other fiber alignment using rotating mandrel. However, although there was significant alignment of the fibers along the circumference of the axis for the first 15mins, the fiber alignment starts to go astray after that [Katta et. al. (2004)].

The loss in alignment was attributed to accumulation of charge on the deposited fibers on the drum. Since the drum was rotating at such a low speed, the aligning of the fibers may not be due to the mechanical winding of the fibers around its circumference. The electric field profile created by the parallel wires with gaps between them may play a part in aligning the fibers. Section 3.8.4 of this chapter discussed a technique which made used of parallel electrode with a gap between them to encourage fiber alignment. This may be the main factor contributing to the alignment of the fibers.

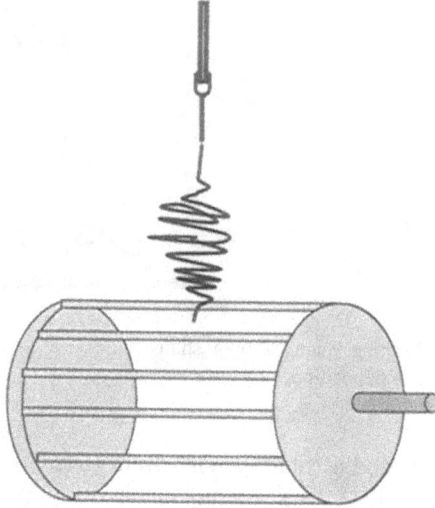

Fig. 3.22. Electrospinning on wire drum rotating at 1rpm.

3.8.2. A Knife Edge Disk

From the previous example, it was shown that controlling the path of the electrospinning jet by altering the profile of the electrostatic field from the needle to the collector is able to obtain well aligned fibers. By using a knife-edge rotating, it is possible to direct the electrostatic field lines to the knife-edge of the disk. Coupled with the rotation of the disk, excellent alignment of the nanofibers can be achieved [Theron et. al. (2001)]. Due to the concentration of the field lines towards the knife-edge, the electrospinning jet will tend to follow the direction of the field lines and converges towards the knife-edge. As the jet travel towards the knife-edge, the rotation of the disk will continuously wind the fibers along the knife-edge. It has been demonstrated that polyethylene oxide nanofibers of diameter ranging from 100nm to 400nm with a pitch of 1 to 2μm can be aligned on the knife-edge.

The pitch is caused by the repulsion of the fibers as they are deposited on the knife-edge. The typical rotation speed of the disk to obtain aligned fibers was about 1000rpm. By using a small rotatable collector attached to the rim of the disk, it is possible to obtain patterned nanofiber mesh. A layer of aligned fibers was initially deposited on the rim of the disk across the table. The angle of the table to that of the knife-edge was then varied before collecting another layer of aligned fibers. Thus a 2D architecture of fibers aligned at specific angle can be achieved as shown in Fig. 3.23.

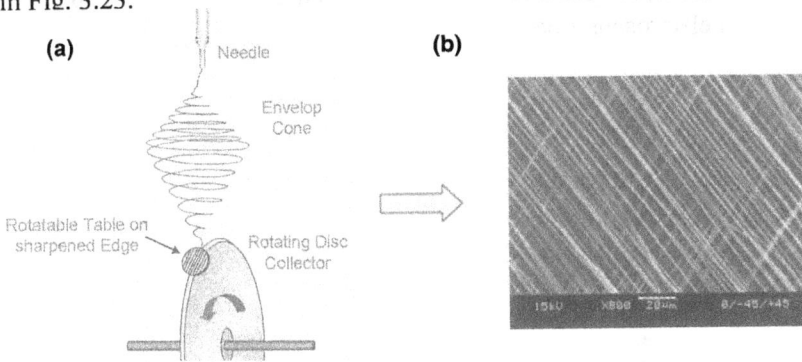

Fig. 3.23 Schematic drawing of a disk collector which has rotatable table at the rim of the disk (a) [Theron et. al. (2001)] and an obtained 2D patterned polycaprolactone (PCL) nanofiber mesh [Courtesy of Inai, Kotaki and Ramakrishna, National University of Singapore].

3.8.3. An Auxiliary Electrode/Electrical Field

In electrospinning, it is possible to control the path of the electrospinning jet by manipulating the electric field. To modify the electric field, an auxiliary electrode made of parallel conducting strips is used [Bornat (1987)] as shown in Fig. 3.24. This auxiliary electrode was given a negative voltage such that the electric field extends from the positively charged needle to the auxiliary electrodes. The parallel conducting strips has the effect of concentrating the electric field along the orientation of the parallel strips.

Thus the electrospinning jet has a greater tendency to spin in the direction of the orientation of the strips. When a non-conducting rotating mandrel is placed between the needle tip and the auxiliary electrode, the fibers can be easily picked up as it accelerates towards the electrode. In this way, aligned fibers along the circumference of the fibers can be obtained at a lower rotation speed of less than 1000rpm as compared to using a rotating mandrel alone [Teo et. al. (2005)]. Smaller diameter tube can be used in this setup as the rotating mandrel to form tubular structure made of electrospun fibers.

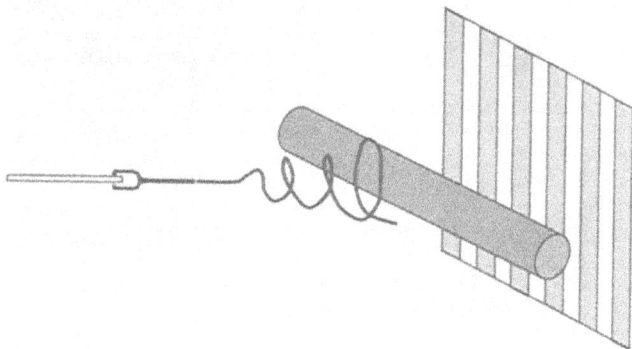

Fig. 3.24. Use of parallel strips as auxiliary electrode to assist in the alignment of the fibers.

To alter the electric field such that more electric field lines converge towards the auxiliary electrode, parallel knife-edged bars can be used instead of strips. These were found to improve on the alignment of the deposited fibers along the tube [Teo et. al. (2005)]. It has also been shown that using an auxiliary electrode to alter the electric field can control electrospinning jet such that the alignment of the electrospun fibers can be altered. [Teo et. al. (2005)] made use of a knife-edged aluminum bar to create an electric field profile such that the fibers that were collected on the tube were aligned in a diagonal direction instead of along the circumference of the tube as shown in Fig. 3.25. This is very useful in creating composite tubes made of electrospun fibers.

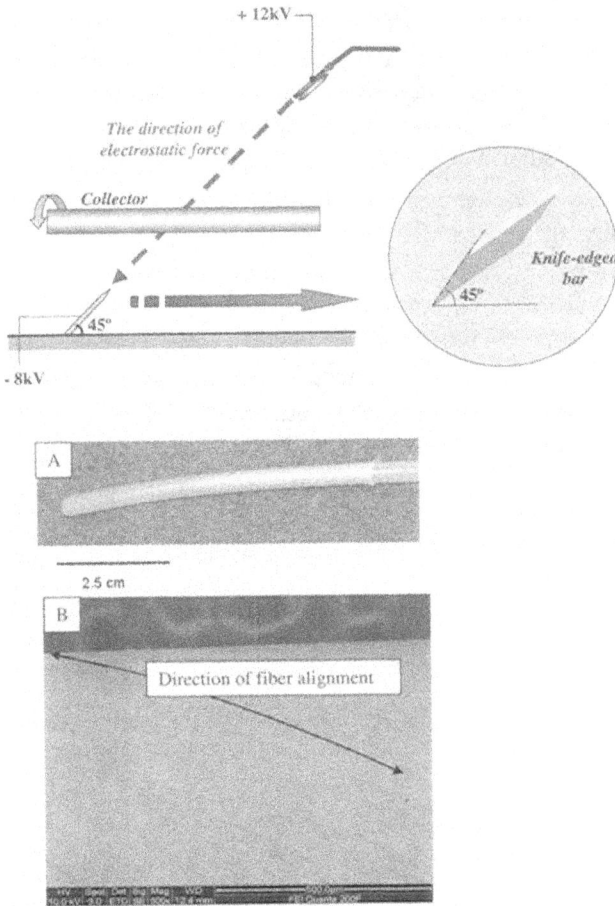

The direction of electrostatic force

Collector

Knife-edged bar

45°

45°

+ 12kV

- 8kV

A

2.5 cm

B

Direction of fiber alignment

Fig. 3.25. Setup to obtain fibers that are aligned diagonally on the tube [Teo et. al. (2005)].

3.8.4. *Parallel Conducting Collector*

Previous mentioned methods of obtaining aligned or patterned fibers involve the use of mechanical rotation of a disk or mandrel. In the following method, the alignment of the electrospun fibers is obtained purely through the behavior of electrospinning jet in an electrostatic field. When two parallel conducting electrodes are placed below the needle with a gap between them as shown in Fig. 3.26, the electric field lines in the vicinity of the parallel electrodes were split into two fractions pointing towards edges of the gap along the electrodes [Li et. al. (2003b)]. Since it is known that the electrospinning jet is influenced by the electrostatic field profile, the jet would stretch itself across the gap as the field lines are attracted towards the electrodes. This results in electrospun fibers aligning itself across the gap between the electrodes. Due to the presence of charges on the electrospun fibers, mutual repulsion between the deposited fibers will enhance the parallel and relatively even distribution of the fibers. One advantage of this setup over previous setups is that the aligned fibers can be easily removed from the collector.

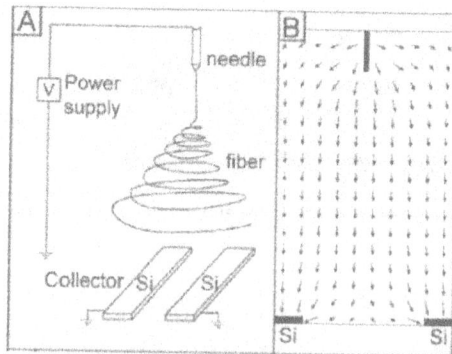

Fig. 3.26.   (A) Electrospinning with two parallel conducting collector. (B) Profile of the electric field [Li et. al. (2003b)].

Making use of the behavior of the electrospinning jet across the gap between electrodes, modification to the setup can be used to achieve more complicated arrangement of the electrospun fibers as shown in Fig. 3.27. However, when the distance between the electrodes was too great, there was a greater tendency for the fibers to be deposited on the electrode than across the gap. Repulsion caused by the charges on the deposited fibers also means that the collection time cannot be more than a few minutes thus this method may not be suitable for collection of a large patterned mesh or thick patterned mesh.

In a similar principle, a frame collector was used to obtain aligned nanofibers. The metal frame collector had the dimension of 20 x 60mm as shown in Fig. 3.28 [Dersch et. al. (2003)]. Aligned poly(lactic acid) (PLA) and polyamide 6 (PA6) nanofibers was achieved using this setup. A fiber orientation parameter (S) was determined to characterize the alignment of the fibers,

$$S = \ <(3\cos2\theta - 1)/2>,$$

where θ is the angle at which the individual fiber forms with the preferred direction controlled by the metal frame. The S value varies in principle from zero for an isotropic orientation to 1 for a perfect alignment. For PLA fibers, the S value obtained was $0.93 + 0.1$.

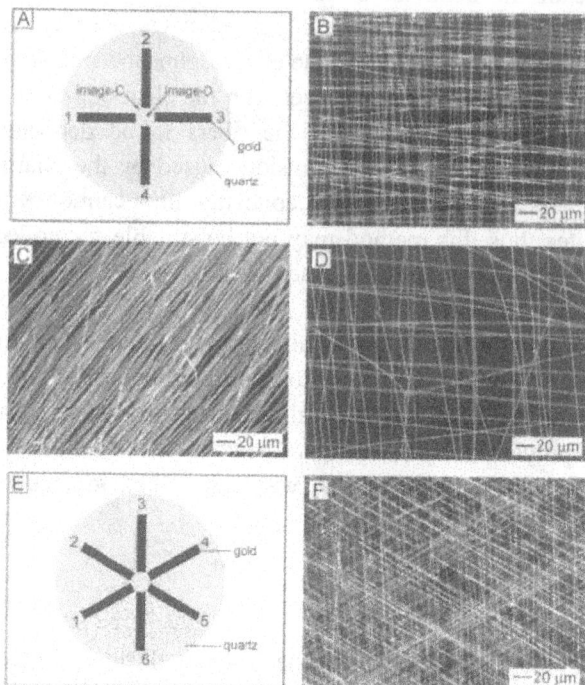

Fig. 3.27. (A,E) Schematic illustrations of test patterns that were composed of four (A) and six (E) electrodes deposited on quartz wafers. B) Optical micrographs of a mesh PVP naofibers collected in the center area of the gold electrodes shownin (A). During collection, the electrode pairs of 1-3 and 2-4 are alternately grounded for abt 5s. C,D) Optical micrographs of PVP nanofibers collected on the four-electrode pattern by grounding all four electrodes at the same time. The region from which these images were captured are indicated in (A). F) Optical micrograph of a tri-layer mesh of PVP nanofibers that was collected in the center area of the gold electrodes shown in (E). The electrode pairs of 1-4, 2-5 and 3-6 were sequentially grounded for 5s to collect alternating layers with the orientations of their fibers rotated by~60° [Li et. al. (2004b)].

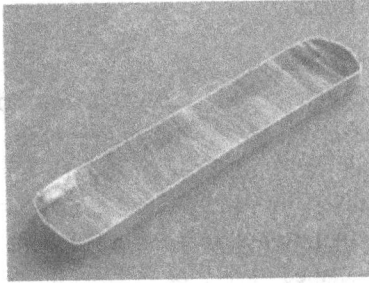

Fig. 3.28. A frame electrode collector for obtaining aligned nanofibers [Dersch et. al. (2003)].

3.9. Fiber Yarn and Textile

When it comes to continuous fiber, most people will think about yarns and textiles. In recent years, linear, 2-dimensional and 3-dimensional textile fabrics have been used beyond traditional apparels and found applications ranging from medical, chemical separation to composite reinforcements and chemical protection. This is due to the unique combination of light weight, flexibility, permeability, strength and toughness of textiles.

However, before fibers can be made into textile, they have to be made into the form of continuous yarns. This has posed a great difficulty for electrospun fibers as they lack sufficient strength to withstand traditional textile performing process. It is also a challenge to control the electrospinning process precisely to obtain yarns with different architecture as most methods of electrospinning were only able to obtain either random fiber mesh or aligned fiber. Nevertheless, the ability to produce yarns made of nanofibers are highly attractive, there are several methods that attempts to address the issue.

3.9.1. Hybrid Fiber Yarns

Since electrospun fibers lack the necessary strength to be processed using traditional textile technology, one recent attempt to create continuous fiber yarn is to make a hybrid consisting of a core filament and electrospun fibers on the surface [Scardino and Balonis (2001); Laurencin and Ko (2004)]. The setup is rather complicated as a vortex air suction with variably controlled air pressure flowing around the orifice, was created in the Y-shaped glass tube to twist the fiber yarn and also to direct the electrospinning jet to the core filament as shown in Fig. 3.29. A rotating drum was used to collect the yarn at the top of the setup. This method of producing yarn combines the strength of the inner core with the high surface area provided by the electrospun fibers.

Despite the advantage of this yarn processing method, the resultant yarn is not made out purely of electrospun fibers. The deposited electrospun fibers may not be tightly bounded to the surface of the filament core and thus the electrospun fibers may come off easily. Optimization of the conditions to deposit the electrospun fibers onto the core filament may be a very tricky and time-consuming process.

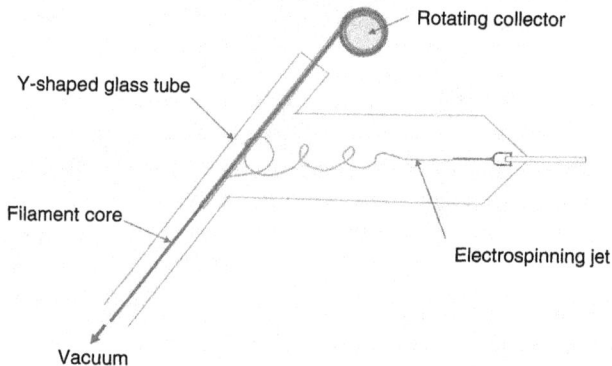

Fig. 3.29. Deposition of electrospun fibers onto a stronger, micron-size filament core.

3.9.2. Electrospun Fiber Yarn

There are a few methods where electrospinning is able to yield electrospun fiber yarn. However, the resultant fiber yarns are of limited length. In section 3.8.2 of this chapter, where a rotating disk was used to collect highly aligned electrospun fibers, a bundle of highly aligned fibers can be taken off the edges of the disk. However, in this case, the length of the yarn is limited by the circumference of the knife-edged disk. In a similar attempt to obtain a continuous yarn made of aligned fibers, fiber bundles are obtained by collecting the electrospun fibers on a rotating drum. The collected nanofibers were linked and twisted into yarns [Fennessey and Farris (2004)]. However, there were no descriptions of how the nanofibers were linked and twisted. Since the nanofibers were linked rather than being long continuous strands, the resultant yarns may be weak at the links.

[Khil et. al. (2005)] used a novel method to producing long, continuous fiber yarn. The method relies on deposition of the electrospun fibers onto a water bath as shown in Fig. 3.30. The deposited fibers were than drawn out of the water as a bundle from the side of the water bath using filament guide bar and collected onto a roller in the form of a yarn. The use of non-solvent such as water also has the advantage of coagulating the fibers especially when less volatile solvents are used to dissolve the polymer [Reneker and Chun (1996)].

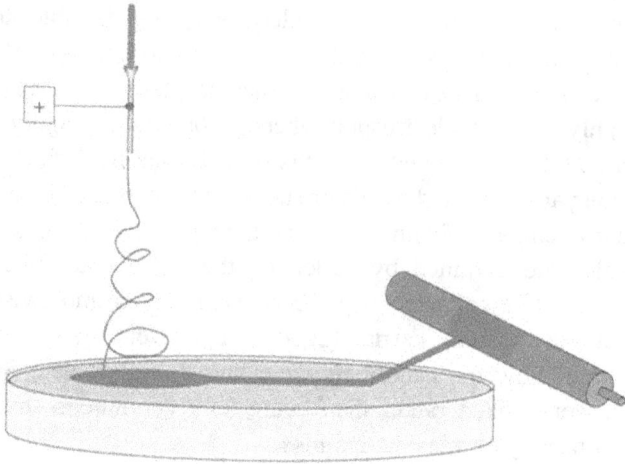

Fig. 3.30 Continuous fiber yarn through deposition of electrospun fibers onto water bath.

3.9.3. Twisted Fiber Yarn

One common method of improving the quality and the strength of the fiber yarn is to twist the fiber yarn. However, as the nanofibers are rather weak, it is not easy to create twisted fibers. One of way creating twisted fibers is to make of the electrospinning process where two parallel electrodes were used as discussed in Section 3.8.4. In this case, the parallel electrodes were made of rings instead of straight bars. During electrospinning, the fibers would be deposited on the conducting rings. As the fibers get deposited, one of the rings was rotated, thereby twisting the aligned fiber bundle as shown in Fig. 3.31.

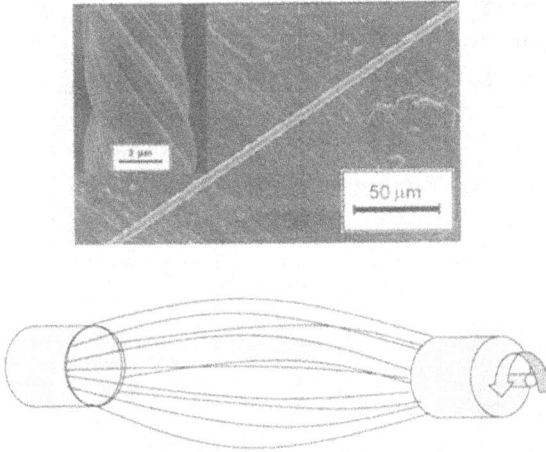

Fig. 3.31. Twisting of aligned electrospun fibers [Dalton et. al. (2005)].

Although by using this method, it is possible to get twisted fiber bundle, a significant drawback is that the length of the twisted fiber bundle was limited to the space between them. Optimizing of the gap between the rings to produce twisted fiber yarn shows that a distance between 40mm and 100mm were able to give favorable twisted yarn. Distance outside the range would results in poor deposition profile of the fibers on the rings with fibers either depositing mainly at the top of the ring or at the bottom of it.

3.10. Variations to Electrospinning

Since fundamentally, electrospinning is the production of fiber strands when the charges within a polymer fluid caused a jet to be ejected from it, several variations to electrospinning have been investigated. These variations come in the form of needleless electrospinning source although they are still based on basic electrospinning principle.

3.10.1. Scanning Tip Electrospinning Source

To carry out electrospinning without the need for a needle, one way is to have a sharp tip such that the charges can be concentrated at the tip. This way, it is possible to increase the charges to a small amount of polymer solution placed at the tip to a level such that a polymer jet will emerge from the tip as shown in Fig. 3.32. To get distinct fibers, the distance between the tip and the collector should be more than 1 cm. At such close distance, it was possible to deposit the nanofibers within a 5mm spot [Kameoka et. al. (2003)].

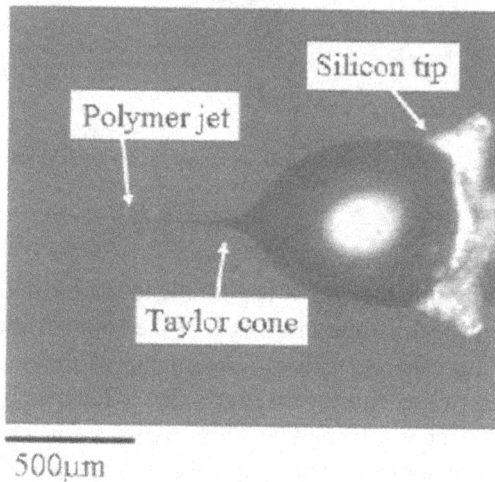

500μm

Fig 3.32. Electrospinning using a silicon tip instead of syringe needle [Kameoka et. al. (2003)].

Although the advantage of this method is that there is no clogging of the syringe needle, the volume of polymer solution that can be electrospun is very small. Electrospinning was only observed for 5-10s. Moreover, as the polymer solution placed at the tip was exposed to air, it may solidify quickly if the voltage was not applied to the solution immediately, further reducing the amount of fibers that can be produced.

3.10.2. Nanofiber Interconnections Between Microscale Features

In a technique which is similar to using scanning tip electrospinning source, nanofibers are formed between micro-electrodes. Rather then using a needle, a droplet of polymer solution was placed on a micro-electrode. At reduced dimensions, the electrodes create a high charge density on the small droplet on them. This causes polymer jet to form between electrodes of opposing polarity as shown in Fig. 3.33.

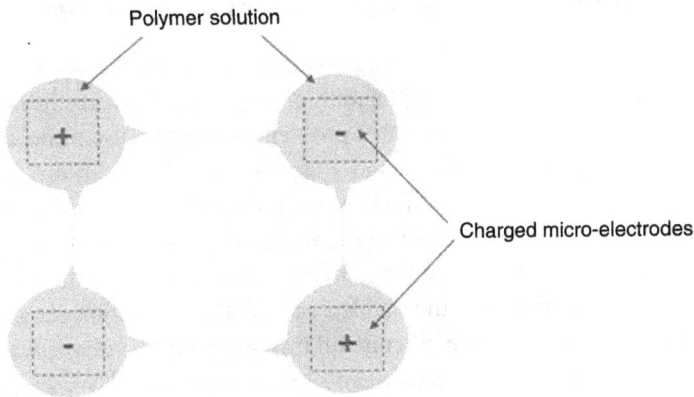

Fig. 3.33. Nanofibers formed between micro-electrodes.

Two methods were suggested to introduce charged polymer solution to the electrodes. The simplest method was to airbrush microscale droplets onto the surface of the electrode. A potential difference was applied between two electrodes and fibers were formed between the electrodes. In a second method, both negatively and positively charged droplets where alternatively sprayed onto an insulating substrate using electrospray ionization by switching the polarity of the high voltage supply. Nanoscale fibers were formed spontaneously between oppositely charged droplets without the need to apply an external potential [Kessick and Tepper (2003)].

Although this is a very interesting method of producing electrospun fibers to link two micro-elecrodes, its application may be very limited. A potential application is in creating electrical connection between micro-chips. However, the nanofiber that links the electrodes may not have uniform dimension throughout its length. This may significantly reduce its usefulness as a nano-wire. Nevertheless, this technique shows the versatility of electrospinning.

3.10.3. *Mass Production Through Needleless Electrospinning*

Although the use of multiple syringes can increase the production rate of electrospinning, there is a complication of clogging of the needles during the spinning. One way to avoid that problem is to make use of needleless electrospinning method [Yarin and Zussman (2004)] as shown in Fig. 3.34. In this case, magnetic fluids were prepared by mixing magnetite with silicone oil. A permanent magnet was used to generate the magnetic field to induce the formation of numerous spikes on the fluid surface. The polymer solution was then added such that the solution remains on the surface of the magnetite-silicone oil mixture. A high voltage was then applied to the mixture. An electrically grounded metal saw was used to collect the fibers. Quasi-steady electrospinning can be carried out with a huge increase in the production of electrospun fiber.

(A)

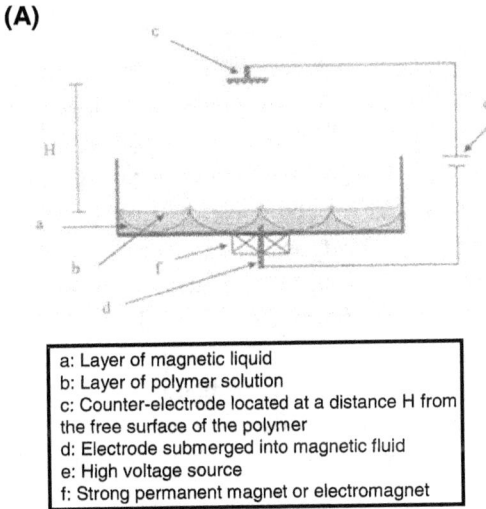

a: Layer of magnetic liquid
b: Layer of polymer solution
c: Counter-electrode located at a distance H from the free surface of the polymer
d: Electrode submerged into magnetic fluid
e: High voltage source
f: Strong permanent magnet or electromagnet

(B)

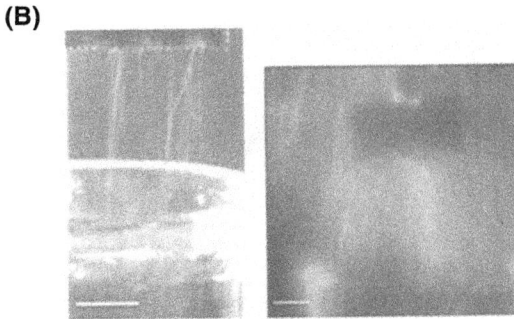

Fig. 3.34.   (A) Schematic drawing of upward needless electrospinning (B) and view of multiple jets attracted to a piece of a metal saw used as a counter-electrode [Yarin and Zussman (2004)].

3.11. Conclusions

In order to control any process, it is important to investigate the parameters that influence it. By varying the parameters, electrospun fibers of different diameters, fiber surface porosity, beaded and non-beaded fibers can be produced. Depending on the application, fibers with different morphology and structures can be produced by using specific parameters and setups. The type of fiber mesh produced can also be varied by using different setups. Tubular scaffold and yarns, made of both aligned and random fibers can be produced till date. Structures made of fibers arranged in specific patterns may also be produced through electrospinning. All these are possible by manipulation of the external electric field and using different types of collector. With greater understanding of electrospinning, we may be able to have a greater control over the behavior of the electrospinning jet and its resultant fibers.

Chapter 4

Modeling of the Electrospinning Process

4.0. Nomenclature

Unless otherwise specified, the symbols used in this chapter adhere to the following nomenclature.

α = coefficient of surface tension

δ = radial perturbation

E = vertical component of electric field or Young's modulus

e = charge

ε = strain

f = force

G = elastic modulus

g = gravitational acceleration

γ = shear

h = distance from pendent drop to ground collector

I = constant total current in the jet or second moment of area

K = liquid conductivity

k = jet curvature

$$L = \frac{e}{R_0} \sqrt{\frac{1}{\pi G}} = \text{length scale}$$

l = length of ideal rectilinear jet

λ = perturbation wave length

M = linear momentum

m = equivalent mass

μ =viscosity

n = flow index of non-Newtonian fluid

v = kinematic viscosity

P = position coordinate

p = liquid pressure

R = jet radius

R' =slope of jet surface with respect to jet axis.

R_0 = initial jet radius

r = radial distance from jet axis

ρ = density

σ =surface charge density or normal stress

σ_V = viscoelastic stress

T = surface tension

t = time

τ = shear stress

V_0 = applied voltage

v = jet velocity

ω = perturbation frequency

z = downward vertical distance from pendant droplet

4.1. Introduction

The importance of electrospinning process modeling is significant. Although knowledge on the effect of various processing parameters and constituent material properties can be obtained experimentally, theoretical models offer in-depth scientific understanding which can be useful to shed light on the contributing factors that cannot be fully measured experimentally. Results from modeling also elucidate how processing parameters and fluid behavior lead to the nanofiber of desired properties. The term "properties" refers to basic properties (such as fiber diameter, surface roughness, fiber connectivity, etc), physical properties (such as stiffness, toughness, thermal conductivity, electrical resistivity, thermal expansion coefficient, opaqueness, density, etc) and specialized properties (such as biocompatibility, degradation curve, etc for biomedical applications). In addition to electrospinning of polymer solution for obtaining nanofibers, other variations of electrospinning exists, namely the combined electrospinning and sol-gel process [Choi et al. (2003a & 2004d)] and the melt-electrospinning processes [Khurana et al. (2003); Lyons et al. (2003 & 2004)]. Since this chapter deals with the electrospun nanofibers, we herein consider the modeling aspects of simple electrospinning that uses polymer solution.

Needles to say, the various methods formulated by numerous research groups are highly motivated by different intended applications of the nanofibers, as well as their academic familiarity. It would be sufficient, in this chapter, to briefly describe the different methods adopted, observe the similarities and discrepancies in all their approaches. Then, we review some results that can be useful to bear in mind for those who intend to begin the research and development of nanofibers using the electrospinning process.

4.2. Preliminaries

The syringe, when charged either negative or positive, experiences an increase or decrease respectively of electrons in the polymer solution (see Fig. 4.1).

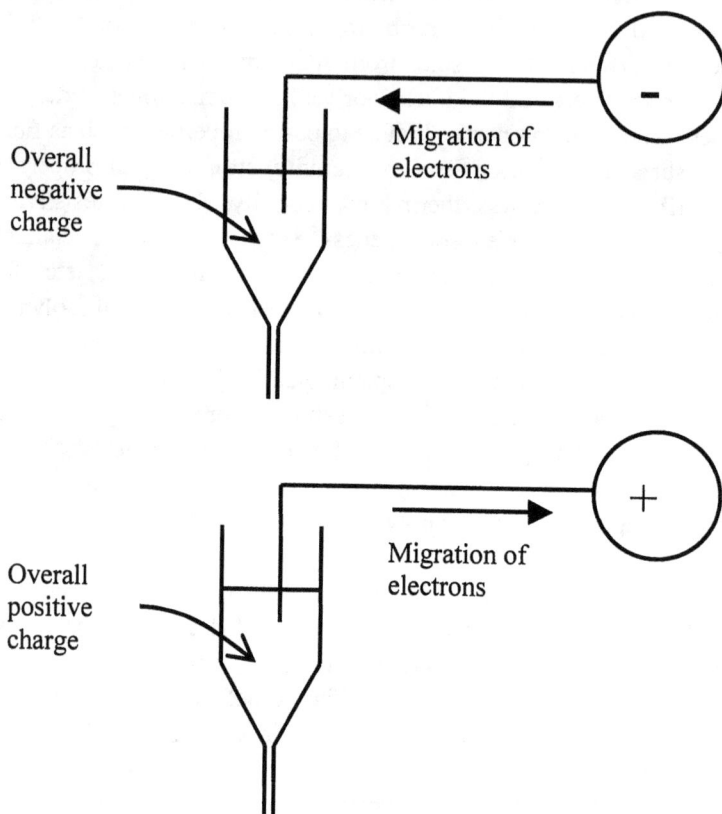

Fig. 4.1.   Flow of free electrons due to applied voltage causes charge build-up.

Even though the main parameter of charge build-up is primarily correlated to the magnitude of applied voltage, the other parameters such as the density of free electrons in an aromatic structure and existence of ions play their role in increasing the charge build-up. While the connection to a positive charge simply decreases the number of free electrons in the polymer solution, the connection to negative charge increases the number of free electrons in the polymer solution – the former due to outflow of aromatic free electrons such as those of aromatic compounds.

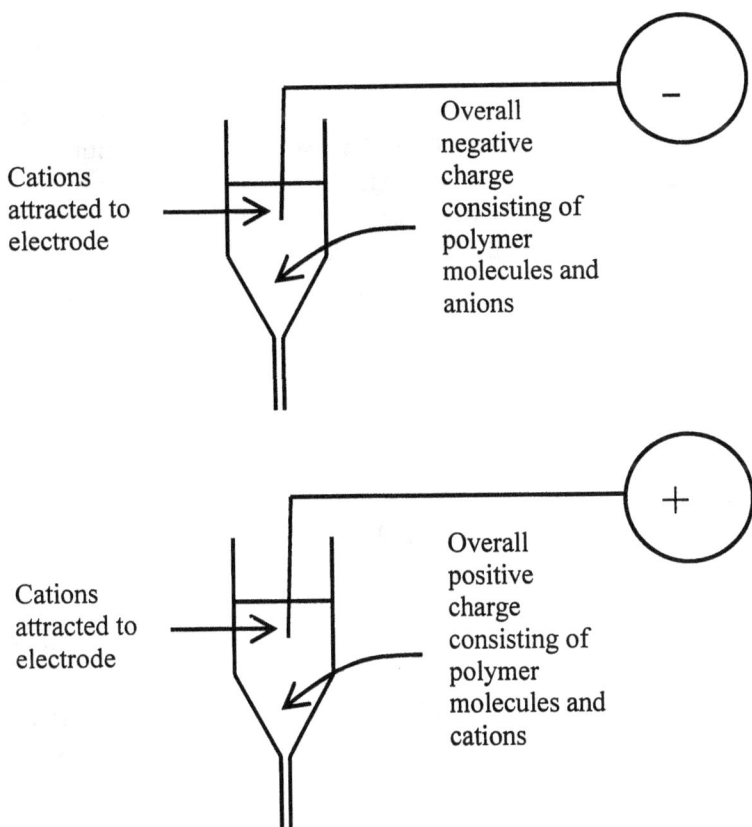

Fig. 4.2. Attraction of oppositely charged ions to the electrode caused charge build-up.

More importantly, the existence of ions in the overall neutral polymer solution helps in the charge build-up. When the electrode is charged negative or positive, cations and anions, respectively, are attracted to the electrode – thereby causing a build-up of negative or positive charge on the polymeric-solvent liquid as a result of excessive anions and cations respectively (see Fig. 4.2).

There exists a force, as a result of charge build-up, acting upon the droplet coming out from the syringe needle pointing toward the collecting plate which can be either grounded or oppositely charged. Furthermore, similar charges within the droplet promote jet initiation due to their repulsive forces. Nevertheless, surface tension and other hydrostatic forces inhibit the jet initiation because the total energy of a droplet is lower than that of a thin jet of equal volume upon consideration of surface energy. When the forces that aid jet initiation (such as electric field and Coulombic) overcome the opposing forces (such as surface tension and gravitational), the droplet accelerates toward the collecting plate. This forms a jet of very small diameter. Other than initiating jet flow, the electric field and Coulombic forces tend to stretch the jet, thereby contributing towards the thinning effect of the resulting nanofibers.

Depending on the processing parameters (such as applied voltage, volum flow rate, etc) and the fluid properties (such as surface tension, viscosity, etc) as many as 10 modes of electrohydrodynamically driven liquid jet have been identified [Jaworek and Krupa (1999)]. The scope of jet modes is highly abbreviated in this chapter because most electrospinning processes that lead to nanofibers consist of only two modes, the straight jet portion and the spiraling (or whipping) jet portion. Insofar as electrospinning process modeling is involved, the following classification indicates the considered modes or portion of the electrospinning jet.

1. Modeling the criteria for jet initiation from the droplet [Senador et al. (2001); Yarin et al. (2001b)];
2. Modeling the straight jet portion [Feng (2002 & 2003); Spivak et al. (1998 & 2000)]; and

3. Modeling the entire jet [Reneker et al. (2000); Yarin et al. (2001a); Hohman et al. (2001a & 2001b)].

The reader is reminded that in general modeling approaches are less than complete from the actual case, and that the level of incompleteness does not necessarily denote any form of inferiority. Instead, the attainment of a complete and "perfect" model would be impractical if simple models that require 10% of the "perfect" model's calculation time can adequately predict an acceptable level of accuracy, say 90%, of the considered process. It follows that such incompleteness would help to focus and deal with the detailed calculation of the specific electrospinning portion of interest.

4.3. Assumptions

Just as in any other process modeling, a set of assumptions are required for the following reasons:
To furnish industry-based applications whereby speed of calculation, but not accuracy, is critical, and
To simplify – hence enabling checkpoints to be made before more detailed models can proceed.
For enabling the formulations to be practically traceable.

4.3.1. Jet Representation

The first assumption to be considered as far as electrospinning is concerned is conceptualizing the jet itself. Even though the most appropriate view of a jet flow is that of a liquid continuum, the use of nodes connected in series by certain elements that constitute rheological properties has proven successful (e.g. [Reneker et al. (2000); Yarin et al. (2001a)]. The second assumption is the fluid constitutive properties. In the discrete node model (e.g. [Reneker et al. (2000)]), the nodes are connected in series by a Maxwell unit, i.e. a spring and dashpot in series, for quantifying the viscoelastic properties.

The use of a spring and dashpot in series is highly suitable for a viscoelestic liquid, since a spring and dashpot in parallel be more suited for a viscoelastic solid. This is further elaborated in the next section by considering of a few simple viscoelastic models.

4.3.2. Modeling Viscoelastic Behavior

Viscoelastic behavior is rate-dependent. Such substances exhibit combined elastic and time-dependent response, and are described by some functional relation between stress and its time derivatives and strain and similar time derivatives. In analyzing viscoelastic models, we apply two types of elements: the dashpot element (Fig. 4.3) which describes the force as being in proportion to the velocity (recall friction), and the spring element (Fig. 4.4) which describes the force as being in proportion to elongation. One can then develop viscoelastic models using combinations of these elements.

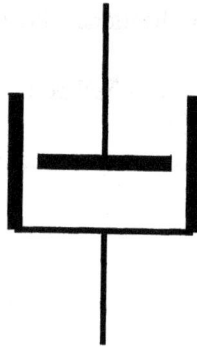

Fig. 4.3. Dashpot element describes friction or viscosity.

Fig. 4.4. Spring element describes elasticity.

In order to obtain the overall force and strain, we note the following:
Forces are common for elements arranged in series,
Forces are summative for elements arranged in parallel,
Elongation is summative for elements arranged in series, and
Elongation is common for elements arranged in parallel.

With these guidelines, the following viscoelastic models can be obtained:

(a) Newtonian dashpot.

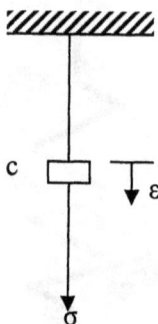

Fig. 4.5. A highly idealized inviscid liquid.

For the Newtonian dashpot, we have

$$\sigma = c\frac{d\varepsilon}{dt} = c\dot{\varepsilon} \, .$$

(4.1)

(b) Kelvin-Voigt solid featuring a parallel spring and dashpot.

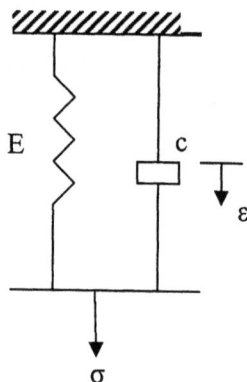

Fig. 4.6. Kelvin-Voigt solid.

For a Kelvin-Voigt solid, we have

$$\sigma = E\varepsilon + c\dot{\varepsilon} \tag{4.2}$$

based on statements (ii) and (iv).

(c) Maxwell model (a fluid), with a spring and dashpot in series.

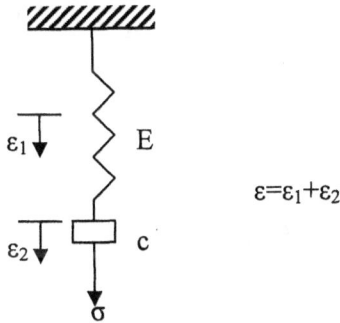

Fig. 4.7. Maxwell fluid.

For a Maxwell model, we apply statements (i) and (iii)

$$\varepsilon = \varepsilon_1 + \varepsilon_2 \tag{4.3}$$

Substituting

$$\dot{\sigma} = E\dot{\varepsilon}_1 \tag{4.4}$$

$$\sigma = c\dot{\varepsilon}_2 \tag{4.5}$$

into Equation (4.3), we obtain

$$\sigma + \frac{c\dot{\sigma}}{E} = c\dot{\varepsilon} \tag{4.6}$$

(d) The standard linear solid which combines the Kelvin-Voigt and the Maxwell solids

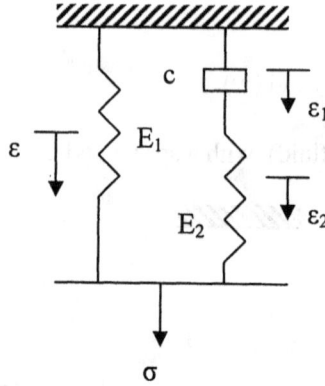

Fig. 4.8.   Standard linear solid.

Since

$$\varepsilon = \varepsilon_1 + \varepsilon_2 \tag{4.7}$$

and

$$E_2\varepsilon_2 = c\,\dot{\varepsilon}_1 \tag{4.8}$$

for this standard model, we have

$$\sigma + \frac{c}{E_2}(\dot{\sigma}) = E_1\varepsilon + \left(\frac{c}{E_2}\right)(E_1 + E_2)\dot{\varepsilon} \tag{4.9}$$

It is clear that both the Newtonian (Fig. 4.5) and the Kelvin-Voigt (Fig. 4.6) are highly unsuitable due to the absence of elastic component and permanent elongation respectively. The Maxwell unit (Fig. 4.7) was selected by [Reneker et al. (2000)] due to its suitability for liquid jet as well as its simplicity. Other models are either unsuitable for liquid jet (such as the standard linear model displayed in Fig. 4.8) or too detailed.

In the continuum model a power law can be used for describing the liquid behavior under shear flow

$$\tau = \mu \left(\overset{\bullet}{\gamma} \right)^{n} \tag{4.10}$$

where

τ = shear stress

μ =viscosity

γ = shear

n = non-Newtonian flow index such that $0 < n < 1$ ($n = 1$ refers to Newtonian fluid). [Spivak et al. (1988 & 2000)] used this power law, written as

$$\tau = \mu \left[tr \left(\overset{\bullet}{\gamma} \right)^{2} \right]^{\frac{n-1}{2}} \overset{\bullet}{\gamma}, \tag{4.11}$$

where

n = flow index of Oswald deWaele Law

r = radial distance from jet axis

t = constant

for describing the jet flow. At this juncture, we note that the power law is characterized from a shear flow, whilst the jet flow in electrospinning undergoes elongational flow.

4.3.3. Coordinate System

The method for coordinate system selection in electrospinning process is similar to other process modeling – the system that best bring out the results by (i) allowing the computation to be performed in the most convenient manner and, more importantly, (ii) enabling the computation results to be accurate. In view of the linear jet portion during the initial first stage of the jet, the spherical coordinate system is eliminated. Assuming the second stage of the jet to be perfectly spiraling, due to bending of the jet, the cylindrical coordinate system wound be ideal. However, experimental results have shown that the bending instability portion of the jet is not perfectly expending spiral. Hence the Cartesian coordinate system, which is the most common of all coordinate system, is adopted.

4.3.4. Liquid Incompressibility

One generalization of fluid is its compressibility, which translates into the variation of density as a result of change in environmental pressure and temperature. Whilst fluid density variation must be accounted for gases as well as liquid in pipes and pressure vessels, in some instances it can be assumed to be constant for liquid flow that is unbounded – such as liquid jet. Hence the assumption of constant density for liquid that exits the syringe is a reasonable one.

4.4. Conservation Relations

One of the major considerations to be imposed upon all calculations is the conservation relations. Most electrospinning models apply three conservation quantities, namely that of the mass, the momentum, and the charge.

4.4.1. Conservation of Mass

Except for nuclear reaction, this law requires that no mass is created nor destroyed, but merely transported from one location to another. At steady state the mass entering a control volume is equal to the mass that exits. Hence, for the case of unsteady state, the difference in the rate of mass flow would contribute towards the rate of change of mass in that control volume. With reference to Fig. 4.9, the mass in the jet bounded by the cross-sections z and $z + dz$ (denoted by the shaded region) is

$$m = \rho \pi R^2 dz \tag{4.12}$$

where ρ = liquid density and R = jet radius. Over an infinitesimal time of dt, the flow of mass into and out from the control volume are

$$m_{in} = \rho \pi R^2 v dt \big|_z \tag{4.13}$$

and

$$m_{out} = \rho \pi R^2 v dt \big|_{z+dz} \tag{4.14}$$

respectively where v is the liquid velocity.

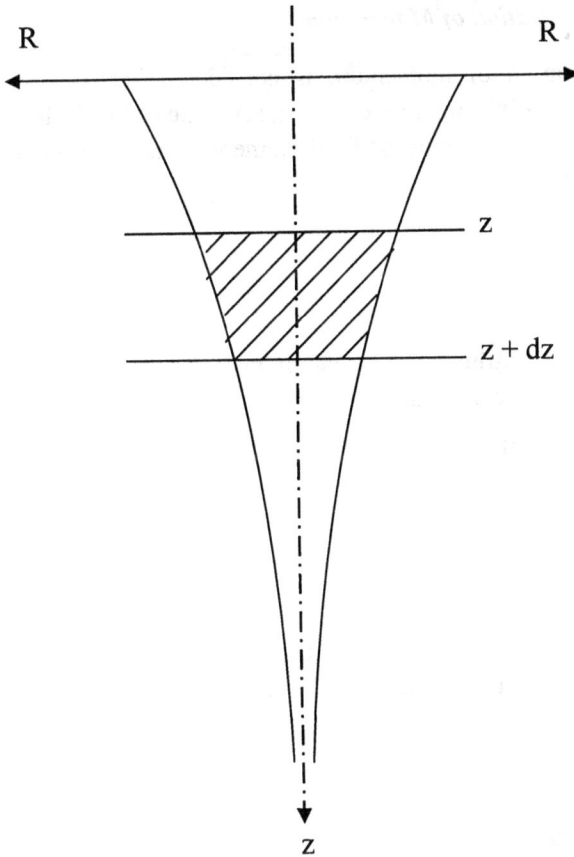

Fig. 4.9.   A segment of jet flow for analysis.

Taking mass balance

$$\Delta_t m = m_{in} - m_{out} \, ,$$ (4.15)

we have

$$\frac{\partial R^2}{\partial t} + \frac{\partial v R^2}{\partial z} = 0 \, .$$ (4.16)

4.4.2. Conservation of Momentum

The conservation of momentum is considered herein as would any process that involves motion, except quasi-static ones. With reference to Fig. 4.9, the momentum of the fluid segment bounded by sections z and $z + dz$ is

$$M = \rho \pi R^2 v dz .$$ (4.17)

Bearing in mind that

$$dz = v dt ,$$ (4.18)

the influx and outflux of momentum into and out from the bounded volume within time interval dt can be written as

$$M_{in} = \rho \pi R^2 v^2 dt \big|_z$$ (4.19)

and

$$M_{in} = \rho \pi R^2 v^2 dt \big|_{z+dz}$$ (4.20)

respectively. Additionally, due to liquid pressure exerted on the cross-section z and $z + dz$ (see Fig. 4.10), the resulting momentum is taken from the integral of cross-sectional force over time, i.e.

$$M \big|_z = p \pi R^2 dt \big|_z$$ (4.21)

and

$$M \big|_{z+dz} = p \pi R^2 dt \big|_{z+dz}$$ (4.22)

where p = liquid pressure.

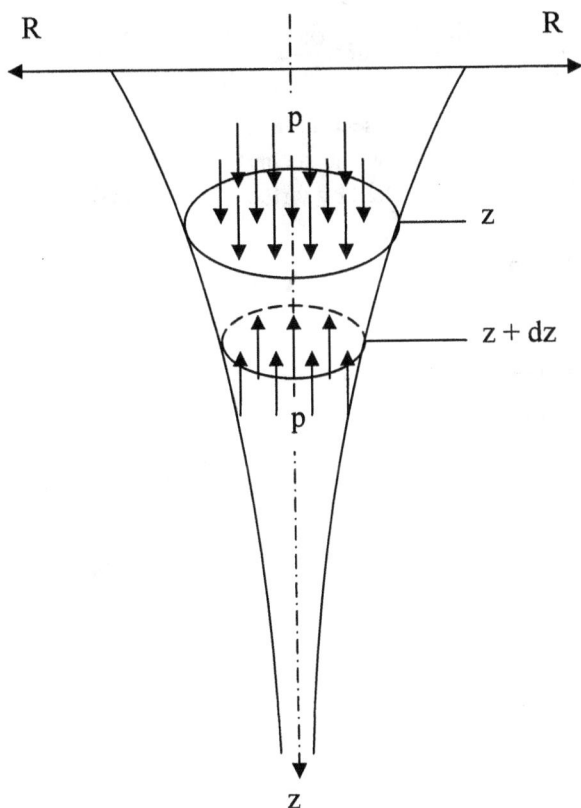

Fig. 4.10 Hydrostatic pressure on fluid segment.

Taking momentum balance

$$\Delta_t M = M_{in} - M_{out} + M\big|_z - M\big|_{z+dz},$$ (4.23)

we arrive at

$$\frac{\partial}{\partial t}\left(vR^2\right) + \frac{\partial}{\partial z}\left(v^2 R^2\right) = -\frac{1}{\rho}\frac{\partial}{\partial z}\left(pR^2\right)$$ (4.24)

in which the viscous terms are excluded. Writing the term $(v^2 R^2)$ as a product of v and vR^2 the LHS of Equation (4.24) can be expanded as

$$R^2 \frac{\partial v}{\partial t} + v \frac{\partial R^2}{\partial t} + v \frac{\partial v R^2}{\partial z} + v R^2 \frac{\partial v}{\partial z} = -\frac{1}{\rho} \frac{\partial p R^2}{\partial z}. \tag{4.25}$$

The purpose of splitting the second term on the LHS of Equation (4.24) as above is intended to allow the second and third terms on the LHS of Equation (4.25) to be cancelled by virtue of Equation (4.16). Hence

$$\frac{\partial v}{\partial t} + v \frac{\partial v}{\partial z} = -\frac{1}{\rho R^2} \frac{\partial p R^2}{\partial z} \tag{4.26}$$

A more detailed momentum balance equation was developed by [Feng (2002)] as follows:

$$\frac{d}{dz}\left(\pi R^2 \rho v\right) = \pi R^2 \rho g + \frac{d}{dz}\left[\pi R^2\left(-p + \tau_{zz}\right)\right] + \frac{T}{R} 2\pi RR' + 2\pi R\left(t_t^e - t_n^e R'\right) \tag{4.27}$$

where

ρ = liquid density

g = gravitational acceleration

p = liquid pressure

T = surface tension

t_t^e, t_n^e =tangential and normal tractions on jet surface due to electricity

R' =slope of jet surface with respect to jet axis.

Further understanding on the hydrodynamics of jet flow can be obtained from [Yarin (1993)].

4.4.3. *Conservation of Charge*

With regard to the electrified nature of the jet, the consideration of charge would differentiate the modeling of electrospinning from most fluid flow. An example of the conservation of charge equation was given as [Feng (2002)]

$$I = \pi R^2 KE + 2\pi R v \sigma \tag{4.28}$$

where
K = liquid conductivity
E = vertical component of electric field
I = constant total current in the jet
σ =surface charge density

Here, the second and third terms on the RHS of Equation (4.28) refer to the current flowing across the jet cross-section and perimeter, respectively, in the direction parallel to the jet axis.

4.5. Consideration of Forces

In the flow path modeling, we recall the Newton's Second Law of motion

$$m\frac{d^2P}{dt^2} = \sum f \tag{4.29}$$

where
m = equivalent mass
and the various forces are summed as

$$\sum f = f_C + f_E + f_V + f_S + f_A + f_G + \cdots \tag{4.30}$$

in which subscripts C, E, V, S, A and G correspond to the Coulombic, electric field, viscoelastic, surface tension, air drag and gravitational forces respectively. A description of each of these forces based on the literature is summarized in Table 4.1. Here,

e = charge

l = length of ideal rectilinear jet

V_0 = applied voltage

h = distance from pendent drop to ground collector

σ_V = viscoelastic stress

G = elastic modulus

μ = viscosity

α = coefficient of surface tension

k = jet curvature

ρ = density

ν = kinematic viscosity

Table 4.1. Description of itemized forces or terms related to them.

Forces or stresses	Discrete nodes (e.g. [Reneker et al., 2000])
Coulombic	$$f_C = \frac{e^2}{l^2}$$
Electric field	$$f_E = -\frac{eV_0}{h}$$
Viscoelastic	$$\frac{d\sigma_V}{dt} = \frac{G}{l}\frac{dl}{dt} - \frac{G}{\mu}\sigma_V$$
Surface tension	$$f_S = \frac{\alpha\pi R^2 k}{\sqrt{x_i^2 + y_i^2}}\left[\mathbf{i}|x|sign(x) + \mathbf{j}|y|sign(y)\right]$$
Air drag	$$f_A = 0.65\pi R\rho_{air}v^2\left(\frac{2\nu R}{\nu_{air}}\right)^{-0.81}$$
Gravitational	$$f_G = \rho g\pi R^2$$

4.6. Instability

In the absence of perturbation, the jet flow considered so far would theoretically lead to straight jet since the models adopt axisymmetrical jet profile. In actual situation, these straight jets eventually change in mode – resulting in bending instability. The sources of instability are numerous and varied such that no reasonably complete nor consistent list appear thereof.

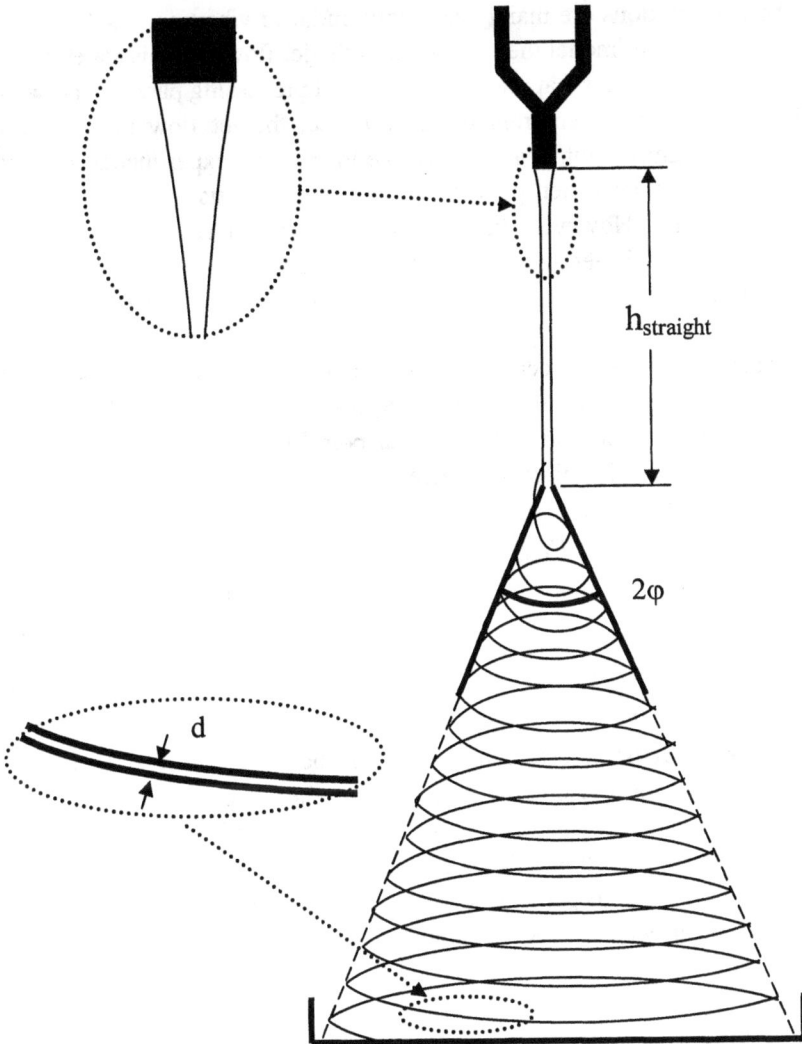

$h_{straight}$

2φ

d

Fig. 4.11. Conservative characteristics of an electrospinning process.

The perturbations are mainly externally induced vibration. As a result it is impossible to model the exact path of the jet flow and this statement is justified by the fact that under exactly equal processing parameters, same liquid and similar environmental influence, the jet flow path are not identical. Simply put, it is not possible to tract the experimental jet path and the simulated jet path on a one-to-one basis due to different perturbations. However, the conservative characteristics such as the length of straight jet portion, half angle of conical envelope and average nanofiber diameter, etc can be reasonably modeled. See Fig. 4.11.

Whilst perturbations occur everywhere along the jet path, the instability occurs only at a certain distance away form the syringe tip. One of the major factors that lead to the critical point is the decrease of bending stiffness as the jet diameter decreases along the flow path. Since the bending stiffness is,

Bending Stiffness $= EI$ (4.31)

where E is the Young's modulus whilst the second moment of area can be written as

$$I = \frac{\pi}{64}d^4 = \frac{\pi}{4}R^4$$ (4.32)

assuming circular cross-section, then the bending stiffness for a fixed polymer solution decreases rapidly with the drop in jet radius. The bending instability can be explained on the basis of a perturbed jet (see Fig. 4.12(a)) as follows. With reference to Fig. 4.12(b), when node Y moves to Y' due to perturbation, the two Coulomb forces from X and Z pushes it further away [Reneker et al. (2000)]. Since each Coulomb force is

$$F = \left(\frac{e}{l}\right)^2$$ (4.33)

along the lines XY' and ZY', therefore the horizontal force component from both coulomb forces can be resolved as

$$F_1 = 2F\cos\theta = 2\frac{e^2\delta}{l^3}.$$ (4.34)

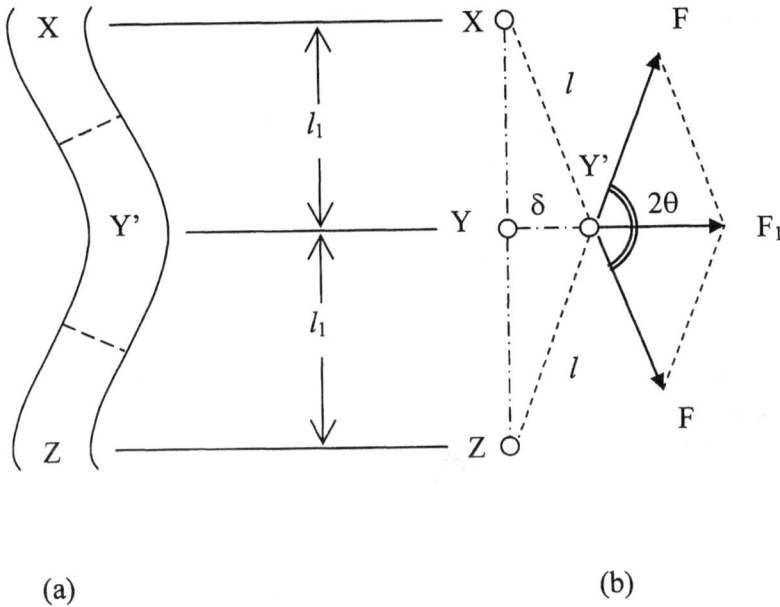

(a)   (b)

Fig. 4.12. Coulombic forces exerted by fluid segments X and Z on Y' [Reneker et al. 2000]: (a) liquid jet, (b) idealized nodes.

However, surface tension counteracts the bending instability because bending always causes an increase of the jet surface area. As such, instability begins when the instability force becomes greater than the surface tension force, while the continued presence of the latter resists the development of too large a curvature by the perturbation XYZ. It follows that perturbation is needed to be induced on the jet flow such that the value of e, δ and l would be enough to produce bending instability, which defines the critical stage. Although two types of perturbation has been proposed in the node approach [Reneker et al. (2000)] for the sole purpose of inducing the bending instability, we herein consider time-dependant perturbation

$$\begin{Bmatrix} x \\ y \end{Bmatrix} = 10^{-3} L \begin{Bmatrix} \sin(\omega t) \\ \cos(\omega t) \end{Bmatrix} \tag{4.35}$$

at syringe tip, i.e. at $z = 0$, where

ω = perturbation frequency
t = time

$$L = \frac{e}{R_0} \sqrt{\frac{1}{\pi G}} = \text{length scale}$$

λ = perturbation wave length,
instead of the spatial perturbation due to the better comparison with experimental observation.

The factor 10^{-3} was arbitrarily chosen in order to induce as small a perturbation as possible, such that the off axis component would be accumulated along the straight jet axis until a stage whereby the jet reaches its critical state that translates into bending instability. With perusal to Fig. 4.13, the circular perturbation at syringe tip leads to a realistic jet flow path.

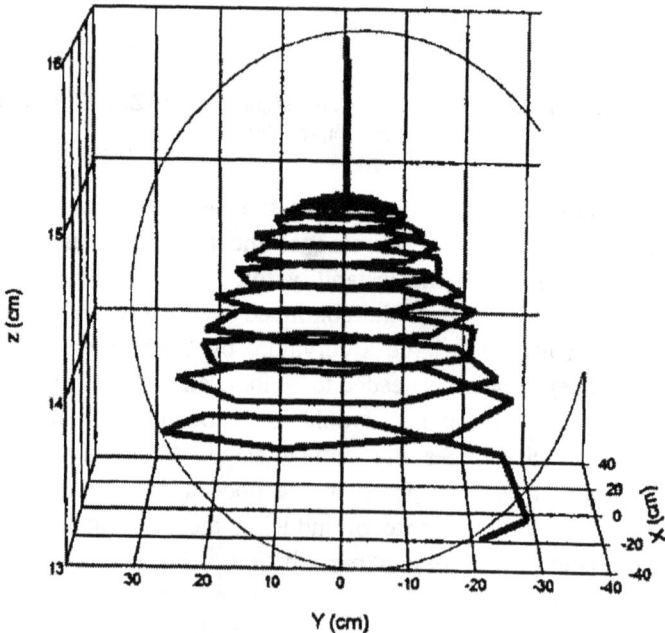

Fig. 4.13. Simulated jet flow as a result of (a) time-dependent perturbations by [Reneker et al. (2000)].

In the model by [Hohman et al. (2001a)], 3 different sources of instabilities were identified: (i) axisymmetric Rayleigh instability, (ii) electric field induced axisymmetric instability, and (iii) bending instability. This approach can be said to be comprehensive in that all instabilities imaginable were accounted for. Perturbation has been incorporated into the jet radius, jet velocity, electric field and surface charge density by introducing a complex number $\exp[\omega t + ikx]$ [Hohman et al. (2001a)]. However, the bending jet flow path was not simulated.

4.7. Results

The simulated results were understandably influenced by the purpose which the modeling works were done. The theoretical treatment by [Feng (2002 & 2003)] allowed the description of jet radius as a function of downward vertical axis (z), Reynolds number (Re), Electric Peclet number (Pe), ratio of electrostatic force to inertia (Fe/I), Deborah number (De) and flow index (n) among others. This flow index, n, is defined from the flow constitutive equation described in Equation (4.10) (see Fig. 4.14 (a)-(d)).

Despite of the discrepancy observed between experiment and theory [Reneker et al. (2000)], the theoretical foundation is nevertheless sound. This can be inferred from the fact that when evaporation and solidification aretaken into consideration, the jet flow path is shown to be more realistic, as evident in Fig. 4.15 [Yarin et al. (2001a)].

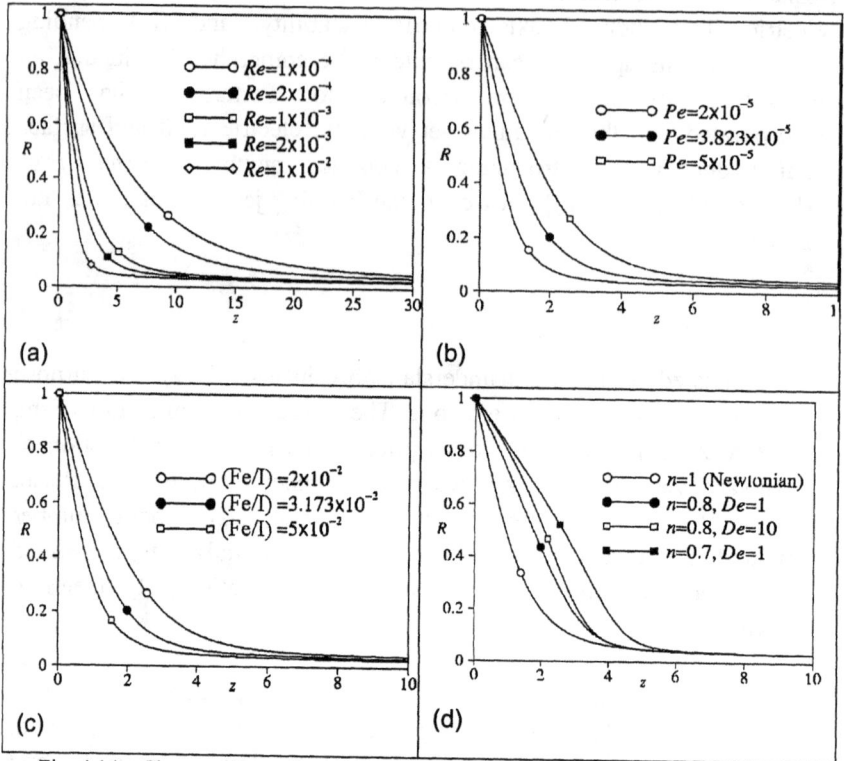

Fig. 4.14. Change of jet radius with distance from the syringe tip as functions of (a) Reynolds number, (b) Electric Peclet number, (c) ratio of electrostatic force to inertia, and (d) Deborah number and flow index, *n*. [Feng (2002)].

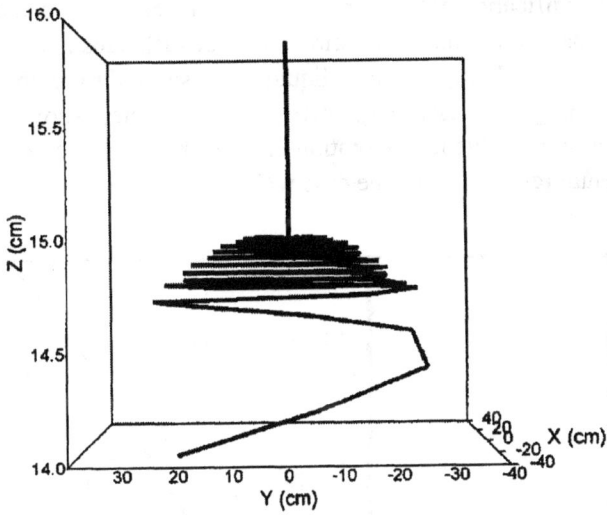

(a)

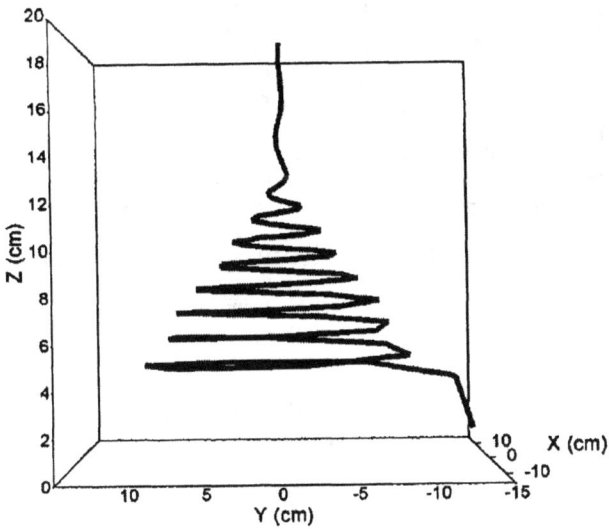

(b)

Fig. 4.15. Jet path calculated (a) without considering evaporation and solidification, and
(b) considering evaporation and solidification [Yarin et al. (2001a)].

The consideration of solvent evaporation is highly important because it implied solidification which, in turn, leads to changes in the material properties as pointed out by [Yarin et al. (2001a)]. This is not surprising in view of the change from a liquid to a solid during the event of electrospinning process. Fig. 4.16, for example, shows that the consideration of solvent evaporation enables the theory to coincide with experimental results, unlike the case without evaporation.

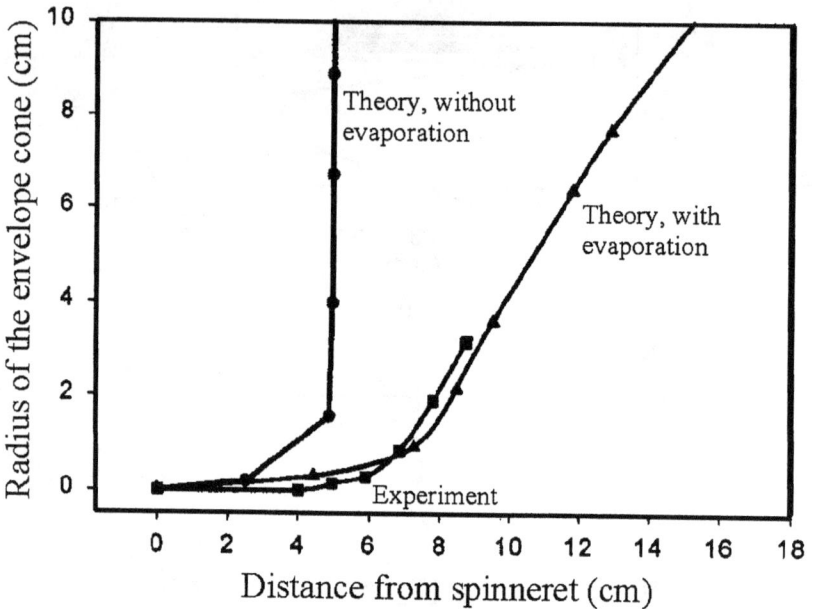

Fig. 4.16.　Jet envelope of a typical electrospinning profile (Adapted from [Yarin et al. (2001a)]).

Hence the trace of beads (such as that shown in Fig. 4.17) is therefore reliable. Consideration of volume constancy, we conceptually have

$$\pi R_0^2 l_0 (C_p + C_s) = \pi R_f^2 l_f (C_p) + \pi R_f^2 l_f (C_s) \qquad (4.36)$$

where

C_p = polymer concentration

C_s = solvent concentration

R_0 = initial jet radius

R_f = final fiber radius

l_0 = initial length of jet segment

l_f = final length of fiber segment.

Practically, the second term on the RHS of Equation (4.36) refers to the volume of solvent that evaporates. Hence the final fiber's segment volume is smaller than its corresponding jet volume, and that, by canceling the evaporated term, the ratio of final-to-initial volume would simply give the polymer concentration

$$C_p = \frac{l_f}{l_0}\left(\frac{R_f}{R_0}\right)^2.$$

(4.37)

By the knowledge of the polymer concentration during experimental preparation C_p, syringe inner radius R_0 as determined by the equipment set-up, and the prescribed initial distance between nodes l_0, one can obtain the fiber radius R_f upon obtaining the computed final distance between nodes l_f.

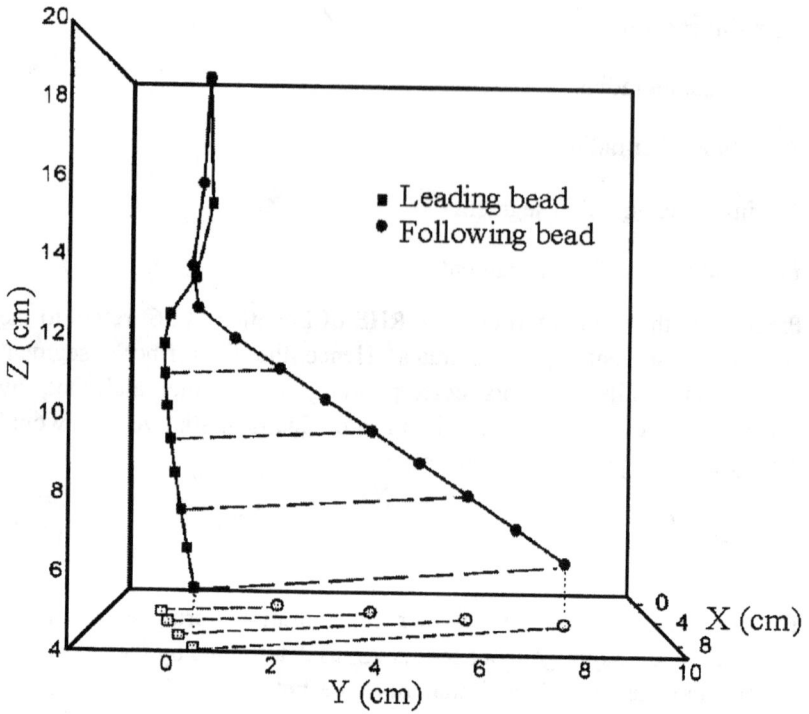

Fig. 4.17. Trace of two nodes [Yarin et al. (2001a)].

It follows that the theoretical results for draw ratio, as furnished in Fig. 4.18, gives a reasonable prediction of the experimental results. Without the consideration of evaporation, the draw ratio unrealistically increases.

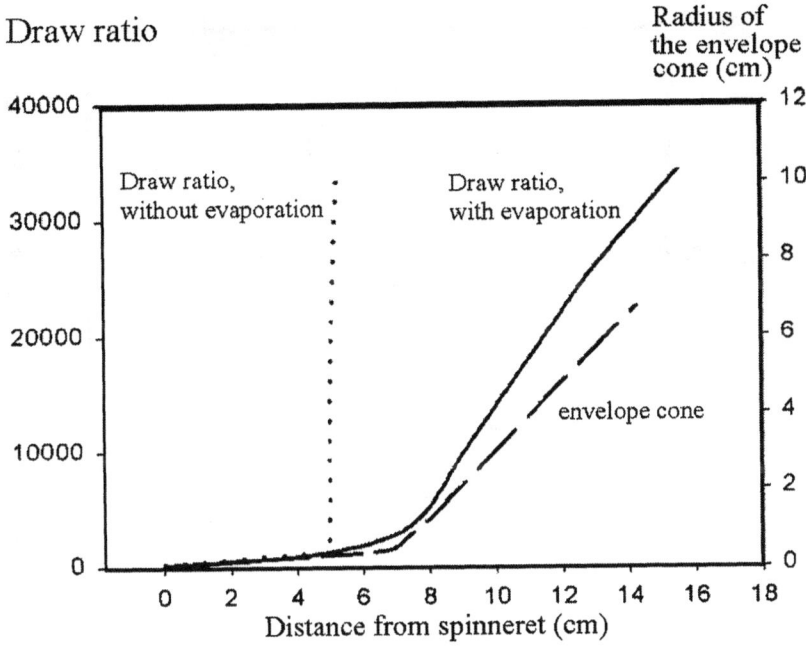

Fig. 4.18. Draw ratio during the event of electrospinning (Adapted from Yarin et al. [2001a]).

It is interesting to note that the viscosity of the jet changes abruptly between the straight jet portion and the bending instability portion shown in Fig. 4.19, thereby implying the almost sudden change from a liquid to solid at the point of instability. If, on the other hand, the rise of viscosity of very gradual such that the jet reaches the collecting late before the viscosity attains that of the solid, then one obtains the interconnected fibrous mesh.

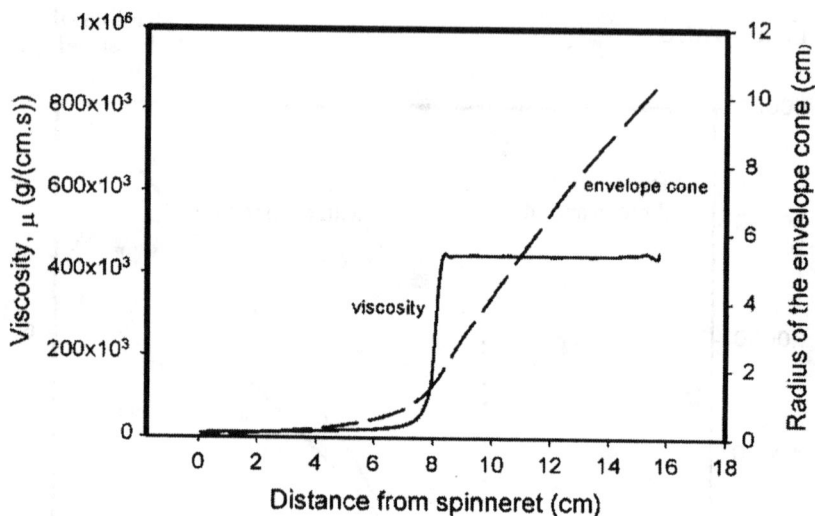

Fig. 4.19. Change in viscosity during the event of electrospinning (Adapted from [Yarin et al. (2001a)]).

An operating diagram showing the states of stability and instability is shown in Figs. 4.20 (a) and (b) for glycerol and polyethylene oxide (PEO) respectively [Hohman et al. (2001b)]. The data points refer to experimental measurements of the instability thresholds for these parameters. The shaded portion on the lower left hand side denotes the occurrence of varicose instability while the shaded region at the upper portion of the diagram refers to bending instability. These theoretical plots are reasonably verified by the experimental data points.

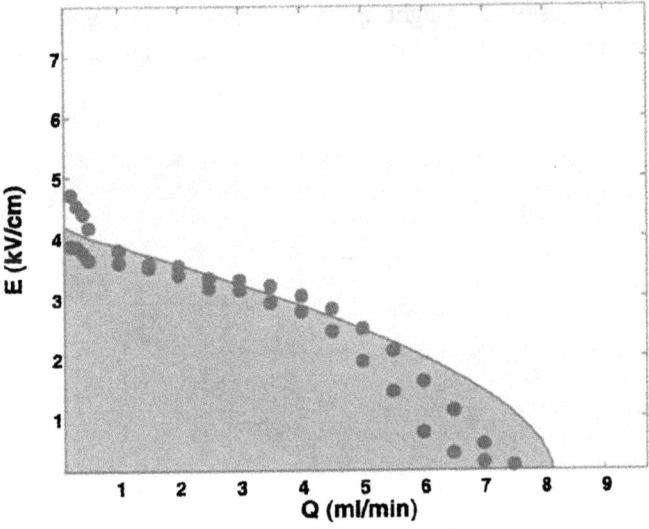

(a)

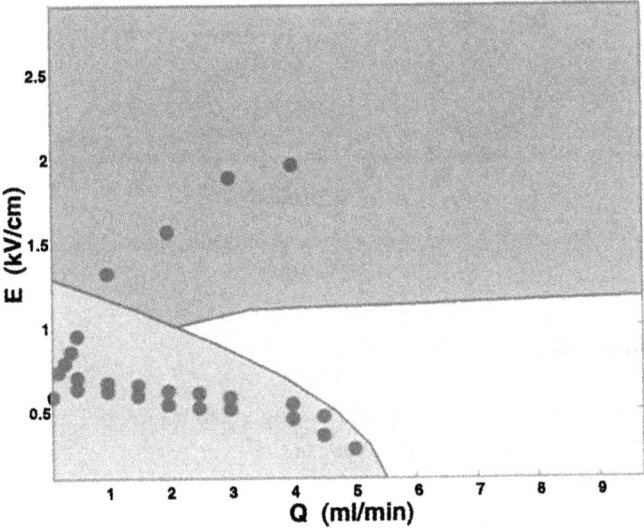

(b)

Fig. 4.20.   Instability regions for the case of (a) glycerol, and (b) PEO [Hohman et al. (2001b)].

When the viscosity is brought up by 10 times as compared to PEO, an operating diagram with all instability modes are displayed, as shown in Fig. 4.21.

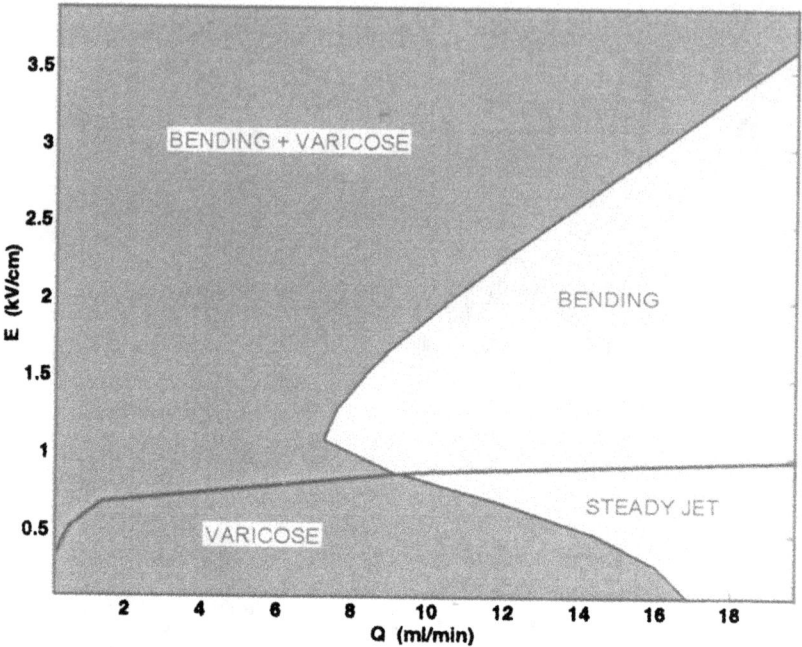

Fig. 4.21. An operating diagram displaying all instability modes by [Hohman et al. (2001b)].

4.8. Future Trends and Challenges

Some issues remain to be addressed in regard to the modeling techniques of the electrospinning process. These are:
1. What other forces are there to be included?
2. The best perturbation method.
3. The most realistic perturbation magnitude.
4. Whether the sign of applied charge at the syringe tip poses any significance.

With the advancement of the electrospinning process and application of the electrospun nanofibers, fabrication of advanced nanofibers are being attempted by means of novel electrospinning processes. Towards this end, the challenges ahead would be of the following:

4.8.1. *Jet Flow With Particles*

Electrospinning of polymer solution that contains solid particulates arises due to the need to embed the nanofibers with fillers in order to cater for higher strength, for development of sensor devices and for incorporating medicine and nutrients. The existence of particulates will inevitably alter the overall liquid viscoelastic properties. Additionally, the presence of nano-scale particulates will require new understanding on their influence to the effective properties and performance of the electrospun nanofibers.

4.8.2. *Core-shell Flow*

This electrospinning of concentric jet solution process was introduced in order to produce core-shell nanofibers and tubular nanofibers. The former is important for the use of nanofibers to deliver time-delayed drugs whilst the latter for nano-scale transport. Tubular nanofibers are obtained when the core material degrades with the passage of time. With the flow of two concentrically distinct liquid, it is expected that the overall jet properties are changed. A challenge for the core-shell and tubular nanofibers is the attainment of consistent shell thickness (or functionally graded shell thickness along the longitude if it so desires) as well as the prevention of shell collapse for the tubular nanofibers.

4.8.3. Field-assisted Flow

So far the process modeling has been confined to cases whereby there nanofibers are collected in a random network. Where alignment of nanofibers are required, the use of rotating disk and pair conductors as collector will definitely induce stretching and change in electric field respectively. Along this direction of thought, the use of electric field to control the jet path for attaining patterned nanofibrous structure has been shown to be technically feasible [Li et al. (2004b)].

4.8.4. Multi-jet Flow

In the multi-jet electrospinning, multi-spinnerets are used to increase the number of jets for high volume production of electrospun nanofibers. Alternatively multi-jet flow can also be performed by needleless approach whereby the polymer solution is electrically forced towards the upward direction. The formation of several Taylor cones on the surface of the polymer container determines the number of simultaneous jet flow. The use of multi-spinneret, however, gives greater controllability as the number of simultaneous jet flow is determined by the number of spinerettes. The mechanics of multi-spinneret electrospinning will differ from that of single syringe electrospinning due to changes in the electric field and jet overlap.

4.8.5. Gas-assisted flow

In the gas-assisted flow process, gas flows parallel to the straight portion of the jet in the region concentrically surrounding the jet. It is believed that the jet flow will be more controllable and that the locus of nanofibrous membrane on the collection plate will be more defined. The influence of aerodynamic forces could be more significant and, if so, the mechanics of the jet flow will be altered.

4.9. Conclusions

As a consequence of the different assumptions and methods, we appreciate the fact that all these models are useful in one way or another. The node model and bending instability [Reneker et al. (2000); Yarin et al. (2001a)] is able to give the jet path and, therefore, would be recommended for experimentalist to design their experiments to control the collection area of electrospun nanofibers. The combined varicose and bending instability model [Hohman et al. (2001a & 2001b)] has been demonstrated to be reliable for predicting the instability modes of the jet using fluid properties and process parameters. It follows, then, that the jet stability can be controlled by refining the process parameters and carefully choosing the polymer for the desired fluid electro-rheological properties.

A selection of modeling methods for the electrospinning process has been briefly surveyed and their corresponding results have been given. Currently, two schools of thought are applicable on the basis of the jet conceptualization – continuous media and discrete nodes. In summary, the main concepts adopted are those of conservation laws (such as the conservation of mass, charge and momentum) and Newton's Second Law of motion. The latter concept enables the acceleration of jet can be obtained by considering the various forces acting on the jet, such as Coulombic, electric field, surface tension, viscoelastic forces, etc. There is also a need to assume the viscoelastic properties of the polymer solution in order to apply its constitutive relations.

Chapter 5

Characterization

Electrospun polymer nanofibers in the form of membrane have many potential applications in the field of bioengineering, environmental engineering & biotechnology, energy & electronics, and defense & security, as introduced in chapter 7. The explanation of specific characterization on each application is omitted here. However, it must be borne in mind that the basic properties of nanofibrous membranes are (1) morphology, (2) molecular structure and (3) mechanical properties. The morphology of single fiber such as average fiber diameter, pores on the fiber surface as well as the morphology of nanofibrous membrane such as porosity, are basic properties of the deposited nanofibers on the collector. Hydrophobic property is also related to the morphology of the nanofiber and nanofibrous membrane. The molecular structure of a nanofiber bears influence on the optical, thermal and mechanical behavior of the nanofibrous membranes. If these nanofibrous membranes are chemically treated after the spinning process, we are interested in the change of chemical structure of the polymer molecules. Mechanical properties of single nanofiber as well as nanofibrous membranes are important because without mechanical stability, the nanofibrous membranes cannot be sued in the abovementioned applications. This chapter introduces to the readers the various characterization techniques for obtaining each physical property.

5.1. Morphology

The Morphology of electrospun polymer nanofibers can be characterized by scanning electron microscope (SEM), field emission SEM (FE-SEM) and transmission electron microscope (TEM). Since nanofiber membranes have porous structure, morphological properties include pore geometry and density. In this section, methodologies to characterize morphology of electrospun polymer nanofibers are furnished. A typical sample preparation and sample observation precautions are given.

5.1.1. Fiber Diameter

The diameter of an electrospun polymer nanofiber can be examined under scanning electron microscope (SEM). The principle layout of SEM is shown in Fig. 5.1 [Campbell et. al. (2000a)]. The electron beam is accelerated by holding the tungsten filament at a large negative potential between 1kV and 50kV and whilst the specimen is grounded. When the electron beam impinges on the material surface, backscattered electrons (BE), secondary electron (SE) and X-rays escape from the material surface. In the display of SEM system, SE is captured by the detector for producing the images (SEI).

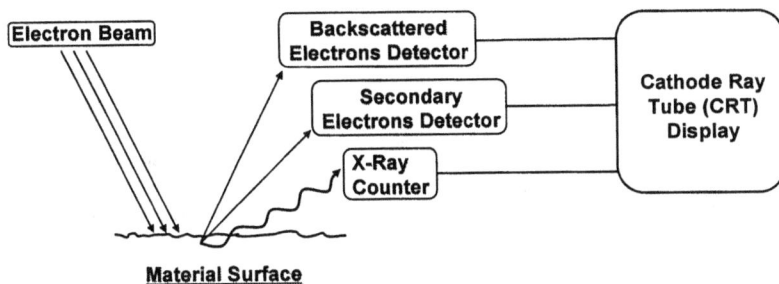

Fig. 5.1. The principle layout of Scanning Electron Microscope (SEM).

If the SEM is equipped with an X-ray detector, particular element on material surface can be mapped with SEM image. SEM sample preparation of electrospun polymer nanofibers can be conducted, as shown in Fig. 5.2. In the electrospinning process, polymer solution is stretched by electrical charge difference between a needle and the earthed collector. During polymer jet is traveling to the collector, solvent is evaporated. After electrospinning, residual solvent may still exist on the nanofibers. Hence, electrospun nanofibers are dried at least one night under vacuum condition. From a completely dried nanofiber membrane, an area of 1cm x 1cm is cut by scissors and attached by means of carbon tape to a copper stub. It is very important at this juncture to ensure that direct adhesion of nanofibers is not recommended since adhesive of carbon tape may damage the nanofibers. This is especially so if the biodegradable polymer nanofibers are treated. Hence, the solution is that spinning is directly done on conductive aluminum sheet at the collector and then the nanofibers on aluminum sheet are attached by carbon tape.

Since polymer nanofibers require conductive coating, gold was selected as coating material due to its ease to vapor deposit and on bombardment with high energy electrons it gives a high secondary yield.

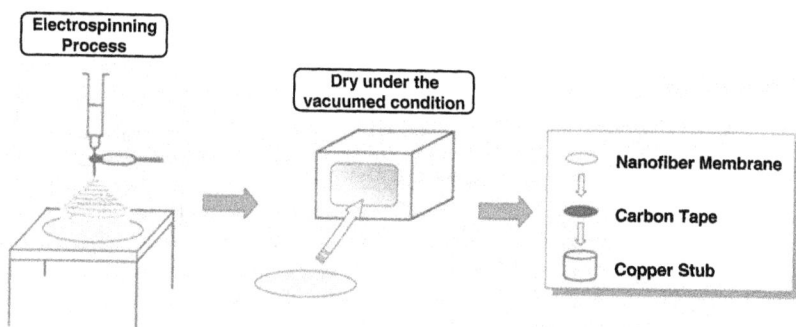

Fig. 5.2. SEM sample preparation of electrospun polymer nanofibers.

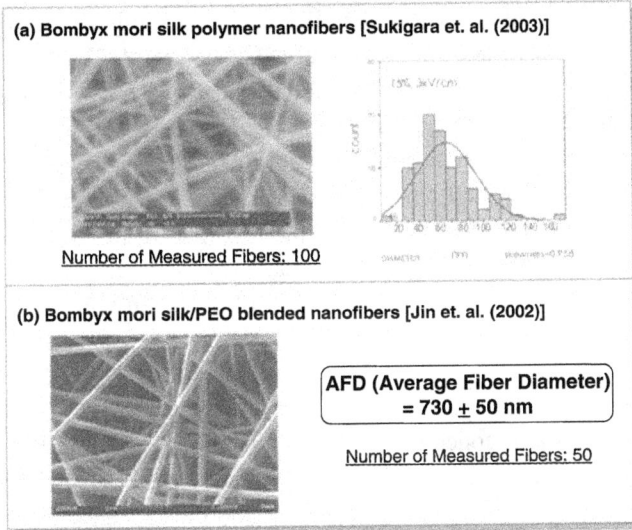

(a) Bombyx mori silk polymer nanofibers [Sukigara et. al. (2003)]

Number of Measured Fibers: 100

(b) Bombyx mori silk/PEO blended nanofibers [Jin et. al. (2002)]

AFD (Average Fiber Diameter)
= 730 ± 50 nm

Number of Measured Fibers: 50

Fig. 5.3.   Examples of fiber diameter measurement of electrospun nanofibers under FE-SEM.

Polymer nanofibers whose diameter is 200nm ~ 1000nm, are observed at around x15,000 magnification with 10 ~ 20kV acceleration voltage under SEM. Basically, each diameter of 50 ~ 100 nanofibers is examined using image analyzer and average fiber diameter and fiber distribution are determined, as shown in Fig. 5.3.   Here, it must be noted that if fiber observation is conducted at extremely high magnification, such as above x15,000 magnification, fiber damage by energetic impinging of electrons takes place.   It is known that a significant temperature rise (tens of degrees) occurs when a material surface is bombarded by an energetic electron beam [Campbell et. al. (2000a)]. Hence, when ultra-fine nanofibers with less than 200nm diameter are observed, the accuracy of measured value is doubtful. If biodegradable polymer nanofibers and biomimetic collagen nanofibers with poor heat resistance are observed, precaution must be taken to prevent fiber damage.

In this regard, Field-Emission SEM (FE-SEM) is highly recommended to observe electrospun polymer nanofibers [Casper et. al. (2004); Demir et. al. (2004); Huang et. al. (2000); Huang et. al. (2001a); Huang et. al. (2001b); Jin et. al. (2002); Kim et. al. (2004e); McKee et. al. (2003); Megelski et. al. (2002); Nagapudi et. al. (2002); Nah et. al. (2003); Ohgo et. al. (2003); Pedicini and Farris (2003); Sukigara et. al. (2003); Zarkoob et. al. (2004)]. The feature of FE-SEM is that high resolution images can be obtained with low acceleration voltage. Fig. 5.4 shows photographs of electrospun polymer nanofibers taken under FE-SEM. It was shown that image of nanofibers are clearly captured with low accelerated voltage.

Another important concern in observing ultra-fine nanofibers is the thickness of the conductive gold coating. Generally, the thickness of gold coating is around 25nm. If the ultra-fine nanofibers are examined under SEM, coating thickness interrupts the accuracy of diameter measurement. To avoid the coating influence, the nanofiber diameter is measured under the Transmission Electron Microscopy (TEM). TEM consists of electron source, condenser lens, objective lens, intermediate lens, projector lens and CRT display, as shown in Fig. 5.5.

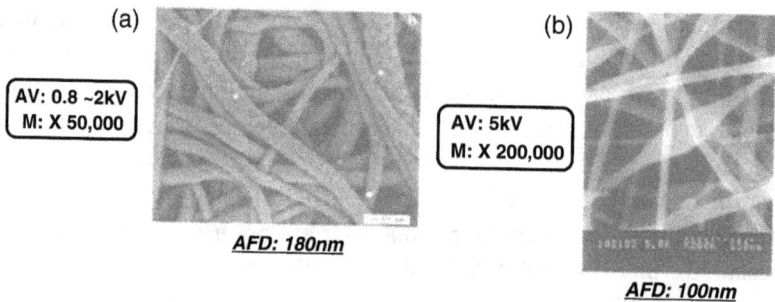

Fig. 5.4. FE-SEM photographs of (a) polyacrylonitrile - acrylic acid (PAN-AA) copolymer nanofibers [Demir et. al. (2004)] and (b) B mori silk nanofibers [Zarkoob et. al. (2004)]. AV, M and AFD stands for accelerated voltage, magnification and average fiber diameter, respectively.

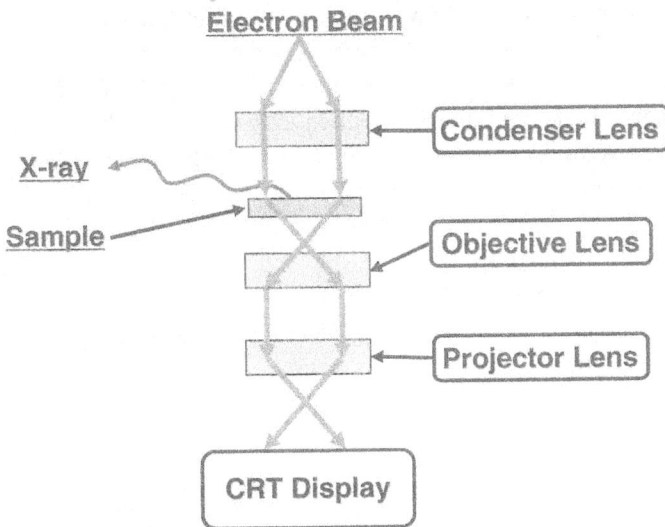

Fig. 5.5. Principle of transmission electron microscope (TEM).

In a TEM, the electron source is generally tungsten filament heated with a low voltage source. The filament is held at a large negative potential and the electrons are accelerated towards specimen with less than 100nm thick. Similar to SEM, X-ray escapes from material surface and the detected X-ray supplies the information of particular element of the sample. After the electron beam passes through the sample, transmitted beams accordingly passes through the other lenses and finally an image is produced at the CRT display.

Fig. 5.6 shows a typical TEM sample preparation method using a metal mesh with 3mm diameter. A metal mesh is subjected to FORMVAR coating and fine supporting polymer film is placed on the metal mesh. Then, carbon coating is further applied to the metal mesh and nanofibers are electrospun on the mesh. Thus, gold coating is not necessary for a TEM sample. [Hou et. al. (2002)] compared the diameter of poly(L-lactide) (PLA) nanofibers using SEM and TEM. The diameter of gold coated PLA nanofibers under SEM was above 10nm while that of uncoated PLA nanofibers was as low as 5nm. This example reveals that TEM observation is a useful methodology to accurately measure the diameter of ultra-fine polymer nanofibers.

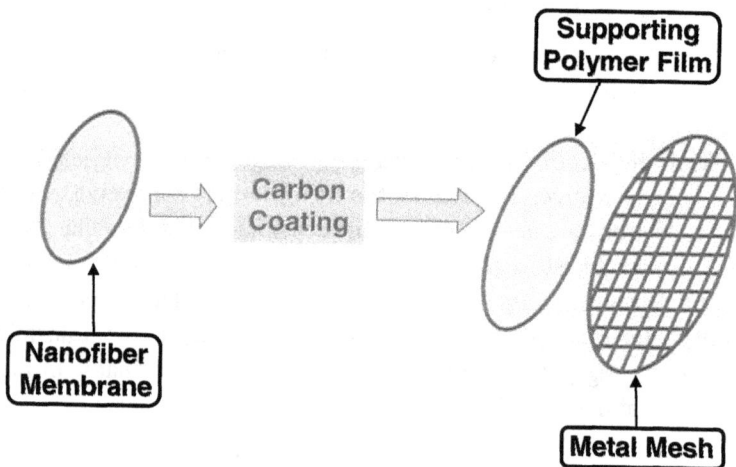

Fig. 5.6. TEM sample preparation method using metal mesh.

5.1.2. Pore Size and Porosity

Pores play an important role in determining the physical and chemical properties of porous substrates and have a deterministic effect on the performance of membranes, catalysts, adsorbents etc. To design electrospun nanofibrous substrates for such applications, it is necessary to analyze the pore-size, its distribution and porosity. In electrospun nanofibrous substrates, two types of pores can be identified: (1) pores on/within each fiber and (2) pores (between fibers) on a nanofibrous membrane. Pore size can be measured by direct or indirect methods. Direct methods involve the use of techniques such as electron microscopy (SEM and TEM) or atomic force microscopy (AFM). Both these techniques can provide detailed and magnified 'images' of the electrospun nanofibers from which pore size and its distribution can be determined. Indirect methods include bubble point measurements, solute retention challenge, molecular resolution porosimetry, extrusion porosimetry and instrusion porosimetry.

Pores on the Surface of Electrospun Fibers

Fig. 5.7 shows poly(L-lactic acid) (PLLA) electrospun fibers having a porous surface structure. These features may also extend to the interior of the fiber (see in Fig. 5.8), as observed by the cross-sectional micrograph of electrospun polysulfone (PSU) fibers. Such structures are very unique and may be useful for drug release applications. As an example, we consider a core region of the nanofiber being filled with a drug element while the shell region of the nanofiber has a porous structure. In this arrangement, the nanofiber becomes a drug-carrying component, whereby the drugs can be released to the surroundings through the porous shell region. As such, drug release rate of nanofibers is definitely influenced by the surface pore size as well as porosity of the fiber. Hence, pore size, its distribution and porosity on single fibers are important parameters.

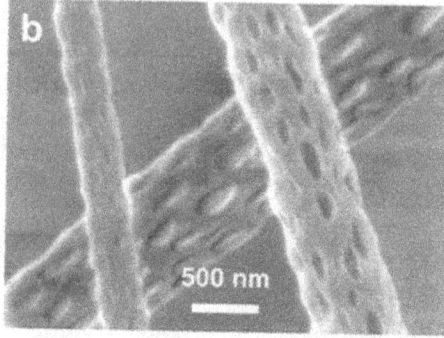

Fig. 5.7.   SEM photograph of porous PLLA nanofibers [Bognitzki et. al. (2001)].

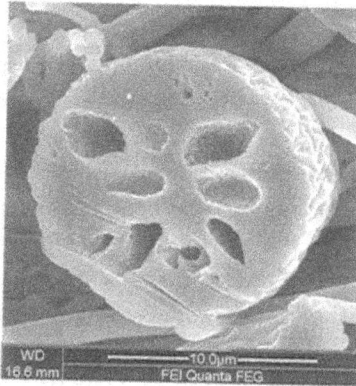

Fig. 5.8.   SEM image of electrospun polysulfone fiber showing porous structure within [Courtesy of Gopal and Ramakrishna at National University of Singapore].

The Atomic Force Microscope (AFM) described in Fig. 5.9 is a useful tool to investigate the various pore sizes on the fiber surface. The AFM consists of a probe (tip), laser, cantilever and detector. It operates by measuring the attractive or repulsive forces between a tip and the sample surface. Based on the sample surface-tip interaction, there are 3 established modes of operation: Contact mode, Non-contact mode and Tapping mode. In contact mode, the tip is in constant contact with the sample surface and scans across the surface. Contact mode can generate image with high resolution, determined by the tip radius. The drawback is a possible damage of both probe and material surface because of their direct contact. This is significant when soft materials are analyzed. On the other hand, although such damage can be avoided using non-contact mode, resolution level is diminished. Hence, most common measurement mode is the tapping mode which consists of both the contact and non-contact modes. The tapping mode overcomes the problems associated with friction, adhesion, electrostatic forces, and other difficulties that can plague conventional AFM scanning methods by alternately placing the tip in contact with the surface to provide high resolution and then lifting the tip off the surface to avoid dragging the tip across the surface.

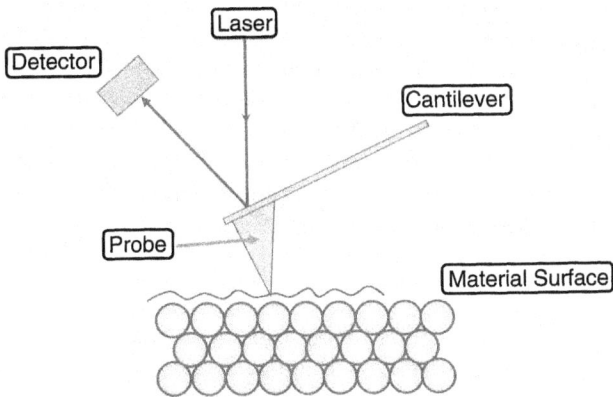

Fig. 5.9. Principle of atomic force microscope (AFM).

Humidity: 60~72%

Range of pore diameters: 200 ~ 1800nm

Average pore diameter: 350nm

(a)

(b)

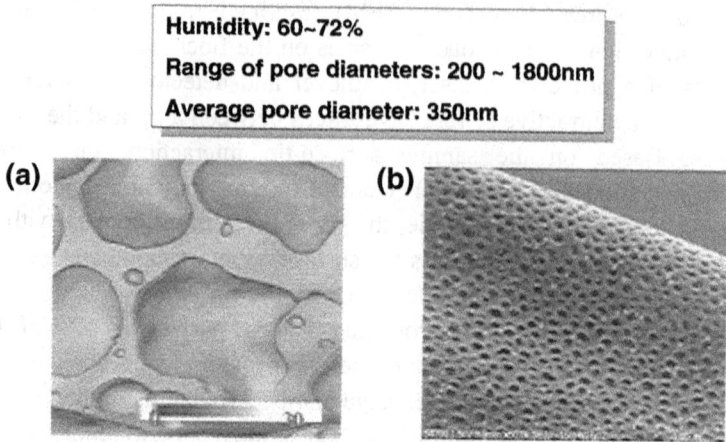

Fig. 5.10.   (a) AFM image and (b) FESEM image of porous polystyrene nanofibers [Casper et. al. (2004)].

In the tapping mode, a probe vibrates with low frequency and softly touches the material surface. Fig. 5.10 shows an example of AFM image of porous polystyrene nanofibers using tapping mode. Polystyrene nanofibers were electrospun with the variation of spinning humidity [Casper et. al. (2004)]. As humidity changes, the evaporation rate of the solvent varies and alters the pore size on the surface of the electrospun fibers. Using an image analyzer, pore distribution and average pore size on the surface of a polystyrene nanofiber were determined to be around 200 ~ 1800nm and 350nm respectively [Casper et. al. (2004)]. The advantage of AFM is that is can measure the pore depth conventional SEM image analysis can only supply the data of pore-size and its distribution.

Another possible method for pore characterization is the use of a molecular resolution porosimetry. Solid surfaces have been characterized by the molecular adsorption theory widely, especially in the fields of catalysis and adsorbents [Kaneko (1994)]. The most common molecule used is N_2 [Gregg and Sing (1982)].

When a surface is exposed to a gaseous environment, gas molecules get adsorbed onto the surface due to the weak van der Waal's type interaction between the surface and the molecules. At constant temperature, the amount of adsorbed gas on a surface depends on the pressure of the gas. Measurement of the amount of adsorption as a function of pressure can give information on the pore structure. The Brunauer Emmett and Teller (BET) theory of physical adsorption is normally used for analysis of adsorption data [Brunauer et. al. (1938)]. The advantage of using a BET analyzer is that it can determine the porosity and the surface area of the substrate on top of pore size and its distribution. To date such methods have not been employed on surface characterization of electrospun fibers. It should be possible to use this technique. The only limitation of this method is that the measured maximum pore size is around 200 nm. Hence it is only suitable for fiber surface analysis and not for membranes.

Pores on Electrospun Nanofibrous Membranes

Fig. 5.11 shows that electrospun nanofibrous membranes possess interstitial spaces between fibers. These are analogous to the pores present in membranes. Thus through electrospinning a highly porous membrane can be generated. In separation technology, the knowledge of pore size and porosity are critical in determining the membrane's performance. Pore size discriminates between the type (size or molecular weight) of species than can permeate through and which will be retained; whereas porosity determines the flux or flow across the membrane. Just like in surface pore analysis, SEM can be used to determine the pore size of electrospun nanofibrous membranes. However, porosity cannot be measured by SEM. Further disadvantage is that only pores on the top surface can be analyzed while pores beneath the membrane surface cannot be directly determined.

The most common methodology adopted for membrane pore characterization is using mercury porosimetry. The principle of mercury porosimeter is shown in Fig. 5.12.

Fig. 5.11. Electrospun polysulfon fibrous membrane, depicting interstitial spaces (pores) between fibers [Courtesy of Gopal and Ramakrishna at National University of Singapore].

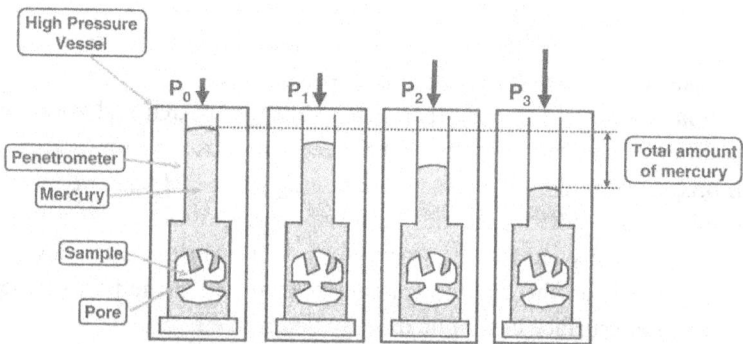

Fig. 5.12. Principle of mercury porosimeter to measure various pore sizes of a sample.

A sample is placed in a cell filled by mercury from a capillary. Since mercury is a non-wetting liquid, it does not instantaneously fill the pores of the sample. However upon application of pressure, the mercury penetrates into pores of a sample. The largest pores get filled first and as the pressure is increased, smaller pores get filled. Pore size (D) is calculated using the following equation:

$$D = -4\,\gamma\,\cos\theta\,/\,P,$$

where γ is the surface tension of mercury, θ is a surface contact angle of mercury on capillary surface and P is the applied pressure.

The advantage of using a mercury porosimeter is that in addition to the pore size and its distribution, the total pore volume and the total pore area can also be determined. [Li et. al. (2002b)] used a mercury porosimeter to analyze poly(D,L-lactide-co-glycolide) (PLGA) nanofiber membrane 1mm thick x 2cm width x 5cm length having a fiber diameter of 500 ~ 800nm. It was revealed that PLGA nanofiber membranes have large porosity structure (92%). Total pore volume was 9.7mL/g and the total pore area was 24m^2/g.

A major drawback of using mercury is that generally very high pressures are required as the pore size diminishes. When thin sections are analyzed, there is a high possibility that the membrane can be destroyed at higher pressures. This is especially true for electrospun nanofibrous membranes. The pores in electrospun membrane are 'dynamic' in nature; they are not rigid pores and self-supporting structures. For successful mercury porosimetry analysis, thick samples are necessary. When samples less than 100μm are used, they are destroyed by the pressure. The other drawbacks of using mercury are its cost and toxicity. It is also possible to use a lower surface tension liquid in place of mercury. In principle, any liquid that does not wet the membrane spontaneously can be used and obeys the abovementioned equation.

By using such liquids, intrusion into pores can be accomplished at much lower pressures than those required with mercury. This greatly reduces the risk of the sample being crushed at high pressures.

Alternatively, instead of intrusion porosimetry, extrusion porosimetry can be employed. In such systems, the sample pores are filled with a wetting liquid. The differential pressure of a non-reacting gas is increased on the sample to displace the liquid from the pores. The liquid displace from the pores and the sample is weighed. Using the aforementioned equation, the pore diameter can be determined. Based on this principle, [Gibson and Gibson (2002)] used a capillary flow porometer, which uses a special low tension fluid, Galden perfluorinated liquid HT 230, to determine the pore- size of their electrospun Estane® 58237. In their study, membranes with only 100μm thickness were successfully characterized. Table 5.1 summarizes the various methods available for pore-size analysis.

5.1.3. *Surface Contact Angle Measurement*

The hydrophobic (or hydrophilic) nature of a substrate has a direct impact on the avenue of its usage. For example, in membrane distillation, where water must not enter the membrane, a hydrophobic membrane has to be used. In separation of protein solutions, hydrophilic membranes are preferred to minimize protein adsorption. Likewise for tissue-engineering scaffolds hydrophilic scaffolds are preferred. The most direct method of measuring these characteristics is via contact angle measurements.

Surface contact angle of electrospun nanofibrous membranes is simply examined by a water contact angle (WCA) machine. A distilled water pendent droplet is injected from a syringe onto the membrane surface. The image of the droplet on the membrane is visualized through the image analyzer and the angle between the water droplet and the surface is measured. Hydrophilic materials show low contact angle (spreading of water across surface) while hydrophobic materials show high contact angle (minimal contact between water droplet and surface).

Table 5.1. Comparison of several analysis techniques to characterize the pore structure of polymer nano fibrous membranes.

Analysis Technique	Principle	Pore Size on Nanofiber Surface	Pore Size on Nanofiber Membrane	Porosity Rate
Electron Microscopy (SEM, TEM)	Uses a beam of electrons to scan the surface of a sample to build a three-dimensional image of the specimen.	Yes (limited)	Yes, but on top surface only	No
AFM	A sharp probe scans a sample surface at a distance over which atomic forces act. The forces between tip and sample cause the cantilever to deflect. A photo detector measures the cantilever deflection and from this information a map of the sample topography can be created	Yes	No (ability decreases with increasing pore diameter)	No
Molecular Resolution Porosimetry (BET Analyzers)	Performed by the addition of a known volume of gas, typically nitrogen, to a solid material in a sample vessel at cryogenic temperatures weak molecular attractive forces will cause the gas molecules to adsorb onto a solid material. Surface area is computed from the amount of vapor adsorbed at a pressure much less than the equilibrium vapor pressure. Pore volume and diameter of pores are obtained from the amount of vapor condensed in pores as a function of vapor pressure.	Yes	No (only pores up to 200nm can be detected	Yes (only fiber porosity)
Intrusion Porosimetry (Mercury Porosimetry)	Pore information is obtained by forcing liquid mercury into pores by increasing the external pressure. This information, as well as information concerning the contact angle, is used to calculate the pore structures.	No (requires through pores)	Yes	Yes
Extrusion Porosimetry (Capillary Flow Porometer)	Pores of a sample are filled with a wetting liquid. Differential pressure of a non-reacting gas is increased to displace the liquid from the pores. Differential pressure yields pore diameter and weight of liquid expelled yields pore volume.	No (requires through pores)	Yes	Yes

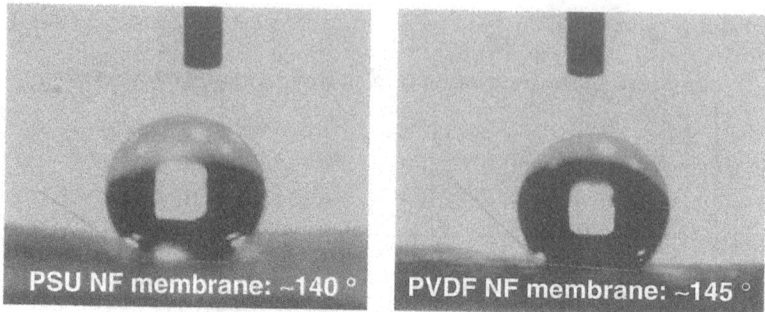

Fig. 5.13. Static Advancing Contact angle measurements on electrospun polysulfon (PSU) and polyvinilidenfluoride (PVDF) membranes [Courtesy of Gopal and Ramakrishna at National University of Singapore].

The most common angle measured is the Advancing Contact Angle. Fig. 5.13 shows the contact angle micrographs of electrospun polysulfone (PSU) and polyvinilidenfluoride (PVDF) membranes. It is interesting to note that compared to PSU films which have a contact angle around 70-90°, the PSU electrospun membranes are exhibiting higher contact angles of around 140°. Thus the fibrous architecture is imparting enhanced hydrophobic characteristics to the membrane. [Acatay et. al. (2004)] had also demonstrated that electrospun architecture can achieve tunable, superhydrophobically stable polymeric fabrics.

Conversely, to reduce the hydrophobicity and make a surface more hydrophilic, surface modification techniques have to be employed. [Fujihara et. al. (2005)] tried to apply hydrophobic polycaprolactone (PCL) nano fibrous scaffolds as bone grafting material. However, the problem was hydrophobic scaffolds were not suitable for osteoblast attachment and proliferation. Hence, air-plasma treatment was adopted to enhance the hydrophilicity of PCL nano fibrous scaffolds. The plasma treatment for 10 minutes decreased the surface contact angle of membranes to zero value. This was possible because plasma treatment introduced polar chemical groups onto the scaffold surface, increasing the surface energy of polymer and thereby decreasing in surface contact angle.

5.1.4. Others

When nanofibers are made of polymer blends or copolymer, component morphology of singe nanofiber is discussed by TEM observation. Fig. 5.14 shows a TEM photograph of bi-component electrospun nanofibers made by co-axial spinning method. TEM observation reveals that core and shell components of single nanofiber are respectively polyethyleneoxide (PEO) and poly(dodecylthiophene) (PDT). The diameter of the core region is measured as about 200nm. Fig. 5.15 is another example of single nanofibers made of styrene-butadiene-styrene triblock copolymer. The nanofibers were stained in osmium tetroxide (OsO$_4$) which can react with polybutadiene phase. It was identified that the dark regions are polybutadiene domains while the lighter shaded domains are polystyrene. Thus, TEM is a powerful tool to identify each material region of nanofibers.

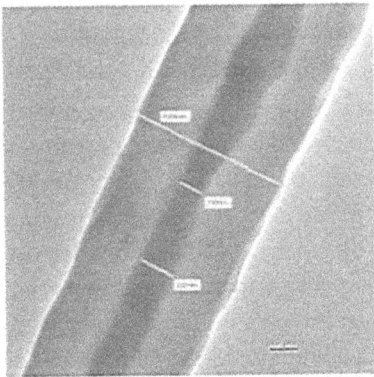

Fig. 5.14. TEM photograph of bi-component electrospun nanofibers. Core component is polyethyleneoxide (PEO) and shell component is poly(dodecylthiophene) (PDT) [Sun et. al. (2003)].

Fig. 5.15. TEM photograph of styrene-butadiene-styrene triblock copolymer nanofibers. [Fong and Reneker (1999)].

5.2. *Molecular Structure*

5.2.1. *Crystalline Structure*

The crystalline structure of electrospun polymer nanofibers has been investigated by many researchers by using X-ray Diffraction (XRD), Differential Scanning Calorimeter (DSC) and Transmission Electron Microscope (TEM) [Bergshoef and Vancso (1999); Bhattarai et. al. (2003); Buchko et. al. (1999); Chen et. al. (2001); Dersch et. al. (2003); Ding et. al. (2002); Fennessey and Farris (2004); Fong and Reneker (1999); Inai et. al. (2005); Lee et. al. (2004); Lee et. al. (2003b); Madhugiri et. al. (2003); Son et. al. (2004c); Zarkoob et. al. (2004); Zhao et. al. (2003); Zhao et. al. (2004); Zong et. al. (2002); Zong et. al. (2003)]. Amongst conventional research works, a few papers discussed crystalline structure of electrospun polymer nanofibers from the viewpoints of size effect, processing parameters and solution properties, as shown in Fig. 5.16. With respect to size effect, when polymer fibers shrink from micrometer to nanometer scale, a question arises on how the crystalline structure is created in nanofibers and how different are they in comparison with the crystalline structure of bulk polymers. Another concern is how processing parameters influence the crystalline structure of the nanofibers. The higher applied voltage may lead to high electrical force which pulls out polymer jets during electrospinning and there must be the difference on crystalline formation as compared to the case where lower voltage is applied during electrospinning. If the nanofibers are collected by a rotating drum or disk, higher rotation speed may lead to higher elongation of polymer jet and alignment of polymer chains along longitudinal direction of a fiber could be expected.

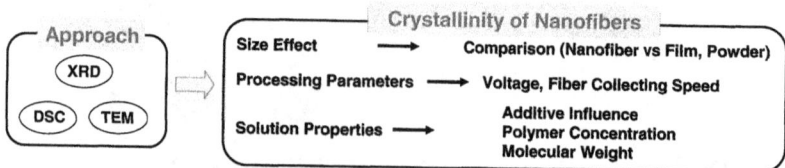

Fig. 5.16. Discussed issues about crystallinity of electrospun polymer nanofibers.

Curiosity has been also placed on the solution properties such as additive influence, polymer concentration and molecular weight. In this section, the principle of methodology to investigate nanofiber crystallinity is briefly described and the discussion shown in Fig. 5.16 is introduced.

The most common arrangement of XRD apparatus is illustrated in Fig. 5.17 [Campbell et. al. (2000b)]. The surface of sample is mounted to the stage with parallel. An incident X-ray beam impinges on the sample which is slowly rotated. At any given position, a multiplicity of Bragg reflections is excited with a crystalline sample and is recorded by a counter. The counter gives a signal proportional to the X-ray intensity. The signal is amplified and displayed on a meter and provide a continuous trace of intensity versus 2θ (θ: incident beam angle). In each semi-crystalline polymer, there is a corresponding intensity peak which shows the crystalline structure. For electrospun polymer nanofibers, the sample preparation was conducted as follows. On the fiber collector table, 13mm diameter coverslips were placed and nanofibers were electrospun. A coverslip is extracted after the spinning and the required sample thickness is around 100μm. Nanofiber membrane on a coverslip is placed on the stage of the diffractometer and characterization is conducted. The description of TEM is omitted as it is already explained in previous section.

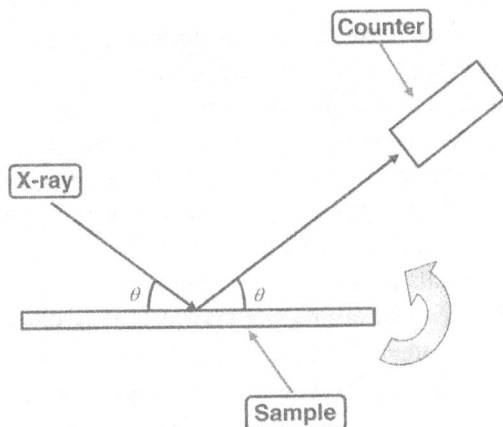

Fig. 5.17. Principle of X-ray Diffraction (XRD).

DSC basically consists of sample and reference holders, heat sink and thermocouples (see Fig. 5.18). Heat flow is delivered from heat sink to sample and reference holders to increase or decrease temperature. At that time, DSC records the required energy to maintain the temperature of both the sample and reference. DSC supplies glass transition temperature and melting point as well as information of crystalline structure.

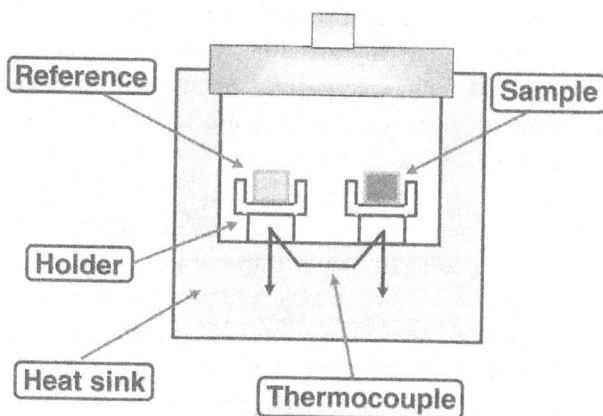

Fig. 5.18. Principle of Differential Scanning Calorimeter (DSC).

Table 5.2 shows the works which discussed the nanofiber crystalline structure from the viewpoint of size effect. There is much curiosity about the difference on crystalline structure between bulk polymer and electrospun polymer nanofibers. As seen in Fig. 5.19(a), [Lee et. al. (2003b)] compared the crystallinity of polycaprolactone (PCL) nanofibers and PCL film. In XRD profile, there was a significant decrease in intensity ratio (I_{110}/I_{200}) of nanofibers, which can be seen during uni-axial tensile deformation process of orthorhombic semi-crystalline polymers. This implies that chains and crystal in nanofibers has some orientation to the same direction of fiber axis during electrospinning process. However, as the retardation of crystallization occurred in nanofibers, crystallinity of PCL nanofibers was lower than that of PCL film.

Table 5.2. Crystallinity of electrospun polymer nanofibers discussed from size effect viewpoint. In each case, nanofibers were collected by rotating drum and fiber orientation of nanofiber membranes was random.

Reference	Material	Fiber Diameter	Discussed Comparison	Method
Lee et. al. (2003b)	Polycaprolactone (Mn = 80,000)	200 ~ 400 nm (50%)	1. Casting Film 2. Nanofibers	XRD
Bhattarai et. al. (2003)	PPDO/PLLA-b-PEG Copolymer (Mw = 42,000)	380 nm	1. Polymer Powder 2. Annealed Nanofibers 3. Dried Nanofibers	XRD
Zong et al (2002)	PLLA (Mw = 100,000)	Unknown	1. Polymer As-received 2. Dried Nanofibers	XRD & DSC

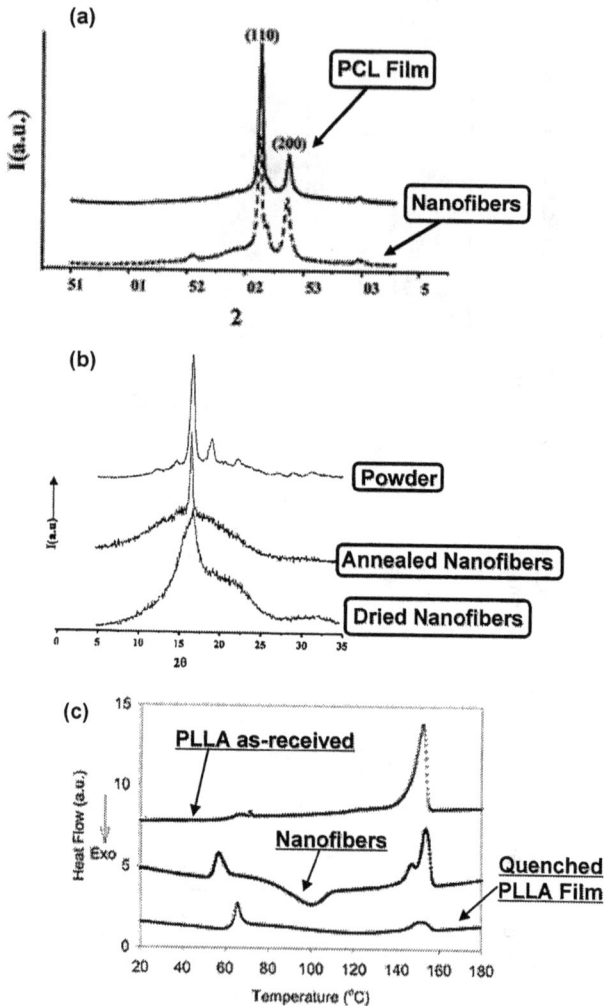

Fig. 5.19.   Crystallinity structure comparison between electrospun polymer nanofibers and bulk polymer: (a) Comparison between PCL film and nanofibers [Lee et. al. (2003b)], (b) Comparison among copolymer powder, annealed nanofibers 50°C for 2 hours and dried nanofibers at 23°C [Bhattarai et. al. (2003)] and (c) Comparison among PLLA as-received, nanofibers and quenched PLLA film [Zong et. al. (2002)].

In the case of [Bhattarai et. al. (2003)] (Fig. 5.19(b)), XRD characterization was conducted on copolymer powder (PPDO:PLLA:PEG = 20:70:10) and electrospun copolymer nanofibers. An XRD curve of copolymer powder revealed the characteristic crystalline peaks at 2θ = 16.4 and 18.7° corresponding to the PLLA segments whereas the other segments of the copolymer (i.e., PPDO and PEG) were not indicated because of the lower material ratio. However, the sharp diffraction peaks seen in copolymer powder were not observed in dried nanofibers. It is interesting to note that the annealed nanofibers at 50°C for 2 hours showed the clearer peak than dried nanofibers. This implies that there are some orders of polymer chains in electrospun nanofibers. As shown in Fig. 5.19(c), the DSC result of [Zong et. al. (2002)] showed that as-received PLLA exhibited 36% crystallinity degree whereas the electrospun PLLA nanofibers exhibited significant lower value. However, XRD result supported that although the polymer chains are non-crystalline in the nanofibers, they are highly oriented. These three cases summarize that electrospun polymer nanofibers do have order of polymer chains although their crystallinity is not significantly high. During polymer jets are traveling to the fiber collector, the electrical force do affects the polymer chain order on nanofibers. In shown reports, it must be noted that randomly aligned nanofibers are collected using rotating drum which implies that the fiber collecting speed is not high. Hence, it can be considered that the order of polymer chains could be also affected if higher voltage is applied to make nanofibers or fiber collecting speed is much faster. Such curiosity can be known from the results shown in Table 5.3 and each case is briefly introduced as follows.

Fig. 5.20 shows how molecular order of electrospun fibers is affected by the applied voltage although fiber diameter is in sub-micron meter level [Zhao et. al. (2004)]. Ethyl-cyanoethyl cellulose (E-CE)C fibers were electrospun on the plate collector with certain distance and then the applied voltage was varied. The crystallinity of electrospun ethyl-cyanoethyl cellulose (E-CE)C fibers increased with increasing the voltage.

Table 5.3. Crystallinity of electrospun polymer nanofibers discussed from the viewpoint of processing parameters.

Reference	Material	Fiber Diameter	Discussed Processing Parameter	Method
Zhao et. al. (2003)	Ethyl-Cyanoethyl Cellulose (Mn = 97,000)	Sub-Micron	Voltage (30 ~ 60kV)	XRD
Fennessey and Farris (2004)	Polyacrylonitrile (Mn = 200,000)	Less than 800nm	Fiber Collecting Velocity (Rotating Drum: 0 ~ 720 m/min)	XRD
Inai et. al. (2005)	PLLA (Mw = 300,000)	890 nm (63 m/min) 610 nm (630 m/min)	Fiber Collecting Velocity (Rotating Disk: 63 & 630 m/min)	XRD

However, the crystallinity decreased with increased voltage above 50kV. It was considered that the fiber crystallinity is affected by the effect of electrostatic field and the time of crystallization during the flight of jet. When the voltage initially increased from zero, the polymer molecules are more ordered. However, the time of the jet flight from the needle tip to the plate collector is shortened with further voltage increase of the electrostatic field, which means that the time of crystallization of (E-CE)C fibers is shortened. In other words, the higher voltage and the longer distance between needle tip and the fiber collector are key points to obtain the more ordered polymer chain structure in the case of the plate collector.

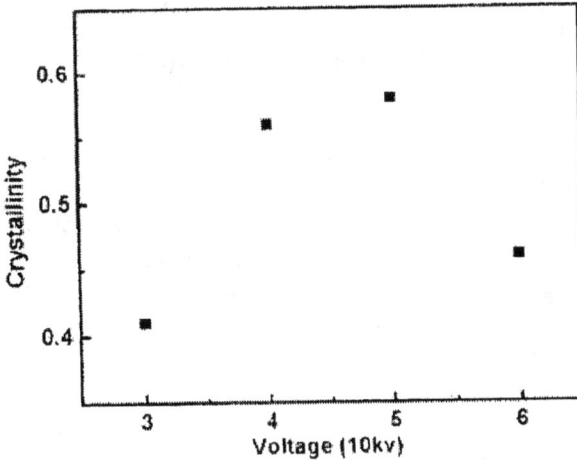

Fig. 5.20. Relationship between crystallinity of electrospun (E-CE)C fibers and the applied voltage [Zhao et. al. (2004)].

The next example is the influence of fiber collecting speed (in the case which uses a rotating drum or a disk) on the crystallinity of nanofibers. If rotation speed of a drum or a disk is faster, polymer jets are strongly stretched which may lead to the more ordered structure. The XRD result shown in Fig. 5.21 reveals this hypothesis.

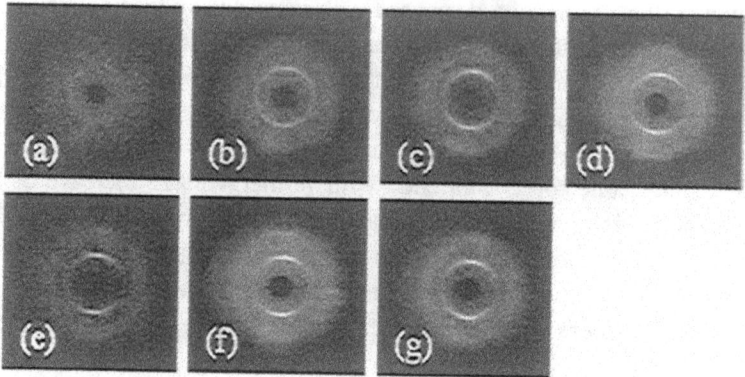

Fig. 5.21 WAXD of electrospun polyacrylonitrile nanofibers collected by a rotating drum with a surface velocity of (a) 0 m/s, (b) 3.5 m/s, (c) 6.1 m/s, (d) 8.6 m/s (e) 9.8 m/s, (f) 11.1 m/s and (g) 12.3 m/s [Fennessey and Farris (2004)].

Polyacrylonitrile nanofibers were collected by a rotating drum with varied speed [Fennessey and Farris (2004)]. XRD diffraction pattern of the nanofibers showed two equatorial peaks; a weak peak at $2\theta = 29.5°$ corresponding to (1120) reflection and a strong peak at $2\theta = 16°$ corresponding to (1010) reflection. The arc width of the strongest equatorial reflection provides an indication of the degree of molecular orientation within the fibers. Hence, it was found that the more ordered molecular orientation is induced when nanofibers are collected by a high-speed rotating drum. [Inai et. al. (2005)] also reported that a speed of rotating disk gives influence to the polymer chain order of nanofibers, as shown in Fig. 5.22. Their XRD result showed that as-spun poly(L-lactic) acid (PLLA) nanofibers did not show any intensity peak, which is seen in other electrospun nanofibers [Bhattarai et. al. (2003); Lee et. al. (2003b); Zong et. al. (2002)]. Once as-spun nanofibers experienced annealing, an intensity peak was seen in fibers collected with both low and high rotation speed. Furthermore, nanofibers collected by a high-speed rotating disk indicated a clearer intensity peak as compared to the nanofibers collected by a low-speed rotating disk. This implies that ordered structure on PLLA nanofibers was more induced in the case of the collection with higher rotating speed.

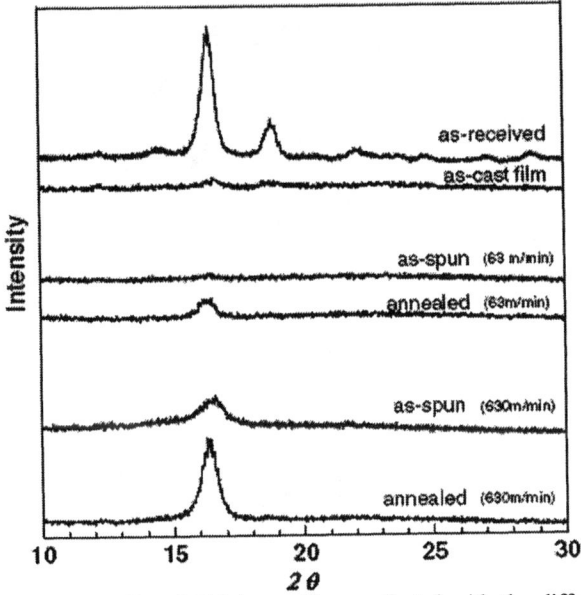

Fig. 5.22 XRD profile of PLLA nanofibers collected with the different disk rotation speed. There are not intensity peaks on as-spun nanofibers. However, as-anneal nanofibers indicated the intensity peak which implies the polymer chain order [Inai et. al. (2005)].

Thus, electrospinning parameters affect the polymer chain order of nanofibers. In order to get ordered chain structure in nanofibers, higher voltage in certain range must be considered with using both flat plate collector and rotating drum or disk collector. Furthermore, fiber collecting speed is also a key point in the case of the rotating collector.

Discussion of solution parameters on nanofiber crystallinity is summarized in Table 5.4. It is intriguing how molecular weight, polymer concentration and additives in polymer solution affect the structure forming of nanofibers. Fig. 5.23 is an example of polyvinylalcohol (PVA) nanofibers crosslinked with the varied amount of glyoxal. The purpose to crosslink PVA nanofibers is to improve waterproof and mechanical property of nanofibers. XRD result showed that (b) neat PVA nanofibers have crystalline structure greater than (a) PVA powder.

Table 5.4. Crystallinity of electrospun polymer nanofibers discussed from the viewpoint of solution properties.

Reference	Material	Fiber Diameter	Discussed Solution Parameter	Method
Ding et. al. (2002)	PVA (Mn = 85,000)	280 nm	Addition of Glyoxal (enhance crosslinking)	XRD & DSC
Inai et. al. (2005)	PLLA (Mw = 100,000)	810 nm (DCM:Pyridine = 50:50) 2000 nm (DCM:Methanol = 80:20)	Addition of Pyridine & MeOH (enhance solution conductivity)	XRD & DSC
Lee et al. (2004)	Atactic PVA	270 ~ 280 nm	Molecular Weight (*Pn = 1,700 & 4,000)	XRD & DSC
Zong et. al. (2003)	PLGA (PLA:PGA = 10:90) (Mw = 75,000)	Beaded Fiber (7.5wt%) 400 nm (10wt%) 1000 nm (15wt%)	Polymer Concentration (7.5, 10, 15wt%)	XRD
Inai et. al (2005)	PLLA (Mw = 100,000)	630 nm (7.5wt%) 810 nm (12.5wt%)	Polymer Concentration (7.5 & 12.5wt%)	XRD & DSC

*Pn: number-average degrees of polymerization

However, PVA nanofibers became amorphous with increasing the amount of glyoxal. This phenomenon was supported by DSC result which indicated no melting point with the degree of crosslinking increased. The addition of pyridine and methanol to enhance solution conductivity is conducted by [Inai et. al. (2005)]. Since conductivity of pyridine is higher than methanol, the more ordered molecular structure was expected on PLLA nanofibers by adding pyridine. As shown in Fig. 5.24, there was no intensity peak on nanofibers made from solutions with both pyridine and methanol. However, once nanofibers experienced annealing, nanofibers made of the solution with pyridine showed higher intensity than nanofibers made of the solution with methanol. Thus, high conductive solution leads to the more oriented molecular order inside of nanofibers.

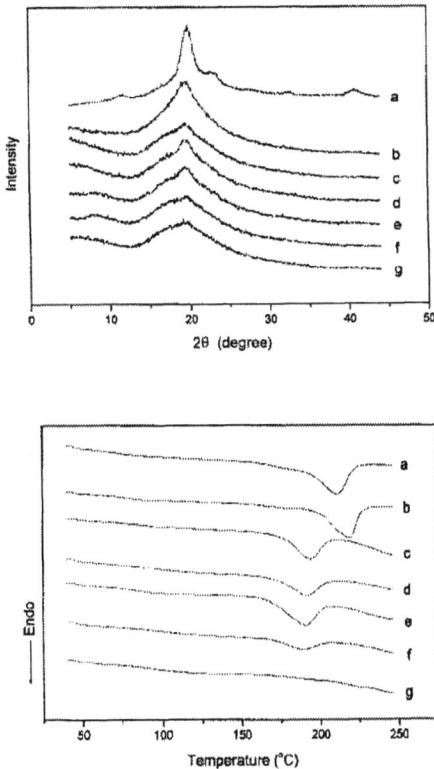

Fig. 5.23. XRD profile and DSC curves of (a) PVA powder and nanofibers crosslinked with (b) 0 wt%, (c) 2 wt%, (d) 4 wt%, (e) 6 wt%, (d) 8 wt% and (e) 10wt% of crosslinking agent (glyoxal) [Ding et. al. (2002)].

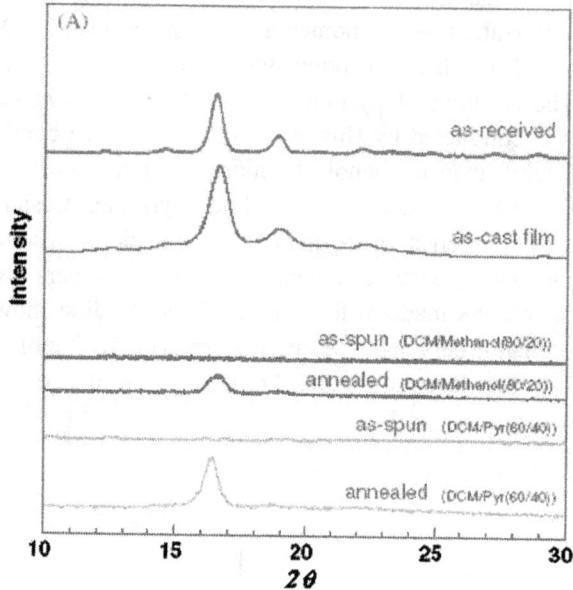

Fig. 5.24. XRD profile of PLLA nanofibers made from the solutions which contain methanol and pyridine [Inai et. al. (2005)].

Fig. 5.25 shows that XRD profile and DSC curves of atactic polyvinylalcohol (PVA) nanofibers [Lee et. al. (2004)]. According to the XRD profile, PVA nanofibers with high molecular weight (P_n=4,000) indicated higher and sharper intensity peaks as compared to nanofibers with low molecular weight (P_n=1,700). Two intensity peaks correspond to (001) and (101) plains of PVA crystalline. It is considered that this is attributed to longer chain length and higher linearity of PVA chain. DSC result also supported high crystallinity of PVA nanofibers (P_n=4,000) since melting temperature of PVA (P_n=4,000) indicated 8°C higher than that of PVA (P_n=1,700). This result reveals that higher molecular weight induces the higher crystallinity on nanofibers. Here, the question arises if polymer concentration is varied with both in low and high molecular weight, what happens to the nanofiber structure.

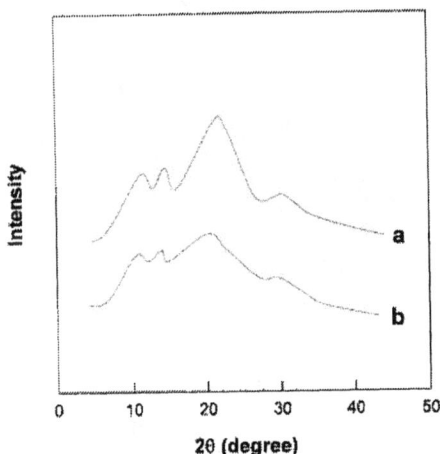

Fig. 5.25. XRD profiles of atactic polyvinylalchohol (PVA) nanofibers with different molecular weights (a) Pn=4,000 and (b) Pn=1,700 [Lee et. al. (2004)]. Pn: number-average degree of polymerization

Although there is not such comparison in the references, two examples may be useful information to address this question. Poly(glycolide-*co*-lactide) (PLGA) nanofibers (PLA:PGA = 10:90) with Mw=75,000 were electrospun with varied polymer concentration [Zong et. al. (2003)]. In their XRD result (see in Fig. 5.26), PLGA nanofibers formed from lower concentration (7.5 and 10wt%) exhibited only amorphous peaks while nanofibers from 15wt% showed two crystalline peaks with an intermediate degree of crystallinity (30%). However, XRD result of PLLA nanofibers (Mw=100,000) showed different tendency [Inai et. al. (2005)] (see in Fig. 5.27). Although as-spun PLLA nanofibers with both high and low concentration did not show intensity peaks, both fibers showed an intensity peak after annealing. However, the intensity peak was higher on nanofibers with low molecular weight. [Inai et. al. (2005)] pointed out that a flow-induced molecular orientation is also related to polymer chain order of nanofibers. In other words, when polymer solution is fed out from a syringe to a needle, mobility of molecular chains is large with decrease of polymer concentration.

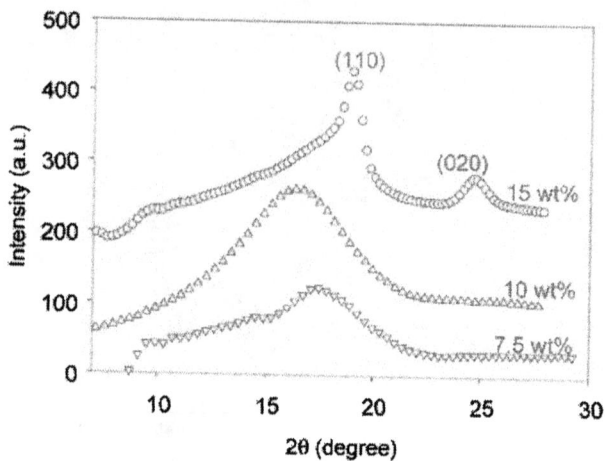

Fig. 5.26. XRD profiles of PLGA nanofibers electrospun from different polymer concentrations [Zong et. al. (2003)].

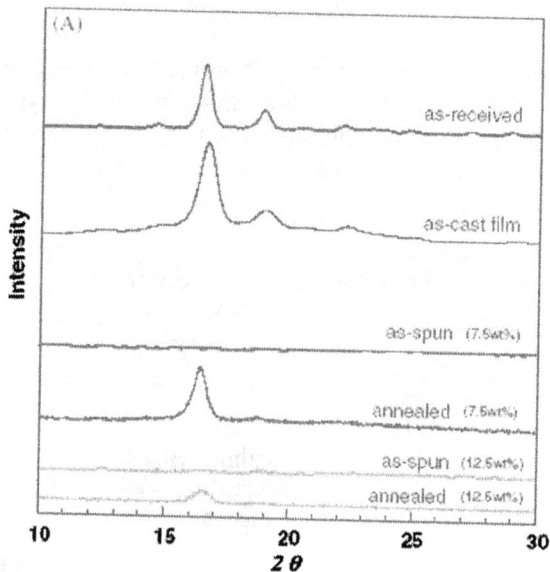

Fig. 5.27. XRD profiles of PLLA nanofibers and annealed nanofibers from different polymer concentrations [Inai et. al. (2005)].

As overall comprehensive, it is likely that electrospun polymer nanofibers have ordered molecular structure although crystallinity of nanofibers is relatively low as compared to bulk polymers. Furthermore, spinning parameters and solution properties do affect the degree of molecular order in nanofibers. It must be noted that the understanding of the molecular order in nanofibers seems to be more difficult if solution parameters are changed as compared to another cases. The reason is that the influence of polymer concentration and molecular weight may be dependent on the chain structure of a polymer and the polymer chain conformation. Those polymer structural parameters also give influence to a flow manner at a fiber spinneret. Thus, further study from polymer chemistry and flow mechanics viewpoints may be required to understand the crystalline structure of electrospun polymer nanofibers from the viewpoint of solution parameters.

5.2.2. Organic Group Detection

For functional group detection on electrospun polymer nanofibers, Fourier Transform Infrared Spectroscopy (FT-IR) which has one moving mirror, one fixed mirror, a beam splitter and a detector, is utilized (see in Fig. 5.28) [Cambell et. al. (2000c)]. As infrared frequency corresponds to molecular frequency, infrared spectroscopy sensitively reflects molecular structure of material. As shown in Table 5.5, two cases (i.e., influence by spinning process and chemical or physical reaction after spinning) are found to investigate chemical functional groups which exist in electrospun polymer nanofibers. As to the influence by spinning process, the concern thing is how chemical structure of polymer is influenced by electrospinning process.

DAM-1 electrospun fibers made of Vitamin E TPGS and DAM-1 powder were examined by FT-IR [Madhugiri et. al. (2003)] and there was no difference on both FT-IR spectrums. It was confirmed that vitamin element exists on nanofibers and original chemical structure was not affected by high voltage.

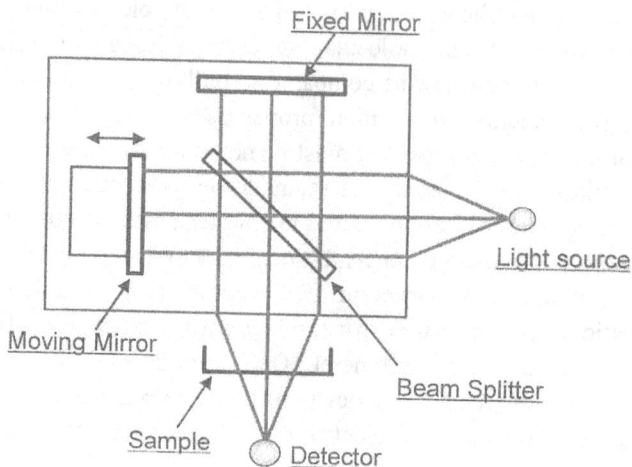

Fig. 5.28. Principle of Fourier Transform Infrared Spectroscopy (FT-IR) apparatus.

On the other hand, chitosan-PEO (chitosan:PEO = 1:1) electrospun fibers prepared by [Duan et. al. (2004)] showed that electrospinning process affects the chemical component. Casting films of original raw materials (i.e., chitosan, PEO and chitosan-PEO) were compared with chitosan-PEO electrospun fibers via FT-IR. FT-IR result showed that C=O-NHR absorption band at 1650 cm^{-1} and –NH$_2$ absorption band at 1550cm^{-1} were found in chitosan film and chitosan-PEO film. However, chitosan-PEO nanofibers did not possess those absorptions while nanofibers possessed the triplet peaks of the C-O-C stretching vibrations appeared at 1148, 1101 and 1062 cm^{-1} seen in PEO film. It was elucidated that chitosan experienced phase separation in chitosan-PEO solution due to the high voltage and shear force of feeding process and no chitosan element is found on nanofibers electrospun from chitosan-PEO solution. As seen in those examples, chemical composition of electrospun polymer nanofibers is occasionally affected under high voltage spinning condition. This is especially so that if highly conductive solvent is chosen to dissolve polymer, there is a possibility that molecular chain is cut or damaged by high electrical charge.

Table 5.5. Fourier Transform Infrared Spectroscopy (FT-IR) examination on electrospun polymer nanofibers.

Reference	Material	Discussed Issue	Comparison
Duan et. al. (2004)	Chitosan : PEO = 1:1	Influence of Spinning Process	Bulk (film) & Nanofibers
Madhugiri et. al. (2003)	DAM-1		Bulk (powder) & Nanofibers
Liu and Hsieh, (2003)	Cellulose		Monomer Grafting Efficacy
Son et. al. (2004c) and Liu and Hsieh, (2002)	Cellulose Acetate	Chemical or Physical Reaction after Spinning	Deacetylation Treatment after Spinning
Jiang et. al. (2004)	Methacrylated Dextran		Cross-linking Degree by UV-light

After making electrospun polymer nanofibers, certain applications may require surface modification to attach chemical function onto the nanofiber surface. For instance, filter media which intends to capture toxic molecular needs to attach such functional group onto the surface of nanofibers. In such cases, the efficacy of surface modification has to be monitored after chemical treatment of nanofibers. [Liu and Hsieh (2003)] conducted monomer grafting with 2-step reactions on cellulose nanofibers made of cellulose acetate. Methyl methacrylate (MMA) was initially grafted to the surface of cellulose nanofiber membranes and further three different monomers were respectively grafted from attached methacrylate groups. Deacetylation of electrospun cellulose acetate nanofibers was also reported by [Son et. al. (2004c); Liu and Hsieh (2002)]. Cellulose acetate nanofibers were deacetylated with several chemical reaction times and the degree of acetylation was monitored by FT-IR spectrum. Another example is methacrylate modified dextrane nanofibers [Jiang et. al. (2004)].

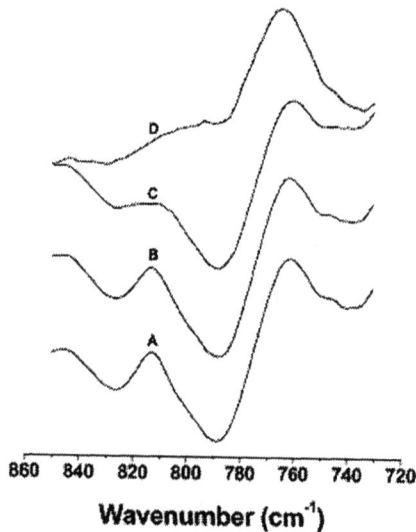

Fig. 5.29.   FT-IR spectrum of methacrylated dextran nanofibers irradiated with UV-light for 0 hours (a), 2 hours (b), 24 hours (c) and 48 hours (d) [Jiang et. al. (2004)].

Photo-cross linking of the modified dextrane nanofibers was encouraged with photoinitiator under UV light. FT-IR spectrum monitored the kinetics of cross-linking with reaction time (Fig. 5.29) and the absorption peaks at 763cm^{-1} and 813 cm^{-1} seen in methacrylated dextran without UV reaction eventually disappeared with reaction time.

5.2.3. *Others*

For electrospun polymer nanofibers, synthesis of polymer or attaching function group to polymer is conducted before electrospinning process. In some cases, nanofibers are chemically modified after spinning process. In addition to the FT-IR analysis, Nuclear Magnetic Resonance (NMR) analysis also provides a means for identifying molecular structure of polymer nanofibers [Huang et. al. (2001a); McKee et. al. (2003); Nagapudi et. al. (2002); Ohgo et. al. (2003)]. Principle of NMR is described in Fig. 5.30. When a sample is placed in non-magnetic field, a nuclear spin of an atom is rotating with various directions (Fig. 5.30(a)).

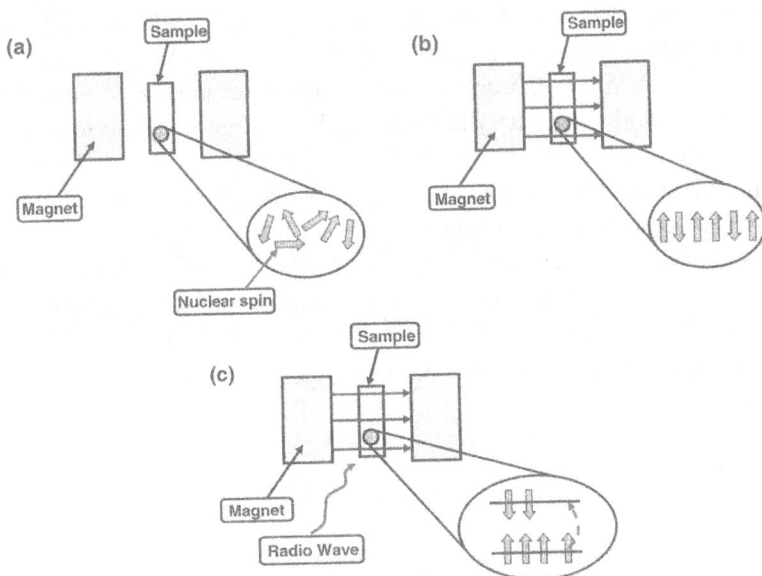

Fig. 5.30. Principle of nuclear magnetic resonance (NMR).

However, once high magnetic field is applied to nuclear spins with certain direction, each direction of nuclear spins is divided to two directions (Fig. 5.30(b)). Since nuclear spins in the opposite direction of magnetic field possess higher energy state, Fig. 5.30(b) can further be described with energy state (see Fig. 5.30(c)). In this condition, when the radio wave is applied to nuclear spins, energy state of nuclear spins jumps to higher level with absorption of electrical magnetic wave. NMR detects this electrical magnetic wave and number of ^{1}H and ^{13}C of a sample can be determined. Moreover, chemical shift is measured from ^{1}H and ^{13}C NMR spectroscopy. As certain value of chemical shift corresponds with the chemical functional group, the chemical structure of polymer chain can be determined. [McKee et. al. (2003)] fabricated nanofiber of linear and branched poly(ethylene terephthalate-co-ethylene isophthalate) synthesized from ethylene glycol, dimethyl terephthalate and dimethyl isophthalate. ^{1}H NMR spectroscopy revealed that synthesized polymers contained diethylene glycol units due to self-condensation of ethylene glycol. The other example is Bombyx mori silk nanofibers treated with methanol after fiber formation [Ohgo et. al. (2003)]. The structure of silk fibroins and residual solvent during the fiber formation process were monitored by ^{13}C NMR spectrum. Fig. 5.31 shows ^{1}H NMR spectrum of elastin-mimetic polymer chemically modified with methacrylic anhydride before fiber formation. Incorporation of the double bonds through peaks at 5.3 (H^a) and 5.6 (H^b) ppm was confirmed. Moreover, acrylate modification produced a 0.25 ppm (H^d) downfield shift of the methylene protons α to the amino group (H^c). Since this elastic-mimetic polymer has a cross-linking feature under visible light irradiation, cross-linking efficiency was also confirmed by ^{13}C NMR spectrum (see in Fig. 5.32). The disappearance of double bonds was seen in the irradiated sample.

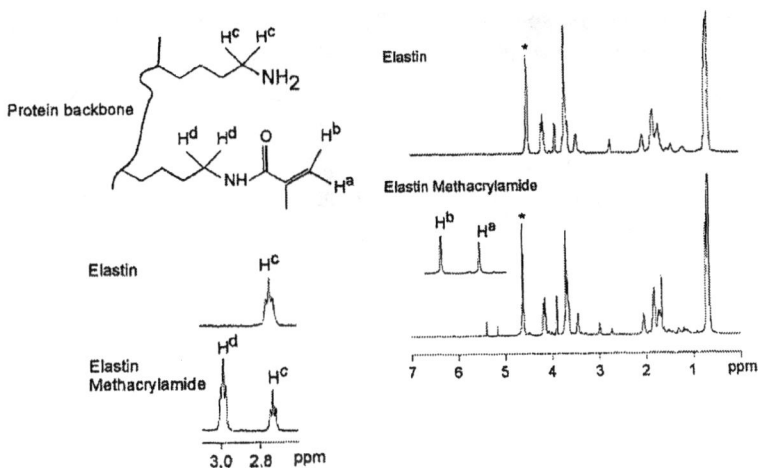

Fig. 5.31. 1H NMR spectra of elastin-mimetic polymer (elastin) and methacrylic modified elastin (elastin methacrylamide) [Nagapudi et. al. (2002)].

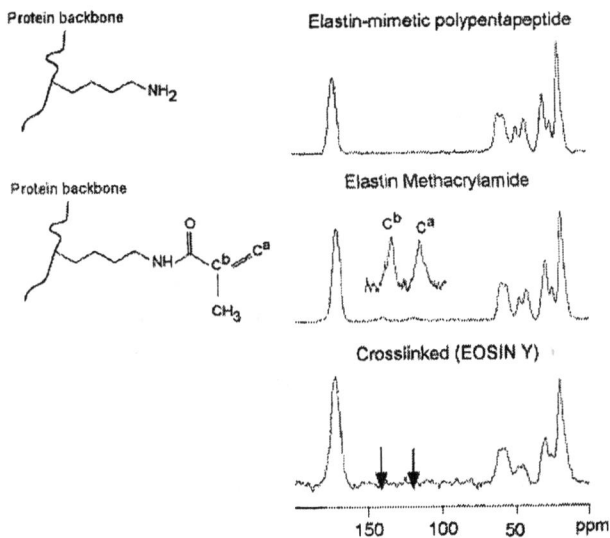

Fig. 5.32. 13C NMR spectra of an elastin-mimetic polymer, methacrylic modified elastin and crosslinked elastin [Nagapudi et. al. (2002)].

X-ray photoelectron spectroscopy (XPS) is another method to investigate the chemical structure of nanofibers [Deitzel et. al. (2002); Duan et. al. (2004); Jin et. al. (2004)]. As shown in Fig. 5.33, when a material is irradiated by soft X-ray, a photoelectron is effused from the material surface due to photoelectric effect. As a photoelectron possesses inherent energy of the atom, chemical element at material surface can be estimated. It is known that the obtained data is arisen from 10nm surface depth.

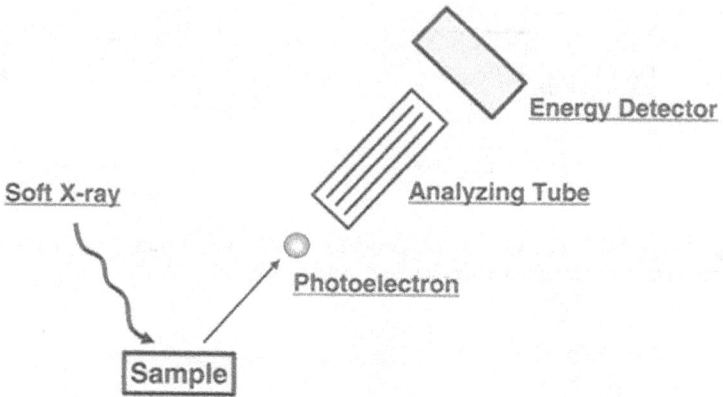

Fig. 5.33.   Principle of X-ray photoelectron spectroscopy.

[Deitzel et. al. (2002)] investigated the surface of random polymethylmethacrylate (PMMA) - tetrahydroperfluorooctylacrylate (TAN) copolymers as film and electrospun conditions. Fluorinated fiber surface was expected for stain resistant and water repellent materials and coatings application. XPS result (see in Fig. 5.34) showed that the atomic percentage of fluorine in the near surface region of the electrospun fibers is about double the atomic percentage of fluorine found in a bulk sample of random copolymer. [Jin et. al. (2002)] electrospun silk/poly(ethylene oxide) (PEO) nanofibers from water solvent with varying weight ratio of silk and PEO. Since silk is relatively hydrophobic, a lower content of silk was expected at the surface of nanofibers.

In order to confirm the hypothesis, XPS was used to estimate the surface composition of nanofibers and it was shown that the composition of silk and PEO at the surface of nanofibers was almost same as blend composition in solution. The other example is blended chitosan/PEO (weight ratio = 50:50) microfibers and nanofibers electrospun from acetic acid solvent. Atomic percentage of C_{1S}, O_{1S} and N_{1S} at the surface of microfibers and nanofibers were investigated and it was shown that chitosan/PEO microfibers only contained C_{1S}, O_{1S}, whereas N_{1S} appeared on the surface of nanofibers. With using FT-IR spectra, XPS result was understood that microfibers mainly contained PEO while nanofibers consisted of chitosan due to phase separation of the chitosan/PEO solutions during electrospinning process.

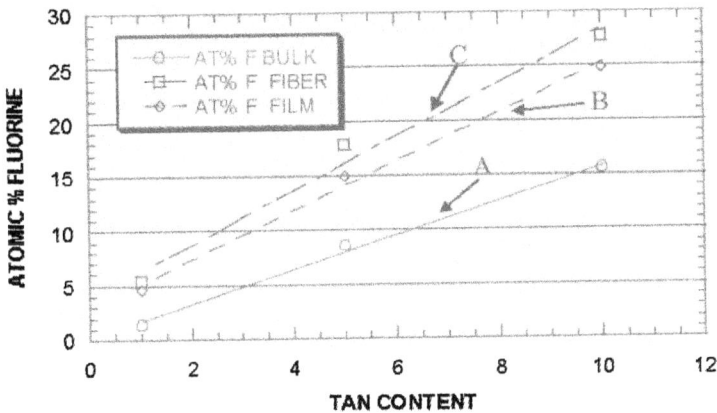

Fig. 5.34. Atomic percent content of fluorine as a function of TAN content; (a) bulk polymer, (b) near surface of thin films and (c) near surface for electrospun fibers [Deitzel et. al. (2002)].

5.3. Mechanical Property

Electrospun polymer nanofibers have several possible applications such as filtration, tissue scaffolds, protective clothing, gas sensors etc. In order to meet with long-life durability in those applications, mechanical properties of nanofibrous membrane as well as a single nanofiber has to be discussed. From the material science view point, the size effect on single fiber property should be an interesting topic. There arise some questions on what happens to the mechanical properties of a single fiber when the fiber diameter shrinks from micron to nano size level. In addition, a question is raised on how fiber diameter shrinks to nano level during fiber processing since crystalline structure of a single fiber is affected by processing condition even in a same diameter of two types of fibers. In terms of the mechanical testing of nanofibrous membranes, traditional tensile testing method is applied. However, the testing of single nanofiber needs much know-how since it is not an easy task to prepare the sample and to apply the loading to a tiny nanofiber. Hence, single fiber testing is limited as compared to nanofibrous membrane testing. In this section, methodologies of mechanical testing on electrospun polymer nanofibers have been introduced and their validity has been discussed.

5.3.1. Single Nanofiber

As summarized in Table 5.6, there are few number of scientific papers discussed about the mechanical property of electrospun single nanofibers. The reasons should be the handling difficulty to extract a single fiber from the electrospun nanofiber web and also non-availability of accurate testing apparatus. In the literatures, three testing apparatus are available to measure tensile and bending properties of a single nanofiber, i.e., 1.Cantilever technique [Buer et. al. (2001)], 2.AFM-based nanoindentation system [Tan and Lim (2004)] and 3.Nano tensile tester [Tan et. al. (2005); Inai et. al. (2005)]. Schematic methodology of cantilever technique is described in Fig. 5.35 [Buer et. al. (2001)].

Table 5.6. Summarized data of mechanical property of single nanofibers.

Material	Fiber Collector	Fiber Diameter	Apparatus	Modulus (GPa)	Strength (MPa)	Reference
PLLA	Plate	270 ~ 410 nm	AFM-based Nanoindentation	0.1~1 (bending)	----	Tan and Lim (2004)
PCL	Frame on the Plate	1000 ~ 1700 nm	Nano Tensile Tester	0.12	40	Tan et. al. (2005)
PLLA	Disk	890 nm (63 m/min) 610 nm (630 m/min) [(·) Disk Rotation Speed to collect Nanofibers]	Nano Tensile Tester	1 2.9	89 183	Inai et. al. (2005)
PAN	Plate	1250nm	Own Setup (Cantilever Technique)	---	308	Buer et. al. (2001)

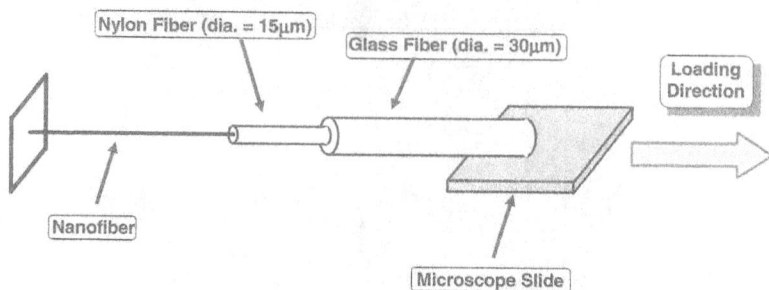

Fig. 5.35. Cantilever technique for tensile testing of single nanofiber [Buer et. al. (2001)].

A 30μm diameter glass fiber, the cantilever, was glued at one end onto a microscope slide and a 15μm nylon fiber was attached at the free end of the glass fiber. Single nanofiber was glued at the free end of the nylon fiber by epoxy adhesive. While the sample fiber was stretched, the deflection of the cantilever was measured with a calibrated eyepiece. The diameter of tested single nanofiber was measured under SEM. The elongation-to-break of electrospun PAN (polyacrylonitrile) fibers was estimated with a caliper. The diameter of tested PAN nanofiber was 1250nm and fiber tensile strength was 300MPa. Using AFM-based nanoindentation system, bending test of PLLA (poly-L-lactic-acid) single nanofiber was conducted by [Tan and Lim (2004)], as shown in Fig. 5.36. A single nanofiber is suspended over the etched groove in a silicon wafer with 4μm wide and 2.5μm deep. An AFM cantilever tip is then used to apply a small deflection at the midway of span length (4μm). Crosshead speed of cantilever tip was 1.8μm/s and the applied maximum load to a single nanofiber was 15nN. Bending modulus of single nanofiber is calculated by beam bending theory with measured deflection and force of a fiber. Fiber diameter of single nanofiber is varied from 270nm to 410nm and it was found that the modulus value decreased with increasing fiber diameter. Nano tensile tester (NanoBionix, MTS, USA) was used to conduct tensile test of single nanofiber and a sample preparation method is respectively described by [Tan and Lim (2004)] and [Inai et. al. (2005)].

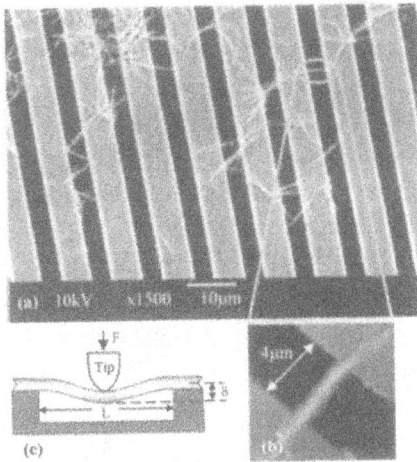

Fig. 5.36. Bending test of PLLA single nanofiber using AFM-based nanoindentation system: (a) PLLA nanofibers deposited onto the silicon wafer, (b) AFM contact mode image of a single nanofiber suspended over an etched groove and (c) Schematic diagram of a nanofiber with mid-span deflected by an AFM tip [Tan and Lim (2004)].

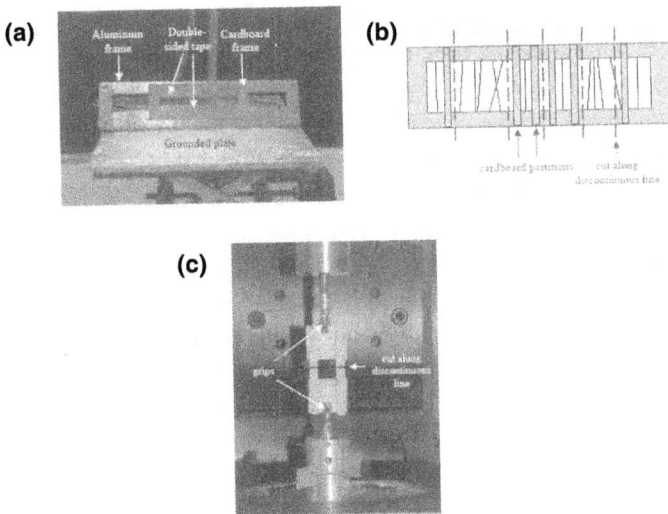

Fig. 5.37. Preparation of tensile specimen of single PCL nanofiber using nano tensile tester; (a) Nanofiber collection way, (b) Extraction method of single nanofiber and (c) Tensile specimen set on nano tensile tester [Tan et. al. (2005)].

As shown in Fig. 5.37, [Tan et. al. (2005)] prepared tensile specimens as follows. First, an aluminum frame with parallel strips was placed at the collector with 60° angel to the horizontal. Then a cardboard frame with 10mm gap between the parallel strips was attached to the aluminum frame. Double sided tape was then pasted on the cardboard frame to attach nanofibers to the frame. An each single fiber is partitioned with cardboard strips and finally extracted from the aluminum frame for tensile test. The gauge length of a specimen was 10mm and the crosshead speed was 1%/s. Polycaprolactone (PCL) fibers with 1000 ~ 1700nm diameter indicated tensile modulus of 120 $\pm$ 30MPa, tensile strength of 40 $\pm$ 10MPa, strain at break of 200 $\pm$ 100%, yield stress of 13 $\pm$ 7MPa and yield strain of 20 $\pm$ 10%.

The sample preparation method by [Inai et. al. (2005)] is described in Fig. 5.38. A disk collector with attachable table was used to collect single nanofibers. In order to get well aligned fibers, conductive plates are placed nearby a paper tab holder. Well - aligned single nanofibers were collected on a paper tab and unnecessary fibers were removed except one single nanofiber located at the center of a paper tab. The prepared sample was examined by nano tensile tester. A PLLA single nanofiber was collected with two different disk rotation speeds (63m/min and 630m/min). In both cases, the average diameter of a single nanofiber was same around 730 ~ 760nm. The gauge length of a specimen is 20mm and the crosshead speed was 0.4%/s. A PLLA nanofiber collected with low speed showed tensile modulus of 1.0 $\pm$ 1.6GPa, tensile strength of 89 $\pm$ 40MPa and strain at break of 1.54 $\pm$ 0.12%. On the other hand, the fiber collected with high speed showed tensile modulus of 2.9 $\pm$ 0.4GPa, tensile strength of 183 $\pm$ 25MPa and strain at break of 0.45 $\pm$ 0.11%.

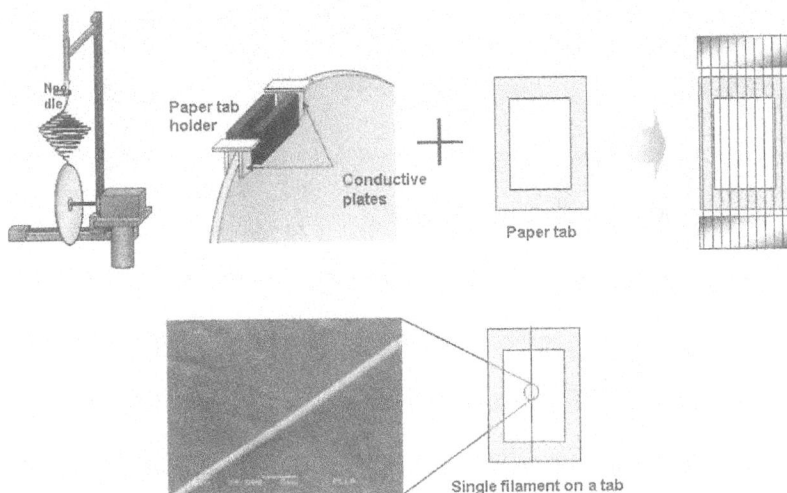

Fig. 5.38.   Preparation method of tensile specimen of PLLA single nanofiber [Inai et. al. (2005)].

In the report by [Buer et. al. (2001)], a question was raised on how a single nanofiber is extracted from electrospun nanofiber web. Another question is the accuracy to detect load and deflection of a single nanofiber. Hence, the obtained tensile strength of a PAN nanofiber may not be a reliable value. On the other hand, testing using AFM-based nanoindentation system and nano tensile tester are useful techniques from the testing accuracy viewpoint. As seen in Fig. 5.39, stress-strain behavior of a single PLLA nanofiber obtained by nano tensile tester possesses high resolution of testing. In terms of the sample preparation way, the nanofiber collection apparatus suggested by [Inai et. al. (2005)] is able to collect a single nanofiber with well-alignment. Furthermore, a single nanofiber can be collected with low and high drawing ratio. In other words it is possible to discuss the tensile property of a single nanofiber with varied crystalline structure using this setup.

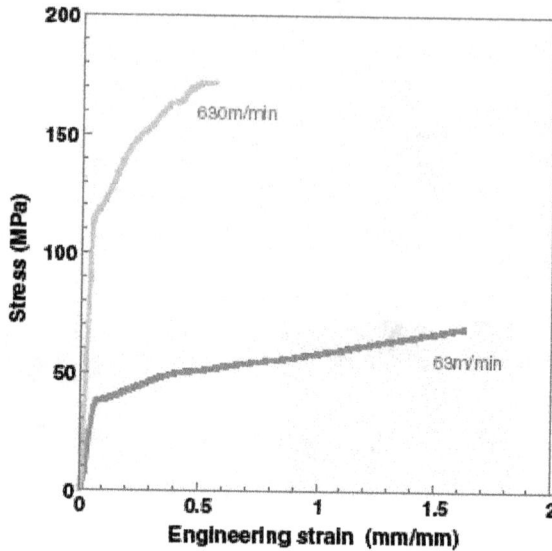

Fig. 5.39.   Stress-strain behavior of a single PLLA nanofiber [Inai et. al. (2005)].

5.3.2. Nanofiber Yarn

The tensile property of PAN nanofiber yarn was examined by [Fennessey and Farris (2004)]. Tensile specimen was prepared as follows. Electrospun PAN nanofibers were firstly collected on the rotating drum with a velocity of 9.8m/s. Unidirectional tows with 32cm x 2cm area extracted from an electrospun nanofibrous sheet were twisted using an electric spinner. The twisted yarns were rinsed in deionized water for 24h and then dried under vacuum at 100°C. Twisting angle of yarns was varied from 1.1 to 17° and the average yarn denier was 446. Prepared nanofiber yarns were examined using an Instron with a crosshead speed of 2mm/min. The gauge length of a specimen was 20mm. Tensile modulus and strength of a PAN nanofiber yarn were respectively 3.8 ± 1.1GPa and 91.1 ± 5.5MPa with a twisting angle of 1.1°. Those values increased gradually to 5.8 ± 0.4GPa and 163 ± 12 MPa with a twisting angle between 9.3 and 11°. However, modulus and strength of a yarn decreased with a twisting angle greater than 11°.

Mechanical property of nanofiber yarns is important as well as that of single nanofibers. The reason is the creation of textile fabrics made of nanofiber yarns which have not been found in recent scientific papers. In order to create nano textile fabrics, nanofiber yarns composing textile structure must be mechanically investigated as nanofiber yarns go through textile performing machine. Nano textile fabrics would have several potentials in medical and chemical separation, and chemical protection [Scardino and Balonis, 2001; Laurencin and Ko, 2004]. Hence, if the potential of nano textile fabrics will be eventually recognized in near future, the discussion about the mechanical property of nanofiber yarns would be hot.

5.3.3. *Nanofiber Membrane*

Tensile properties of electrospun polymer nanofibrous membranes in the literatures are summarized in Table 5.7. For specimen preparation, dumbbell and rectangular shapes are available, as shown in Fig. 5.40. Although rectangular shape specimen is easily prepared, the question is stress concentration vicinity grip part. Hence, there is the possibility to get the exaggerated testing value. On the other hand, dumbbell shape specimen precisely removes the stress concentration and the obtained testing values are cross to true properties. However, preparation of dumbbell shape specimen of nanofiber membranes is costly since sharp dumbbell shape blade must be used. Used tensile apparatus is conventional universal testing machine and testing speed is 10mm/min in 6 cases [Bhattarai et. al. (2003); Lee et. al. (2003b); Wnek et. al. (2003); Lee et. al. (2002a); Khil et. al. (2004); Huang et. al. (2004)]. Testing speed depends on testing purpose, for instance, slow testing speed is suitable to observe membrane deformation manner. As seen in Table 5.7, there are two different fiber collecting ways on discussion of tensile properties of nanofiber membranes.

Table 5.7.   Tensile properties of polymer nano fibrous membranes collected by rotating drum.

Material	Fiber Orientation	Fiber Diameter (nm)	Modulus (MPa)	Strength (MPa)	Reference
Poly(p-dioxanone-co-L-lactide)-block-poly(ethylene glycol)		380	$E_L = 30$ $E_T = 30$	$\sigma_L = 1.2$ $\sigma_T = 1.4$	Bhattarai et. al. (2003)
Polycaprolactone (PCL)		Not specified (between 200 ~ 5500nm)	$E_L = 3.7$ $E_T = 2.7$	$\sigma_L = 1.4$ $\sigma_T = 1.2$	Lee et. al. (2003b)
Poly(vinyl chloride) (PVC)	Aligned	Unknown	$E_L = 7.8$ $E_T = 7.8$	$\sigma_L = 1.8$ $\sigma_T = 1.7$	Lee et. al. (2002a)
Poly(trimethylene terephthalate) (PTT)		400	$E_L = 0.7$ $E_T = 1.1$	$\sigma_L = 2.2$ $\sigma_T = 4.1$	Khil et. al. (2004)
Poly(vinyl alcohol) (PVA) cross-linked by glyoxal (0~10wt%)		280	-------	$\sigma_L = 7 \sim 8.5$ $\sigma_T = 9.5 \sim 10.5$	Ding et. al. (2002)
Acrylate modified Elastomeric Protein (degree of protein cross-linking)		300 ~ 1500	700 (Uncross-linked) 1800 (cross-linked)	16 (Uncross-linked) 43 (cross-linked)	Nagapudi et. al. (2002)
Fibrinogen		700	80	2	Wnek et. al. (2003)

Table 5.7. (continued) Tensile properties of polymer nano fibrous membranes collected by plate.

Material	Fiber Orientation	Fiber Diameter (nm)	Modulus (MPa)	Strength (MPa)	Reference
Bombyx mori	Random	200 ~ 400	0.6	15	Ohgo et. al. (2003)
Poly(lactic-glycolic) acid (PLGA)		400 & 1000	71 (400nm) 87 (1000nm)	6 (400nm) 4.9 (1000nm)	Zong et. al. (2003)
		500 ~ 800	323	23	Li et. al. (2002b)
Gelatin		100 ~ 1900	117 (100nm) 123 (1900nm)	2.9 (100nm) 3.4 (1900nm)	Huang et. al. (2004)
PEO PEO : Colagen = 2 : 1 PEO : Colagen = 1 : 1		100 ~ 150	7 8 12	0.09 0.27 0.37	Huang et. al. (2001a)
Peptide Polymer		450	1800	35	Huang et. al. (2000)
Polyurethane		100 ~ 500	3.7	10	Pedicini and Farris (2003)

(a) Dumbbell shape specimen

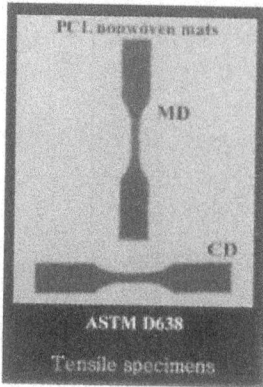

[Lee et. al. (2003b)]

(b) Rectangular shape specimen

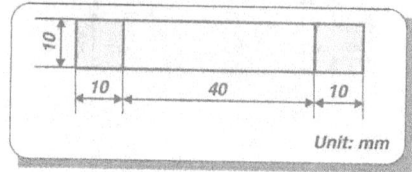

Unit: mm

[Fujihara et. al. (2005)]

Fig. 5.40. (a) dumbbell shape and (b) rectangular shape tensile specimens of electrospun nanofiber membranes.

Randomly-oriented nanofiber membrane was collected on flat plate by [Ohgo et. al. (2003); Zong et. al. (2003); Huang et. al. (2004); Li et. al. (2002b); Huang et. al. (2001a), Huang et. al. (2000); Pedicini and Farris (2003)] while aligned nanofiber membrane was collected by the rotating drum by [Bhattarai et. al. (2003); Lee et. al. (2003b); Wnek et. al. (2003); Lee et. al. (2002a); Khil et. al. (2004); Nagapudi et. al. (2002); Ding et. al. (2002)]. With respect to aligned nanofiber membranes, tensile property in each drum rotating direction and transverse direction is main discussion. As nanofibers are likely to align in drum rotating direction, tensile property of nanofiber specimen could indicate higher value as compared to specimens extracted in transverse direction. However, as seen in Table 5.7, drastic property difference is hardly seen between two fiber orientations. This is attributed to the fiber alignment of collected nanofibers on the rotating drum. Fig. 5.41 shows that even if nanofibers are collected with rotating drum, there is no certainty that the alignment of nanofibers can be unidirectionally oriented.

It is natural that there is only subtle difference of tensile properties. In order to get more precise fiber alignment, the current fiber collection system using rotating drum still needs further improvement.

Fig. 5.41. SEM photograph of poly(trimethylene terephthalate) nanofibers collected with rotating drum [Khil et. al. (2004)]. It can be recognized that fiber alignment is not perfect in MD (drum rotating direction) direction.

5.4 Conclusions

In this chapter, various characterization techniques have been introduced to comprehend (1) Morphology, (2) Molecular Structure and (3) Mechanical Property of electrospun polymer nanofibers. In order to investigate the average fiber diameter of deposited nanofibers, conventional scanning electron microscope (SEM) is available. However, if ultra-fine nanofibers need to be precisely examined, TEM observation which does not need gold coating of 15nm layer is recommended. It has been shown that the molecular structure of nanofibers is greatly affected by electrospinning process. As the order of polymer chains in a nanofiber gives influence to optical, thermal and mechanical response of nanofibrous membranes, the molecular structure of nanofibers need to be understood in specific applications.

This is also applied to the nanofibers which require chemical surface treatment. It has been shown that tensile property of nanofibrous membranes is easily characterized using the conventional testing machine. However, careful sample preparation and assessment are required to understand the mechanical properties of a single nanofiber.

Chapter 6

Functionalization of Polymer Nanofibers

The extensive research activities on electrospun polymer nanofibers are encouraged by great potential of the nanofibers for many applications (See chapter 8). However, most of polymer nanofibers are chemically inert and do not have any specific functions. For the applications to be successful the electrospun polymer nanofibers must be functionalized. For example, for the nanofibers to be used as affinity membrane for IgG purification, protein A or protein G should be covalently bonded on nanofiber surface; for the nanofiber mesh to be used as protective cloth, capture agents need to be immobilized on the nanofiber surface to capture and decompose toxic molecules. In this chapter, principles of polymer surface modification techniques that have been employed or may be employed in the future to functionalize polymer nanofibers will be introduced first, followed by examples of polymer nanofiber functionalization works which are categorized in terms of different applications.

6.1. Polymer Surface Modification

6.1.1. Introduction

Polymer surface modification is an old topic due to the wide application of synthetic polymeric materials in human society in such wide fields as adhesion, membrane filtration, coatings, friction and wear, composites, microelectronic devices, thin-film technology and biomaterials, etc.

247

In general, special surface properties with regard to chemical composition, hydrophilicity, roughness, crystallinity, conductivity, lubricity, and cross-linking density are required for the success of these applications. However, polymers very often do not possess the surface properties needed for these applications. In fact, most of the mechanically strong and chemically stable polymers usually have inert surfaces both chemically and biologically. Vice versa, those polymers having active surfaces usually do not possess excellent mechanical properties which are critical for their successful application.

Due to this dilemma, surface modification of the polymers without changing the bulk properties has been an old classical research topic for many years, and is still receiving extensive studies due to more and more new applications of polymeric materials, especially in the field of biotechnology and bioengineering. Common surface modification techniques include treatments by blending, coating, radiation with electromagnetic wave, electron beam, ion beam [Dong and Bell (1999); Brown (2003)] or atom beams[Chan *et. al.* (1996)], corona or plasmas treatment [Liston *et. al.* (1993); Grace and Gerenser (2003); Chu *et. al.* (2002)], chemical vapor deposition (CVD), gas oxidation, metallization, chemical modifications use wet-treatment and surface grafting polymerization [Uyama *et. al.* (1998); Kato *et. al.* (2003)] etc. Progress in recent years has been summarized by several good reviews [Ikada (1994); Ratner (1995); Desai *et. al.* (2004)].

6.1.2. *Physical Coating or Blending*

Perhaps the easiest and the most strait forward way to modify polymer surface is by blending functional molecules into the bulk polymer or just coating it on the polymer surface. Although the biggest limitation of this technique is the instability of the polymer surface composition caused by the lost of the functional materials from the polymer, it is still a good choice if this lost is too slow to be considered, or not fast enough to affect the application of the materials.

In recent years, physically blending/coating of functional molecules into/onto the polymers was significantly developed by appearance of some new techniques, among which includes "self migration", "self assembly" and "layer by layer" method (Fig. 6.1). In "self migration" method, the functional material with specially designed chemical structure is mixed into the bulk polymer. Driven by tendency to reach the lowest free energy state, the functional materials molecules can automatically migrate towards the polymer surface and finally accumulated onto the polymer surface, changing the properties of the polymer surface significantly [Wang *et. al.* (2000)]. In "self assembly" method, surfactant molecules coated onto substrate surface self assembled into a thin film, driven by hydrophobic interaction, hydrogen bond, electrostatic interaction or chemical reaction [Zhang *et. al.*(1999); Pakalns *et. al.*(1999)]. "Layer by layer" is a special case of the "self assembly" method. In "layer by layer" method the negatively and positively charged macromolecules were alternatively introduced onto the polymer surface through the strong electrostatic interaction [Wang *et. al.* (2004c); Zhu *et. al.* (2003)]. This technique especially makes it much easier to physically immobilize some charged protein or DNA molecules on the polymer surface than to covalently immobilize these biomolecules.

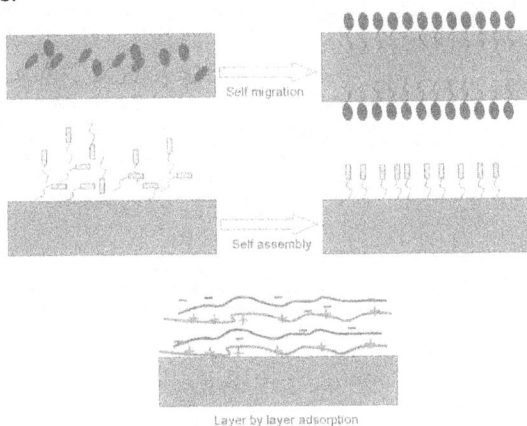

Fig. 6.1. Schematic illustrations of self migration, self assembly and layer by layer adsorption techniques.

6.1.3. Graft Copolymerization

Among the surface modification techniques developed to date, surface graft copolymerization has emerged as a simple, useful, and versatile approach to improve surface properties of polymers for a wide variety of applications. Grafting has several advantages: (a) the ability to modify the polymer surface to have very distinct properties through the choice of different monomers, (b) the ease and controllable introduction of graft chains with a high density and exact localization of graft chains to the surface with the bulk properties unchanged and (c) covalent attachments of graft chains onto a polymer surface assuring long-term chemical stability of introduced chains, in contrast to physically coated polymer chains [Uyama *et. al.*(1998)].

For the occurrence of the graft copolymerization, radicals or groups which can produce radicals like peroxide groups must be introduced onto the polymer surface first. For most of the chemically inert polymers, this can be achieved via irradiation (γ-ray, electron beams, UV, etc), plasma treatment, ozone or H_2O_2 oxidization, or Ce^{4+} oxidization. The general scheme of how the surface of polymer can be grafted is reflected in Fig. 6.2.

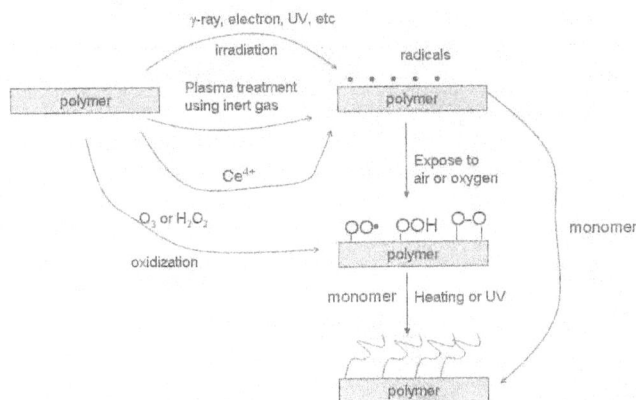

Fig. 6.2. General scheme of graft copolymerization on polymer surfaces.

6.1.3.1. Radiation-Induced Graft Copolymerization

Radiation induced graft co-polymerization is the irradiation of the polymer surfaces with a high energy source (γ-ray, electron beams, UV, etc) and eventually grafting a monomer (or monomers) on the surface. Absorption of high-energy radiation by polymers induces excitation and ionization and these excited and ionized species are the initial chemical reactants for a series of complicated reactions (See Table 6.1) to give free radicals, which can cause monomers to polymerize. The polymer to be surface modified is usually immersed in a monomer solution, so that the radicals produced on the polymer surface can immediately initiate the copolymerization of the monomer.

Table 6.1. Main chemical reactions when polymer receive irradiation.

$P_{(polymer)} \xrightarrow{\text{irradiation}} P^*_{(activated)}$	activation
$P^* \longrightarrow P^{+\cdot} + e_{(electron)}$ $P^* \longrightarrow P\cdot + X\cdot _{(X=H\ or\ Cl,\ etc.)}$ $P^* \longrightarrow R_1\cdot + R_2\cdot$	Radical producing and chain scission reactions
$R\cdot + P \longrightarrow (R\text{-}P)\cdot _{(cross\ linked\ molecule)}$ $P^{+\cdot} + P' \longrightarrow (P\text{-}P')^{+\cdot}_{(cross\ linked\ molecule)}$ $P\cdot + P' \longrightarrow (P\text{-}P')\cdot _{(cross\ linked\ molecule)}$	Cross linking reactions
$P\cdot + O_2 \longrightarrow P\text{-}O\text{-}O\cdot$ $P\text{-}O\text{-}O\cdot + P' \longrightarrow POOH + P'\cdot$ $POOH \longrightarrow R_1' + R_2'$	Formation of peroxide groups
$POOH \longrightarrow PO\cdot + HO\cdot$	Decomposing of peroxide groups producing radicals
$R\cdot \text{(radical)} + M\text{(monomer)} \longrightarrow$ $RM\cdot \xrightarrow{+M} RM_2\cdot \xrightarrow{+M} \cdots \longrightarrow RM_n\cdot$ $PO\cdot \text{(or } HO\cdot) \ M\text{(monomer)} \longrightarrow POM\cdot$ $\xrightarrow{+M} POM_2\cdot \xrightarrow{+M} \cdots \longrightarrow POM_n\cdot$	Graft copolymerization

In addition to producing radicals, exposure of polymers to radiations can also lead to other extensive chemical reactions (See Table 6.1) and physical changes, which may have both detrimental and beneficial consequences in determining the end-uses of the polymers. It is beneficial in the sense that it can cause cross-linking, grafting on the surface of the polymers on the other hand may cause chain scission (breaking of bond) as well thus damaging the polymer [Ghiggino (1989)]. A moderate number of crosslinks can often improve the physical properties of polymers while scission processes usually produce deleterious effects, resulting in materials which are soft and weak. In many cases, crosslinking and scission occurs simultaneously. Chemical nature and morphology of the material determines which of these reactions are predominant [Sangster (1989); O'Donnell (1989); Ivanov (1992); Clough (1991)]. An empirical rule is used to predict the behavior of carbon chain polymers exposed to irradiation. This is reflected in the Table 6.2. Given in Table 6.3 is a list of polymers that undergoes predominantly cross linking or chain-scission as well as polymers which can undergo either cross linking or chain scission depending on appropriate conditions.

Table 6.2. Using empirical formula to predict the behavior of polymer upon irradiation [Ivanov, 1992].

Predominant behavior of polymer upon irradiation	Empirical formula of polymer
Cross-linking	~CH2-CHR~ (as long as there is a H atom on each C)
Degradation/Chain scission	~CH2-CRR'~ ~CX2-CX2~

6.1.3.2. Plasma-Induced Graft Copolymerization

A plasma can be produced from a gas of low pressure (usually <20 pa) if enough energy such as electromagnetic field of radio frequency is added to cause the electrically neutral atoms of the gas to split into positively and negatively charged atoms and electrons.

Table 6.3. List of polymers that undergoes predominantly cross linking or chain-scission [Ivanov, 1992].

Polymers predominantly undergoing cross-linking upon irradiation	Polymers predominantly undergoing chain scission upon irradiation
Polyethylene	Polytetrafluoroethylene
Polypropylene	Polytriflurochloroethylene
Polystyrene	Polyisobutylene
Poly(vinyl chloride)	Poly-α-methyl styrene
Poly(vinylidene fluoride)	Poly(vinylidene chloride)
Poly(vinyl acetate)	Polymethacrylic acid
Poly(vinylalkyl ethers)	Polymethacrylates
Polyacrylic acid	Polymethacrylamide
Polyacrylonitrile	Polymethacrylonitrile
Poly(vinyl pyrrolidone)	Butyl rubber
Polyamides	Polysulphide rubbers
Polyurethanes	Poly(ethylene terephthalate)
Poly(ethylene oxide)	Cellulose and its derivatives
Polyalkysiloxanes	
Polyesters	
Natural rubber	
Synthetic carbon-chain rubbers (except butyl rubber)	
Polymers that can cross-link and degrade depending on appropriate conditions: Poly(vinyl chloride) Polypropylene Poly(ethylene terephthalate)	

The gas molecules are ionized or activated by the applied energy to form an approximately charge neutral mixture of electrons, ions, free radicals, atoms and molecules. Therefore plasma is also called the fourth state of material. The reaction of these species with the polymer can result in free radical in the polymer main chain. The generation of free radicals can happen in two ways [Wertheimer (1999)]: (a) The bombardment of the ions and photons can provide energy higher than the covalent C-C or C-H bonding energy and can break the C-C, or C-H bond to form C free radicals on the midpoint of the polymer main chain, or (b) through elastic or inelastic collision of the electron in the plasma with the polymer, the H can be extracted, resulting in C free radicals.

When these activated polymer are exposed to monomers with unsaturated bond, the radicals on the polymer plays the role of initiator as in conventional polymerization and makes polymerization started. Radicals are usually formed when the polymer are exposed to inert gases (such as argon or nitrogen) plasma. The treatment time usually is from several to several ten seconds, depending on the kind of the polymer. After plasma treatment, the polymer should be immediately immersed into the monomer solution to initiate the graft copolymerization. This is because the radicals produced on the polymer surface by the plasma treatment will react with the oxygen molecules in the air to form a more stable peroxide groups, which can not directly initiate the polymerization. Another strategy, however, is to let the plasma treated polymer exposed to open air or oxygen for more than 10min to transfer all the radicals to the peroxide groups, then using heating or UV to decompose the peroxide groups to initiate the copolymerization. This strategy makes the graft polymerization much easier to perform and the results more reproducible.

The plasma induced graft copolymerization has advantages such as: (i) Modification can be limited to the surface layer, leaving the properties of the bulk material unaltered. Typically, the modification depth is several hundred angstroms. (ii) Nearly all polymer surfaces can be modified by plasma, no matter how the surface is composed. (iii) It is possible to choose the type of chemical modification for the polymer through the choice of gas used in gas plasma. (iv) Gas plasma treatment can avoid the problems encountered in wet chemical techniques such as residual solvent on the surface. (v) Plasma surface treatment is fairly uniform over the whole surface.

6.1.3.3. Oxidization-Induced Graft Copolymerization

Oxidization of polymer surface using O_3 gas [Ying *et. al.* (2004)] or hydrogen peroxide solution [Ma *et. al.* (2002b)] can produce peroxide groups, which can be used for graft copolymerization (Fig. 6.2). Ce (IV) is an oxidization agents often used for graft copolymerization on substrate surface containing hydroxyl groups [Ma *et. al.* (2005a)].

The Ce (IV) and the hydroxyl group form an oxidization-reduction system and react to yield radicals to initiate the copolymerization, as shown in Fig. 6.3.

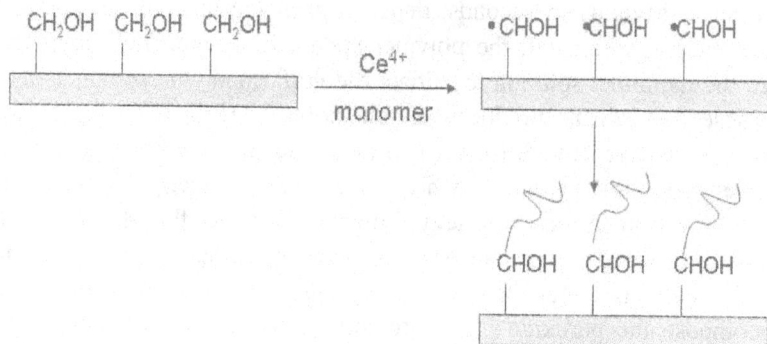

Fig. 6.3. Ce^{4+} induced graft copolymerization on polymer surface containing hydroxyl groups.

6.1.4. Plasma Treatment and Chemical Vapor Deposition

For plasma induced graft polymerization, inert gas is often used to produce radicals on the polymer surface. While if reactive gas such as O_2, SO_2, NH_3 and CO_2 is used, chemical functional groups like hydroxyl, carboxyl, carbonyl and sulfonate etc. can be yielded on the polymer surface directly. However, the functional groups introduced on the polymer surface are not distinct. As aforementioned, plasma consists of a unique mixture of different positively charged or negatively charged ions, electrons, free radicals, atoms and molecules. The variety of the particles in a plasma chamber also cause a variety of the functional groups yielded on the polymer surface. Plasma treatment is usually not expected to introduce a specific functional group, but just to increase the polarity, hydrophilicity and charge of the polymer surface to benefit applications like adhesion, dyeability, blood compatibility, etc.

Another kind of plasma treatment is plasma induced polymer deposition, which is different from the plasma induced graft copolymerization described above. In plasma induced polymer deposition, the monomer gas is directly introduced into the plasma chamber and polymerized. The polymer generated in the gas phase is then deposited onto the substrate surface as a thin film [Mao and Gleason (2004)]. This method is actually one kind of the chemical vapor deposition (CVD) technique and therefore is also called chemical vapor deposition polymerization (CVDP). CVD is a name for a broad of surface modification techniques in which chemical reactions transform gaseous molecules, called precursor, into a solid material, in the form of thin film or powder, on the surface of a substrate.

In a plasma aided CVD process, reactive gases of either high or low temperature are introduced into a plasma chamber and induced to form plasma. The resulted products deposited onto a low temperature substrate to form a thin solid film. Although the CVD process has been mainly used to fabricate semiconductor devices, it is also an important method to modify inert polymer surfaces like TFPE and PE [Lahann *et. al.* (2002)].

6.1.5. Chemical Treatment

Some polymer molecules possess functional groups like hydroxyl, carboxyl, amino and ester, etc. Such kind of polymer can be directly modified by chemical treatment. Chemical modification involves the introduction of one or more chemical species to a given surface so as to produce a surface which has enhanced chemical and physical properties [Mottola (1992)].

Chemical reactions can be carried out at sites that are vulnerable to electrophilic or nucleophilic attack. Structures like benzene rings, hydroxyl groups, double bonds, halogen, ester groups, etc. qualify for such attacks. Polyester like PET, PCL and PLLA can be treated by diamine compounds to introduce amino groups through the aminolysis of the ester groups [Zhu *et. al.* (2002)]. Cellulose surface has plenty of hydroxyl groups and can be modified with many electrophilic agents such as cibacron blue F3GA [Li *et. al.* (2002a)] via the reaction between the hydroxyl groups and the electrophilic agents.

Wet chemical oxidation treatments are also commonly employed to introduce oxygen-containing functional groups (such as carbonyl, hydroxyl, and carboxylic groups) at the surface of the polymer. This can be conducted using gaseous reagents or with solutions of vigorous oxidants. The oxygen-containing functional groups increase the polarity and the ability to hydrogen bond, thus in turn results in the enhancement of wettability and adhesion. Cellulose surface can be oxidized by $NaIO_4$ to transfer hydroxyl groups of the cellulous into aldehyde groups. The aldehyde group has high reactivity to primary amino group in protein molecules to covalently immobilize protein molecules [Boeden *et. al.* (1991)].

Table 6.4. Preparation of functionalized polymer nanofibers.

Applications	Materials	Specific purpose	Methods	References
Affinity membrane	PMMA nanofiber functionalized with - cyclodextrin	Organic waste removal	blending	Kaur et al., 2005
	poly(ether-urethane-urea) nanofiber	IgG purification	Chemical reaction between Succimidyl ester group and protein A	Bamford et al., 1992
	Cellulose nanofiber surface functionalized with cibacron blue	Bilirubin removal and BSA purification	Chemical reaction between the hydroxyl group of cellulose and the cibacron blue F3GA	Ma et al, 2005d

Tissue engineering scaffold	PET nanofiber surface grafted with gelatin	Blood vessel regeneration	graft copolymerized of PMAA, followed by immobilization of gelatin on carboxyl groups in PMAA	Ma et al., 2005a
	PCL nanofiber surface grafted with gelatin	Blood vessel regeneration	Air Plasma treatment to yield carboxyl groups, followed by immobilization of gelatin on the carboxyl groups.	Ma et al, 2005b
	Fibroin nanofiber surface functionalized with Collagen, fibronectin and laminin.	Skin regeneration	coating	Min et al., 2004
	Collagen-coated poly(L-lactic acid)-*co*-poly(ε-caprolactone) P(LLA-CL 70:30) nanofiber	Blood vessel regeneration	coating	He et al., 2005a
	Collagen-blended poly(L-lactic acid)-*co*-poly(ε-caprolactone) P(LLA-CL 70:30) nanofiber	Blood vessel regeneration	blending	He et al., 2005b
	Galactose functionalized poly(e-caprolactone-co-ethyl ethylene phosphate) (PCLEEP)	Artificial liver	graft copolymerized of PMAA, followed by immobilization of 1-O-(60-aminohexyl)-D-galactopyranoside (AHG) on carboxyl groups in PMAA	Chua et al., 2005
Drug delivery	PU nanofibers containing model drug itraconazole and ketanserin	Itraconazole or ketanserin releasing	blending	Verreck et al., 2003a
	PLA-PEG and PLGA with plasmid DND	Plasmid DNA releasing	blending	Luu et al., 2003
Sensor	MoO_3-containning PEO nanofibers	Ammonia or nitroxide sensing	blending	Gouma, 2003
	SnO_2 or TiO_2 coated polyacrylonitrile (PAN) nanofibers	Ammonia or nitroxide sensing	coating	Drew et al., 2003
	poly(acrylic acid)-poly(pyrene methanol) mixed PU	metal ions (Fe^{3+} or Hg^{2+}) or 2,4-dinitrotoluene (DNT) sensing	blending	Wang et al., 2002b

	hydrolyzed poly[2-(3-thienyl) ethanol butoxy carbonyl-methyl urethane] functionalized cellulose acetate nanofibers	viologen (MV^{2+}) and cytochrome c (cyt c) sensing	Layer by layer coating	Wang et al., 2004c
	Aividin functionalized polypyrrole nanofiber	Biotin-DNA detection	blending	Ramanathan et al., 2005
Air filtration & Protective cloth	Polyoxometallate functionalized PU nanofiber	Anti-sulfur mustard protective cloth	blending	Graham et al., 2003
	poly(vinylbenzyl-dimethylcocoammonium chloride - MMA - perfluoro alkyl ethyl acrylate) nanofiber	Anti-bacterial air filtration media	Electrospinning of functional polymer	Acatay et al., 2003
Catalyst support	Lipase functionalized PEO nanofiber	Olive oil hydrolyzing	Blending PEO with the enzymes, followed by crossliking with 4,4_-methylenebis(phenyl diisocyanate) (MDI).	Xie et al., 2003
	R-chymotrypsin immobilized polystyrene nanofiber	Hydrolyzation of n-succinyl-ala-ala-pro-phe p-nitroanilide	Reaction between phenyl nitroxide groups in PS and the protein molecules	Jia et al., 2002

6.2. Functionalization of Nanofibers for Different Applications

6.2.1. Introduction

All the current techniques motioned above may be considered for the surface modification of polymer nanofibers, but with one problem to be noticed: the polymer nanofibers must be protected from being destroyed. Strong reaction conditions like plasma, UV, γ radiation, high temperature, acidic or basic environment may destroy the polymer nanofibers easily, especially when the polymer nanofiber is biodegradable. Biodegradable polymers like PLA and PGA nanofibers may have much faster degradation rate compared with the macro- or micro-scale materials due to the high surface area. Even PET showed considerable degradation rate when it is prepared into nanofibers. Therefore, the reaction condition for the surface modification of polymer nanofibers must be optimized to preserve the nanofibrous morphology. Some surface modification methods like molecular self-assembly [Zhang *et. al.* (1999); Pakalns *et. al.* (1999)] and "layer by layer" electrostatic interaction [Wang *et. al.* (2004c); Zhu *et. al.* (2003)] can be performed under very moderate conditions, thus may be the most potential methods for surface modification of biodegradable polymer nanofibers.

Table 6.4 gives a summarization of works on the functionalization of nanofibers and their respective applications.

6.2.2. Functionalization of Nanofibers for Affinity Membrane Application

One main interest of electrospun polymer nanofiber nonwoven mesh is for affinity membrane applications. Affinity membrane is membrane which bases its separation on the selectivity of the membrane to 'capture' molecules, by immobilizing specific ligands onto the membrane surface. For the nanofiber mesh to be used as affinity membrane, specific ligand molecules need to be immobilized on the nanofiber surface.

The simplest method to introduce ligands on nanofiber surface is to directly mixing the ligand molecules into the polymer solution and then electrospin the polymer solution. An attempt by Kaur et al. has been made to incorporate chemically modified β-CD onto the surface of the nanofiber to target potential applications in organic waste treatment for water purification [Kaur *et. al.* (2005)].

Phenylcarbomylated or azido phenylcarbomylated β-CD was successfully blended with PMMA and electrospun into nanofibrous membrane respectively. The presence of the β-CD derivatives on the surface of the nanofibers was confirmed by ATR-FTIR and XPS (Fig. 6.4).

To determine the functionalized membranes ability to capture small organic molecules, a solution containing phenolphthalein (PHP), a small organic molecule, was used. The functionalized nanofibrous membranes were able to capture the PHP molecules effectively. Thus the functionalized nanofibrous membranes may have the potential to capture similar small organic waste molecules present in water.

A study has been presented on the potentials of electrospun poly(ether-urethane-urea) as an affinity separation membrane [Bamford *et. al.* (1992)]. Succimidyl ester group containing polyurethane has been prepared and the succimidyl groups were reacted with protein A for the functionalization of the membrane surfaces.

Assessments of the capacities of the reactive membranes for covalent coupling of protein A and human immunoglobulin G gave very encouraging data, 4mg/g for protein A and 4mg/g for bounded IgG, respectively.

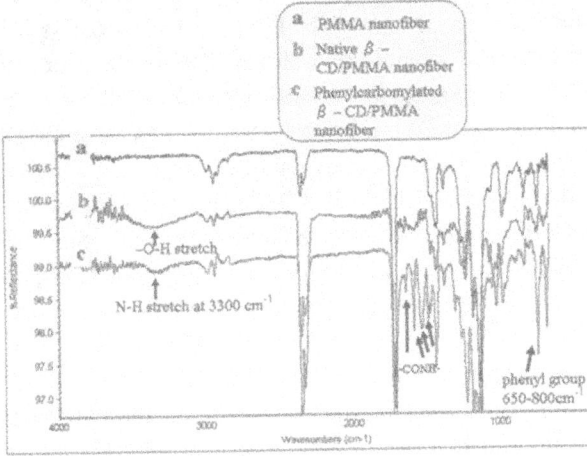

a

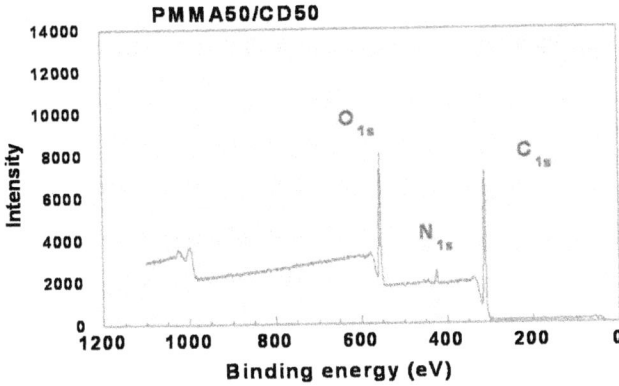

b

Fig. 6.4. (a) ATR-FTIR spectra of the original PMMA nanofiber and PMMA/phenylcarbomylated β-CD nanofiber. The characteristic absorption groups of the phenylcarbomylated β-CD were pointed out. (b) XPS of the PMMA/ azido phenylcarbomylated β-CD (weight ration is 50/50) nanofiber. The nitrogen peak indicated the presence of the azido phenylcarbomylated β-CD on the fiber surface.

Polysulphone has been widely used for membrane preparation due to its excellent mechanical strength and chemical stability in acidic and basic solutions. Nonwoven meshes composed of polysulphone (PSU) ultrafine fibers (diameter, 1~2μm) were fabricated via electrospinning technique and then surface modified towards development of a novel affinity membrane [Ma *et. al.* (2005e)]. After heat treatment of the nanofiber mesh, carboxyl groups were introduced onto the PSU fiber surfaces through grafting co-polymerization of methacrylic acid (MAA) initiated by Ce(IV) after an air plasma treatment of the PSU fiber mesh. Toluidine Blue O (TBO), a dye which can form stable complex with carboxyl groups, was used as a model target molecule to be captured by the PMAA grafted PSU fiber mesh. Further more, the carboxyl groups on the PMAA grafted PSU fiber mesh can be used as coupling sites for immobilization of other protein ligands (Fig. 6.5). Bovine serum albumin (BSA) was chosen as a model protein ligand to be immobilized into the PSU fiber mesh with an encouraging capacity of 17μg/mg, which implies other protein ligands can also be attached using the same method.

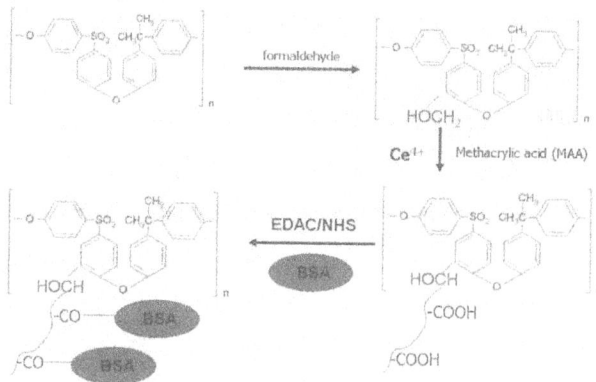

Fig. 6.5. Schematic illustration of the surface modification of polysulphone nanofiber.

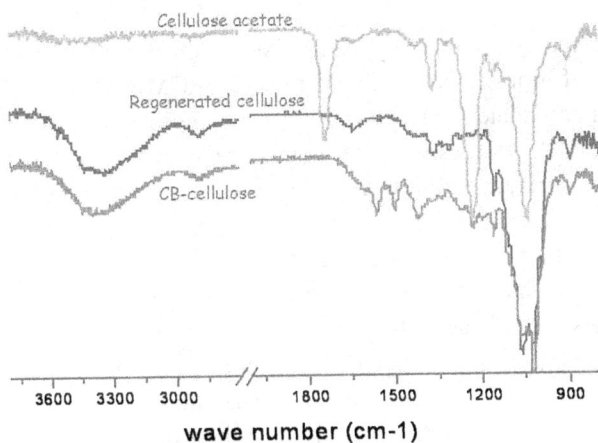

Fig. 6.6. ATR-FTIR spectra of cellulose acetate nanofiber, regenerated cellulose nanofiber and cibacron-blue F3GA (CB) immobilized cellulose nanofiber. The CB immobilized cellulose nanofiber gave characteristic absorption peak of CB at 1565cm^{-1}, indicating its presence .

Cellulose is another important material for membrane preparation. To study the feasibility of applying electrospun cellulose nanofiber mesh as affinity membrane, cellulose acetate (CA) solution (0.16g/ml) in a mixture solvent of acetone/DMF/ trifluoroethylene (3:1:1) was electrospun into nonwoven fiber mesh with the fiber diameter ranging from 200nm to 1μm. The CA nanofiber mesh was heat treated under 208°C for 1h to improve structural integrity and mechanical strength, and then treated in 0.1M NaOH solution in H$_2$O/ethanol (4:1) for 24h to obtain regenerated cellulose (RC) nanofiber mesh, which was used as a novel filtration membrane in this work [Ma *et. al.* (2005d)]. The RC nanofiber membrane was further surface functionalized with Cibacron Blue F3GA (CB), a general affinity dye ligand for separation of many biomolecules. The surface chemical changes were verified by ATR-FTIR spectra (Fig. 6.6). The CB derived RC nanofiber membrane has a CB content of 130μmol/g, and capture capacity of 13mg/g for bovine serum albumin (BSA) and 4mg/g for bilirubin.

6.2.3. *Functionalization of Nanofiber for Tissue Engineering Scaffold Application*

Polymeric nanofiber matrix has similar structure with the nano-scaled nonwoven fibrous extra cellular matrix (ECM) proteins, thus is a wonderful candidate for ECM-mimic tissue engineering scaffold [Ma *et. al.* (2005c)]. A successful tissue engineering scaffold should have cell compatible surfaces to allow cell attachment and proliferation. Most of the surface modification works to improve biocompatibilities of polymeric tissue engineering scaffold center on immobilizations of biomolecules that can be specifically recognized by cells on the biomaterials [Elbert and Hubbell (1996)]. These biomolecules include adhesive proteins like collagen, fibronectin, RGD peptides and growth factors like bFGF, EGF, insulin, etc. The biomolecules can either be covalently attached, electrostatically adsorbed or self-assembled on the biomaterial surfaces.

Collagen, fibronectin and laminin have been coated on electrospun silk fibroin nanofiber surface to promote cell adhesion [Min *et. al.* (2004)]. Collagen-coated poly(L-lactic acid)-*co*-poly(ε-caprolactone) P(LLA-CL 70:30) nanofiber mesh (NFM) with diameter of 470 nm and porosity of 64-67 % was fabricated using electrospinning followed by the plasma treatment and collagen coating. The spatial distribution of the collagen in the NFM was visualized by labeling the collagen with fluorescent dye and it was found that the collagen can be uniformly coated around the polymer nanofiber. Endothelial cells were seeded onto the collagen-coated P(LLA-CL) nanofiber and the results showed that the collagen coated material have higher cell attachment, spreading and viability than the unmodified nanofiber. [He *et. al.* (2005a)].

In another work [He *et. al.* (2005b)], collagen was directly blended into the PLLA-*co*-PCL (70:30) nanofiber. The blended nanofibers were obtained by electrospinning of a mixture of collagen and the copolymer solution. Morphology of the blended nanofibers was investigated by SEM and TEM. ATR-FTIR spectra verified existence of collagen molecules on the surface of nanofibers (Fig. 6.7). The collagen-blended polymer nanofibers preserved the EC's normal phenotype and showed enhanced cell viability, spreading and attachment, indicating a potential application as a tissue engineering vascular graft.

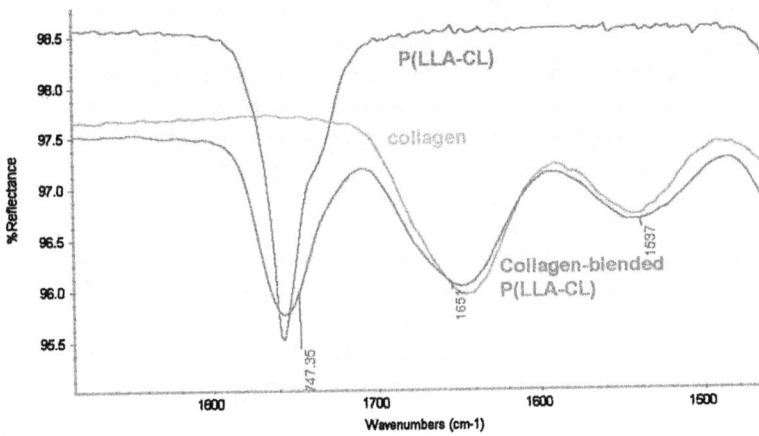

Fig. 6.7. ATR-FTIR of the P(LLA-CL), collagen, and collagen blended P(LLA-CL) nanofiber.

In addition to physically blending or coating, covalently graft ECM protein like collagen and gelatin on nanofiber surfaces is another choice with the ultimate goal of development of bio-mimic tissue engineering scaffolds.

One popular methods to covalently attach protein molecules on polymer surface is to graft poly(methacrylic acid) on the biomaterial surface to introduce carboxyl groups at first, followed by the grafting of the protein molecules using water soluble carbodiimide as coupling reagent [Steffensa *et. al.* (2002); Ma *et. al.* (2002a)]. Non-woven PET nanofiber meshes has been prepared by electrospinning technology and surface grafted with gelatin. The PET nanofiber was first treated in formaldehyde to yield hydroxyl groups on the surface, followed by the grafting polymerization of methacrylic acid (MAA) initiated by Ce (IV). Finally the PMAA grafted PET nanofiber was grafted with gelatin using water-soluble carbodiimide as coupling agent [Ma *et. al.* (2005a)]. For gelatin grafting on PCL nanofiber surfaces, the PCL nanofiber was first treated with air plasma to introduce -COOH groups on the surface followed by covalently grafting of gelatin molecules [Ma *et. al.* (2005b)]. Both the gelatin modified PET and PCL nanofibers showed improved endothelial cell compatibility than the unmodified nanofibers. Moreover, the gelatin modified aligned PCL nanofiber also showed strong abilities to affect endothelial cell orientation.

Chua et. al., used UV radiation induced graft co-polymerization to develop a biofunctional poly(ε-caprolactone-co-ethyl ethylene phosphate) (PCLEEP) nanofiber construct for hepatocyte culture [Chua *et. al.* (2005)]. This was achieved by conjugating hepatocyte-specific galactose ligands onto nanofiber surface. Poly(acrylic acid) was grafted onto the nanofiber scaffold via UV-induced polymerization using acrylic acid (AAc) monomers. The COOH sites on the grafted scaffold was then used as immobilization sites to conjugate1-O- (60-aminohexyl)-D-galactopyranoside (AHG). The functionalized AHG-P(AAc)-grafted nanofiber scaffolds exhibited much higher hepatocyte attachment as compared to unmodified scaffold. Such nanofiber scaffolds would be advantageous in the design of a bioartificial liver-assist device, where the hepatocytes could attach to a scaffold with high surface area immobilized with a cell-specific ligand, maintain their differentiated functions, and remain stable against the perfusion and shear forces in the bioreactor.

6.2.4. *Functionalization of Nanofibers for Sensor Application*

Sensor means devices that respond to physical or chemical stimulus such as biomolecules concentration, gas concentration, thermal energy, electromagnetic energy, acoustic energy, pressure, magnetism, or motion, by producing an easily detectable and measurable signal, usually electrical or optical. Electrospun nanofiber received great research interests for sensor application due to its unique high surface area. High surface area is one of the most desired parameters for the sensitivity of conductimetric sensor film. Conductimetric sensor based on semi-conducting oxides is a kind of low cost detectors for reductive gas. The operating principle of these devices is associated primarily with the adsorption of the gas molecules on the surface of semiconducting oxides inducing electric charge transport between the two materials, that changes the resistance of the oxide. The structure configuration of the metal oxide materials is one of the key parameters controlling the gas senescing process. Now there is an increasing trend in chemical sensing to utilize nanostructured materials as gas sensing elements because the high surface areas and the unique structure features are expected to promote the sensitivity of the metal oxide to the gaseous component.

For the polymer nanofibers to be used as the above described conductimetric sensor, it should be functionalized with metal oxide semiconductors. Gouma produced MoO_3-containning PEO nanofibers by electrospin a mixture of MoO_3 so-gel and PEO solution [Gouma (2003)]. Gas sensing test of the electrospun nanofiber mat was carried out using ammonia and nitroxide as model gas. Electrical resistance of the sensing film as a function of the gas concentration was measured and the results showed that the nanoscaled metal oxide fiber offer high sensitivity and fast response to the harmful chemical gases. Drew coated SnO_2 and TiO_2 on polyacrylonitrile (PAN) nanofibers using a "liquid-phase deposition" technique. The coatings were thin enough to maintain the nanofibrous morphology, thereby retaining the large surface area of the electrospun membrane [Drew *et. al.* (2003)]. Such metal oxide-coated nanofibrous membranes are expected to provide unusual and highly reactive surfaces for improved sensing, catalysis and photoelectric conversion applications.

Single polyprrole nanofiber containing avidin with diameters of 100nm and 200nm were produced by electrospinning avidin-blended polyprrole solution. The single nanofiber was studied as biosensors for detecting biotin labeled biomolecules such as DNA. Exposing of the avidin functionalized single polyprrole nanofiber to the biotin labeled biomolecules will allow the binding between the avidin molecules and the biotin groups, changing the electric resistance of the single nanofiber. Thus the nanofiber was a novel nano-sensor for biomolecule detection [Ramanathan *et. al.* (2005)].

Wang report the first application of electrospun nanofibrous membranes as highly responsive fluorescence quenching-based optical sensors [Wang *et. al.* (2002b)]. A fluorescent polymer, poly(acrylic acid)-poly(pyrene methanol) (PAA-PM), was used as a sensing material. Optical chemical sensors were fabricated by electrospinning PAA-PM and thermally cross-linkable polyurethane latex mixture solutions. The fluorescence can be quenched by metal ions (Fe^{3+} and Hg^{2+}) and 2,4-dinitrotoluene (DNT) so the nanofibrous material can be used as sensor to detect the substances. These sensors showed high sensitivities due to the high surface area-to-volume ratio of the nanofibrous membrane structures.

In another work [Wang *et. al.* (2004c)], a fluorescent probe, hydrolyzed poly[2-(3-thienyl) ethanol butoxy carbonyl-methyl urethane] (H-PURET) was immobilized on the cellulose acetate electrospun nanofibrous membranes using "layer by layer" electrostatically self-assembly method. Also by the quenching mechanism, the nanofiber can by use as optical sensor to detect trace amount of methyl viologen and cytochrome *c* in aqueous solutions. The high sensitivity is attributed to the high surface-area-to-volume ratio of the electrospun membranes and efficient interaction between the fluorescent conjugated polymer and the analytes.

6.2.5. *Functionalization of Nanofiber for Protective Cloth Application*

There exists a great potential in the polymer nanofibers for use as aerosol filters in face masks and in protective clothing against chemical and biological warfare agents [Gibson *et. al.* (1999); Schreuder-Gibson, *et. al.* (2002)]. Polymeric nanofibers have been studied as a carrier for active chemistry that may allow for improvements in chemical protective properties. The nanofibers can provide a huge surface area to be functionalized with chemical groups, which are reactive with toxic gases and chemicals. Graham mixed polyoxometallate, a catalyst for the oxidative degradation of sulfur mustard (a chemical weapon agent) with the PU (EstaneTM 58238) solution and electrospun the mixture into nanofibers. The capability of the catalyst in the electrospun nanofiber was found even higher than the catalyst alone [Graham *et. al.* (2003)].

Acatay synthesized an antibacterial agent (quaternary ammonium salt) containing perfluorinated terpolymer, poly(vinylbenzyl-dimethylcocoammonium chloride - MMA - perfluoro alkyl ethyl acrylate). This terpolymer was electrospun into a fluffy nanofiber mesh with fiber diameter as low as 40nm. The nanofiber was immersed into liquid solutions containing *Escherichia Coli* to test the anti-bacterial ability. Although the antibacterial activity in liquid solution was detected to be weak due to its low surface tension, the material may be used in air filtration applications [Acatay *et. al.* (2003)].

Two of the most common classes of chemical warfare agents include the nerve agents (Organophosphorus compounds) and the blister agents (Mustard gas). The motivation of develop a novel protective cloth using electrospun polymer nanofiber is to impart a functional group into/onto the nanofiber surface in such a way that when the gases come into contact, by chemical reaction on the surface of the nanofibers, their active groups get deactivated into relatively harmless by-products. Additionally these nanofibers would have sufficient resistance to block the bacterial contaminants such as Bacillus Anthracis by virtue of small pore size.

For nerve gases, the reactive moiety is the organophosphorus group. This could be deactivated by hydrolysis. The hydrolyzing agent chosen for this purpose is the oxime moiety (Fig. 6.8). Polymer contained in a solvent would be treated with the appropriate compounds and the oxime group would be attached to the polymer backbone. The functionalized polymer solution is then electrospun into fibers which contain the reactive oxime groups. The oxime functionality would aid in hydrolysis of the organophosphate group into mild organic acids and secondary alcohols. Another approach proposed is to modify the electrospun nanofiber surface with enzymes which catalyze hydrolysis of Organophosphorus agents such as Organophosphorus hydrolase and Acetylcholine esterase. The enzyme can be regenerated by washing the used up fibers in a lean solution containing oxime groups which re-activates the functional agent. However, the oxime modified fibers would possess longer shelf life and activity compared to the enzyme surface modification due to the fact that enzymes denature at a rapid pace.

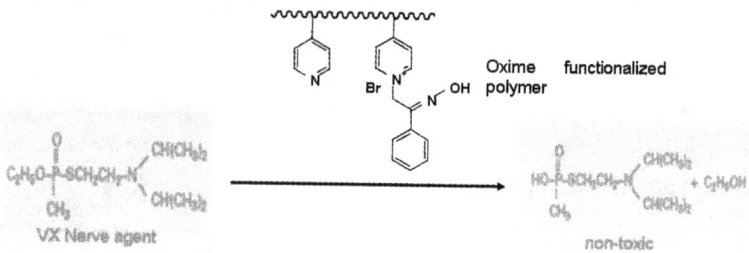

Fig. 6.8. Hydrolyzation of phosphorous organic nerve agent by oxime groups.

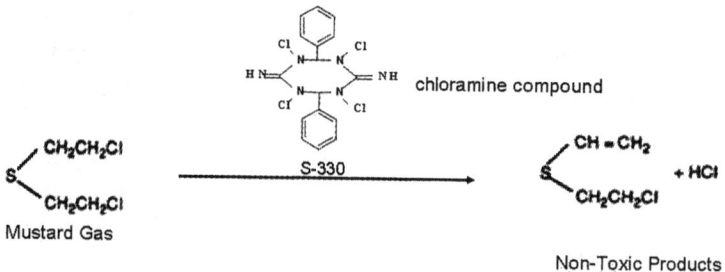

Fig. 6.9. Oxidization of Mustard gas by chloramines compound.

Mustard gas detoxification can be attempted by two pathways, hydrolysis and oxidation. Oxidation pathway is however proposed [McCreery and Michael (1995)] as the hydrolysis reaction leaves behind products that are not entirely safe. Oxidation of mustard gas is by using chloramine functionality (Fig. 6.9). Polymer would be blended with the reagent and by substitution reaction the chloramine functionality would be imparted onto the polymer backbone, this would be electrospun into fibers. When the fibers come into contact with an aerosol containing the mustard gas, it would oxidize the latter into mild non-toxic by products. The oxime functionality and the chloramine functionality also inherently possess bactericidal and fungicidal activity which helps in prevention of fungal growth or entry of bacteria into the mask. The nanofiber filter can be fabricated in such a way that the maximum pore size lies in the sub micron level, which is small compared to the size of bacterial spores. Thereby, bacterial warfare agents also get eliminated.

6.2.6. *Functionalization of Nanofibers for Other Applications*

Due to its higher surface area to volume ratio, nanofiber has been studied for applications as drug delivery carriers and catalyst support. The blending (or mixing) technique is a common choice for the nanofiber functionalization. Drugs, growth factors and genes can be directly mixed into the polymer solution and electrospun to prepare drug carriers with controlled release properties. Verreck prepared PU nanofibers containing model drug itraconazole and ketanserin, using dimethylformide (DMF) and dimethylacetamide (DMAc) as solvent, respectively. The releasing of the drug showed liner relation with the time [Verreck *et. al.* (2003a)]. Luu mixed plasmid with PLA-PEG and PLGA solution in DMF and electrospun the mixture into nanofibers. Release of plasmid DNA from the scaffolds was sustained over a 20-day study period, with maximum release occurring at 2 h [Luu *et. al.* (2003)].

Improvement of catalytic efficiency of immobilized enzymes on nanofiber surface was demonstrated. R-chymotrypsin was chemically attached on polystyrene (PS) nanofibers derived with phenyl nitroxide groups and examined for catalytic efficiency to hydrolyze model compound, *n*-succinyl-ala-ala-pro-phe *p*-nitroanilide. The apparent hydrolytic activity of the nanofibrous enzyme in aqueous solutions was over 65% of that of the native enzyme, indicating a high catalytic efficiency as compared to other forms of immobilized enzymes [Jia *et. al.* (2002)]. In Xie's work [Xie *et. al.* (2003)], the mixture of lipase and poly(ethylene oxide) (PEO) or poly(vinyl alcohol) (PVA) was electrospun into ultra-thin fibrous membranes. The fibrous membranes were rendered insoluble by chemical crosslinking with 4,4_-methylenebis(phenyl diisocyanate) (MDI). The lipase encapsulated in the ultra-fine fibrous membranes exhibits higher catalytic activity towards hydrolyzing olive oil than that in the cast films from the same solution.

6.3. Conclusions

Electrospun polymer nanofiber is receiving an extensive research interest for diverse applications. However, most of the polymer nanofibers do not possess any specific functional groups and they must be specifically functionalized for successful applications. Currently available techniques for polymer surface modification like blending or coating, graft copolymerization and plasma treatment have be used for the functionalization of polymer nanofibers.
The reaction conditions should be controlled properly to prevent the nanofiber morphology from being destroyed.

The most popular and simplest nanofiber modification methods are physical blending and coating. Surface grafting polymerization has also been used for attaching ligands molecules and adhesive proteins on nanofiber surface for application of affinity membrane and tissue engineering scaffold, respectively.

Chapter 7

Potential Applications

7.1. Introduction

As introduced in previous chapters, much effort has been applied to electrospun polymer nanofibers to comprehend the fundamental phenomena of fabrication process as well as the physical and chemical properties from the material science viewpoint. Fig. 7.1 shows that 60% of electrospun polymer nanofiber researches have been devoted to the studies of their processing and characterization. Consequently, nanofiber researchers are able to flexibly control the fiber morphology, fiber diameter and patterning of fiber deposition. These developments have provided much impetus to the dream of realizing the potential applications of nanofibers. As seen in Fig. 7.1, major possible applications are categorized into Bioengineering, Environmental Engineering & Biotechnology, Energy & Electronics and Defense & Security. In all areas, the demand of new novel materials has been aspired and electrospun polymer nanofibers could make the new wave of material research.

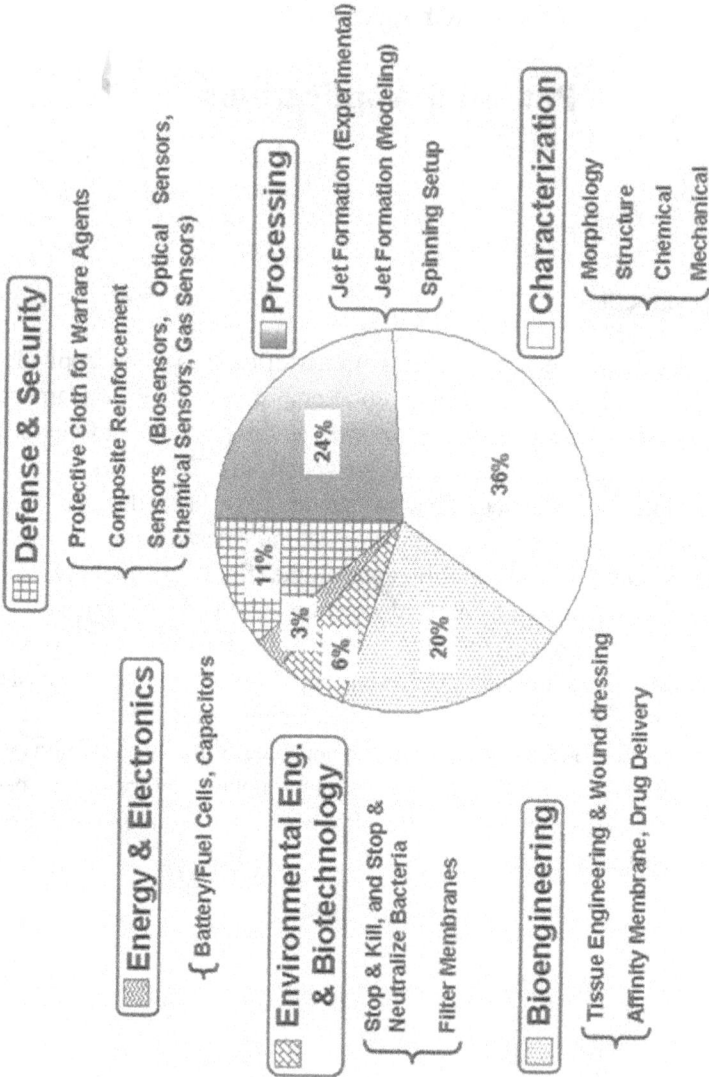

Fig. 7.1. Research category of electrospun polymer nanofibers.

Nanofiber researches in bioengineering consists of tissue engineering and wound dressing, affinity membrane and drug delivery. These fields have gained great interest since the percentage of senior citizens has been projected to increase dramatically in this century. For tissue engineering and wound dressing, electrospun polymer nanofibers are treated as tissue scaffolds which enhance cell growth and proliferation.

The nanofiber scaffolds with seeded cells can be implanted to patient's body to repair the damaged tissues. For drug delivery system, nanofibers are considered as a potential drug carrier. Here, nanofibrous membranes incorporated with drug component can be patched on wound of surgery or encapsulated into pharmaceutical capsules to deliver the drug through digestive system of patient. Affinity membranes have a function to separate the targeted bio-molecule using ligands chemically attached on the nanofiber surface. As compared to conventional filtration column, nanofibrous affinity membranes are expected to realize high efficient separation due to high surface area of nanofibers.

For environmental engineering, the function of nanofibrous membranes is to filter moisture and dusts. Unclean air which contains bacteria can be also purified using nano-size level fibrous membranes. Here, if the surface of nanofibers is chemically modified to kill bacteria, such anti-bacteria filter media is useful to protect people in residence and office.

In view of the increasing level of awareness towards national security at the turn of the millennium, the United Nations (UN) occasionally sends the troop for peace keeping operative (PKO) purpose. In such circumstances, electrospun nanofiber has found its potential application as protective clothing for the soldiers' safety. Such novel protective clothing can capture and neutralize harmful chemical and biological warfare agents. So far there have been no publications which discuss such nanofibrous protective clothing. As such, the background and the concept of nanofibrous protective clothing is introduced in this chapter. Sensors for detecting toxic gases are also possible application of nanofibers for this category of defense.

Nanofibers made of sensing materials may show the excellent sensing behavior within short operating time. It is acknowledged that some of the parts of army constructions as well as transporting aero planes are made of polymer based composite materials. Conventional composite materials possess micron-size carbon, glass and aramid reinforcement fibers. It is considered that electrospun nanofibers with high surface area can supply better efficient reinforcement and may be possible to achieve excellent mechanical performance.

In this century, the economy of developing countries such as China and India has been healthily expanding, thereby causing the energy consumption to increase. Several eco-clean energy resources, such as wind generator, solar generator, hydrogen battery and polymer battery have been investigated so far. With respect polymer battery, researcher's desire has been placed on the battery assembled with polymer nanofibrous membranes as new generation energy.

Thus, in this chapter, abovementioned potential applications of electrospun polymer nanofiber are introduced and the research trend of each topic is briefly analyzed.

7.2. Affinity Membranes

Affinity membrane was developed to permit the purification of molecules based on physical/chemical properties or biological functions rather than molecular weight/size. Rather than operate purely on the sieving mechanism, affinity membrane based its separation on the selectivity of the membrane to 'capture' molecules, by immobilizing specific ligands onto the membrane surface. Affinity membrane reflects technological advances in both fixed-bed liquid chromatography and membrane filtration, and combines both the outstanding selectivity of the chromatography resins and the reduced pressure drops associated with filtration membranes [Klein (2000)]. Compared with the conventional particle–based column chromatography, in which particles or beads need to be packed into a column to form a fixed gel bed, the affinity membrane chromatography have advantages of lower mass transfer limitation, reduced pressure drops and increased flow rate and productivities, as shown in Fig. 7.2.

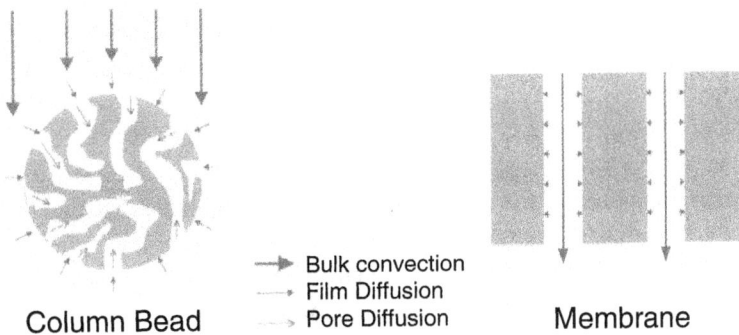

Column Bead

→ Bulk convection
⋯→ Film Diffusion
⋯→ Pore Diffusion

Membrane

Fig. 7.2. Schematic drawing of mass transfer on column bead and affinity membrane. Mass transfer in a column bead is mostly by pore diffusion, and is much slower than mass transfer in affinity membrane, which is mainly by convective flow.

These advantages with regard to rapid reaction kinetics and high productivity lead to an extensive study in affinity membrane, in response to an increasing demand for preparative amounts of proteins, peptides and nucleic acid required by the rapid developments in biotechnologies and the pharmaceutical potential of biomolecules. Affinity membrane chromatography is now an attractive and competitive method for purifying proteins or other biomolecules from biological fluids. In addition to biomolecules, affinity membranes have also been utilized in bioprocess recovery to remove cell debris, colloidal or suspended solids and virus particles from homogenized suspensions of bacterial cells [Roper and Lightfoot (1995)].

Typical geometries of affinity membranes are thin sheet, hollow fiber, polymer rod, spiral wound and membrane stack, as shown in Fig. 7.3 [Roper and Lightfoot (1995)]. In each configuration, membranes possess pore structure and ligands that capture the targeted molecules are chemically or radically coupled on the substrate surface. Much different type of substrates are commercially available which made of polyethylene (PE), polypropylene (PP), polysulfone (PS), polyethersulfone (PES), nylon 6, 66, polyvinylalcohol (PVA) and cellulose esters, and the attachment procedure of ligands is dependent on the substrate type [Klein (2000)]. Ligands are generally classified into 1) amino acids, 2) antigen and antibody ligands, 3) dye ligands, 4) metal affinity ligands, 5) other biological and 6) ion exchange ligands. For instance, Cibacron Blue, the dye reacted with the membrane, is a well known ligand for binding mammalian albumin. The required functions of affinity membranes are high purification effectiveness within short processing time. Furthermore, molecule-capturing property must be uniformly distributed throughout the entire membrane. The idea of using polymer nanofiber as affinity membrane is based on the fact that the polymer nanofibers deposited on the collector during electrospinning assemble into a membrane-like nonwoven fiber mesh.

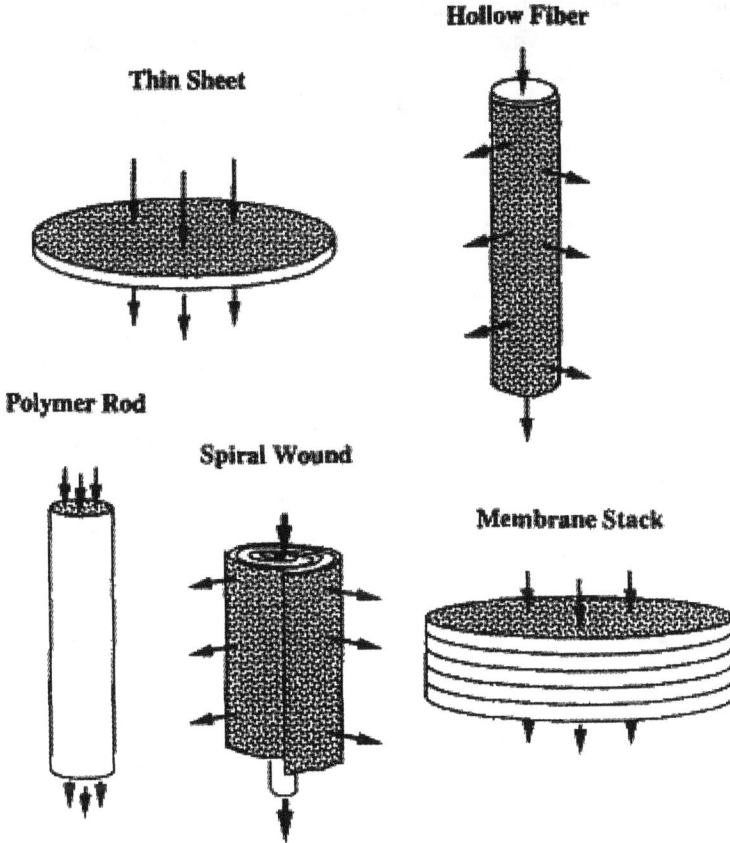

Fig. 7.3. Configurations of affinity membranes commercially available or laboratory prepared geometry [Roper and Lightfoot (1995)].

Electrospun nonwoven mesh possess properties like high porosity, micro-scaled pore size and high interconnectivity of the interstitial space, and above all, the nano-scaled fiber diameter give rise to an increased surface area as compared with the conventional nonwoven filter media composed of currently available textile fibers with typical diameters of several microns. A large surface area to volume ratio is one of the most important requirements for an ideal affinity membrane.

In order to address functions of affinity membranes, polymer nano fibrous membranes which possess large surface area to volume are a promising candidate. The Ramakrishna's group at National University of Singapore has been working on affinity membranes made of electrospun polymer nanofibers and Fig. 7.4 shows three preliminary works, i.e.,

✓ Beta-Cyclodextrin (β-CD) functionalized polymer nanofiber for organic waste removal

✓ Sushi peptide (S3Δ) immobilization on polymer nanofiber surface for lipopolysaccharides (LPS: endotoxin) removal

✓ Bovine Serum Albumin (BSA) functionalized polymer nanofiber for bilirubin removal.

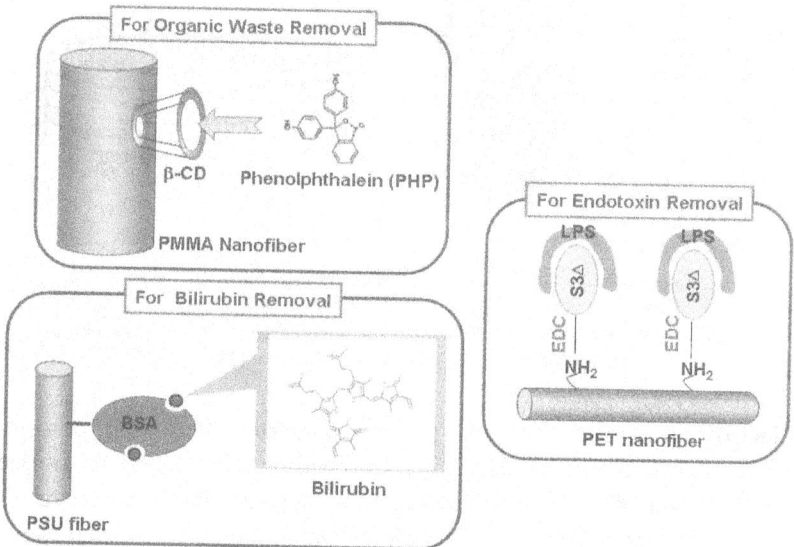

Fig. 7.4. Affinity membrane works using electrospun polymer nanofibers from Ramakrishna's group at National University of Singapore.

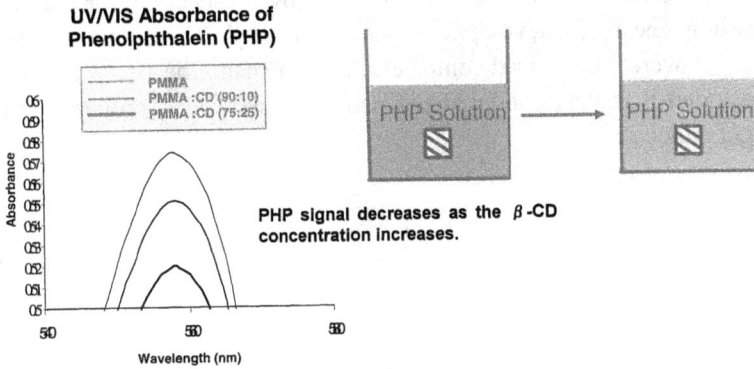

Fig. 7.5. UV/VIS absorbance of PHP on β-CD modified polymethylmethacrylate (PMMA) nanofibrous membranes [Courtesy of Ma and Ramakrishna at National University of Singapore].

For β-CD functionalization, phenolphthalein (PHP) was used as a model of organic waste and the capturing property of polymehtylmethacrylate (PMMA) nanofiber membranes was discussed as shown in Fig. 7.5. It was shown that UV/VIS absorbance of PHP decreased with increasing the amount of β-CD attached on the surface of PMMA nanofibers. Endotoxin (lipopolysaccharide-LPS) is a poisonous substance liberated by gram-negative bacteria. Sever septic shock caused by endotoxin during bacterial infections is responsible of over 100,000 death in US alone. Hence, the surface of polyethyleneterephathalate (PET) nanofibers was modified with sushi peptide (S3Δ) for LPS removal. Confocal microscope photograph shows that the surface modified PET nanofibers effectively captured LPS by means of S3Δ (see in Fig. 7.6). Similarly, polysulfone (PSU) nanofibers functionalized by bovine serum albumin (BSA) showed the capable capturing of bilirubin. Bilirubin is a metabolic waste product in human blood and high concentrated bilirubin is detected in the blood of patients suffering from hyperbilirubinemia.

Hence, the excessive bilirubin must be removed especially for the baby, of whom the brain development can be affected. Thus, three kinds of ligands were introduced onto electrospun nanofiber surface. It is concluded that the functionalized nanofiber showed ability to capture target molecules.

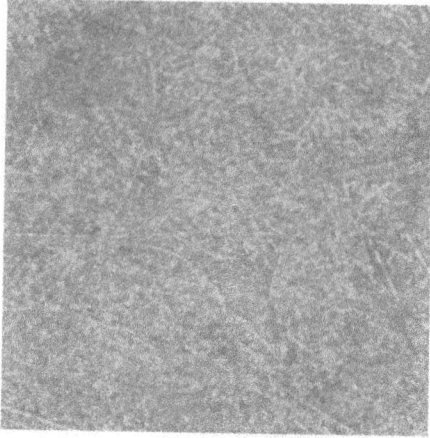

Fig. 7.6. An image of confocal microscope. LPS captured by the S3Δ peptide functionalized PET nanofiber [Courtesy of Ma and Ramakrishna at National University of Singapore].

7.3. Drug Release

In general, patients suffering from diseases take drug orally. Although drug is delivered to the damaged site, the amount of delivered drug decreases against initial drug dose as drugs also spread to the healthy site through digestive organs. Therefore, patients occasionally need to take excessive amount of drug for several times, which may induce undesirable side effect. It is considered that the ideal drug dose is to deliver the minimum required amount of drug to the disease site. Furthermore, those minimized amount of drugs must be effectively absorbed at the disease site. Generally, drug uptake into human body is faster with the smaller size of the drug and its coating material. Hence, drug delivery system has been developed using polymeric materials in the form of nano or micro particles, hydrogels and micelle [Kim et. al. (2004d)]. Although those polymeric drug delivery materials improved therapeutic effect and reduced side effect, there is still need to address how to precisely control the drug releasing rate. Based on such background, researchers have recently focused on the usage of polymer nanofiber membranes which encapsulate medical drugs instead of conventional polymeric materials.

For drug release system of nanofibers, drug chemical component is mixed with polymer solution and then nanofibers are electrospun, as summarized in Table 7.1. The main emphasis is to monitor the drug releasing rate from nano fibrous membranes. For this purpose, UV-Visible spectrophotometer is widely utilized. The principle of UV-Visible spectrophotometer is shown in Fig. 7.7. This apparatus utilizes ultraviolet and visible lights and the wave length of the lights is in the range of 180nm ~ 800nm. The energy of ultraviolet and visible lights corresponds to the difference of electron condition energy in atoms and molecules. Hence the spectrum obtained through the detector shows intensity of absorbed light as well as the information of sample molecules. Using this principle, the drug release profile of a nanofiber sample in testing liquid is determined.

Table 7.1. The list of nanofibers incorporated with several drugs so far. In all cases, electrospinning process was conducted with using polymer solution mixed with drug.

Material of Nanofibers	Drug Type	Drug Function	Reference
Poly(L-lactic acid), PLLA	Triethyl benzyl ammonium chloride (TEBAC)	------	Zeng et al (2003)
	Sodium dodecyl sulphate (SDS)	------	
	PPO-PEO ether (AEO10)	------	
	Rifampin	Anti-tuberculosis	
	Paclitaxel	Anti-cancer	
Polyurethane, PU	Itraconazole	Treatment for Tinea Pedis	Verreck et al (2003a)
	Ketanserin	Treatment for Wound Healing	
Hydroxylpropylmethyl cellulose, HPMC	Itraconazole	Treatment for Tinea Pedis	Verreck et al (2003b)
Poly(lactic acid), PLA	Tetracycline hydrochloride	Treatment for Periodontal Disease	Kenawy et al (2002)
PLA / Poly(ethylene-co-vinyl acetate), PEVA			
Poly(lactide-co-glycolide), PLGA	Mefoxin	Antibiotics	Kim et al (2004d)
PLGA / poly(ethylene glycol)-b-poly(lactide) (PEG-b-PLA)			
Poly(D,L-lactic acid), PDLA	Mefoxin	Antibiotics	Zong et al (2002)

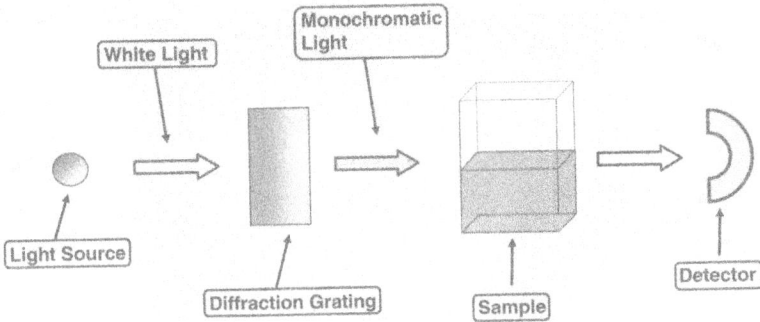

Fig. 7.7. Principle of UV-Visible spectrophotometer.

The first trial of drug release system using nanofibers may well be attributed to the work of [Kenawy et. al. (2002)]. Poly(lactic acid) (PLA), poly(ethylene-co-vinyl acetate) (PEVA) and PLA/PEVA blend (50:50) nanofibers were incorporated with tetracycline hydrochloride drug which is useful for the treatment of periodontal disease. The amount of the drug was 5% in each nanofiber system, respectively. The prepared nanofiber membranes with drug (see in Fig. 7.8) possess 200cm^2 total surface area. According to the release profile of the drug (see in Fig. 7.9), PLA nanofibers exhibited instantaneous release while PEVA and PLA/PEVA blend samples gradually increased the drug release rate within 120 hours. There is an optimum range of drug amount per unit time after patients take medicine. In other words, instantaneous high drug release causes the side effect while too low amount of drug is not effective therapy. Herein, the burst release of drug like PLA sample may not be preferable and steady drug release is suitable. [Verreck et. al. (2003a)] and [Zeng et. al. (2003)] also made an effort to avoid initial burst drug release on their nano fibrous membranes.

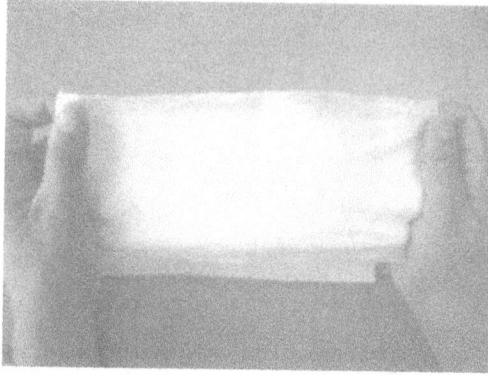

Fig. 7.8.   Poly(ethylene-co-vinyl acetate) (PEVA) nano fibrous membrane which contains tetracycline hydrochloride [Kenawy et. al. (2002)].

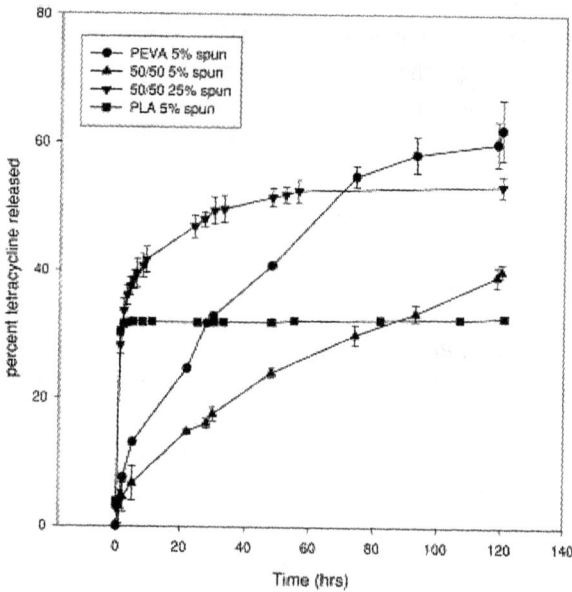

Fig. 7.9.   Drug release profile of 1) Poly(ethylene-co-vinyl acetate) (PEVA), 2) Poly(lactic acid) (PLA) and PLA / PEVA (50:50) nano fibrous membranes [Kenawy et. al. (2002)].

At what factor should the burst release of drug factor be? [Kenawy et. al. (2002)] observed PEVA nanofibers under SEM and found crystals of tetracycline hydrochloride on the fiber surface. Although 5% amount of the drug was mixed with polymer solution, it seems that the drug aggregates existed near the surface of nanofibers. In the case of poly(L-lactic acid) (PLLA) nanofibers with rifampin drug [Zeng et. al. (2003)], drug dispersion in nanofibers was checked by the original red color of rifampin. Except confirming the site of drug inside of nanofibers, the investigation of drug diffusion through the polymer and aqueous pores is an important issue to understand drug release manner from nanofibers [Verreck et. al. (2003a)]. In this case, the distribution of fiber diameter may need to be minimized to supply the constant drug release.

On the other hand, burst drug release is preferred for the type of medical treatment that uses antibiotics. [Zong et. al. (2002)] prepared poly(D,L-lactic acid) (PDLA) nanofibers incorporated with Mefoxin antibiotics and investigated its drug release profile. It was found that burst drug release was recognized in the first 3 hours and reached around 90% release rate up to 50 hours. Furthermore, based on Fick's diffusion law, it was speculated that a large amount of drug molecules were aggregated in the vicinity of fiber surface. [Zong et. al. (2002)] considered that such drug release property is suitable to prevent bacteria infection which generally occurs within the first few hours after surgery. Similar concept is also seen in the work of [Kim et. al. (2004d)] which developed poy(lactide-co-glycolide) (PLGA) nano fibrous membranes incorporated with 5wt% mefoxin. [Kim et. al. (2004d)] further demonstrated the efficacy of the medicated nano fibrous membranes by the incubation test using S. *aureus* bacteria, as shown in Fig. 7.10. Nano fibrous membranes with and without drug were placed in an agar plate and incubated at 37°C for 4 hours. Then, membranes were removed and proliferation of bacteria was allowed overnight. It was found that membranes containing drug inhibited bacteria growth in a much larger area that the membrane size due to the diffusion of drug.

Thus, the amount of instantaneous drug release may be dependent on the targeted medical treatment. However, in any medical cases, the drug release rate must be controlled or adjusted by changing the component of drug amount, minimized distribution of fiber diameter and drug distribution in a nanofiber. In this respect, the work of [Verreck et. al. (2003b)] which utilized pharmaceutical hard gelatin capsule is unique. Hydroxylpropylmethyl cellulose (HPMC) nano fibrous membranes containing drug was folded and placed in hard gelatin capsule. While nanofiber membranes themselves released 100% drug around 4hours, nanofibers filled in the capsule gradually released the drug over 20 hours.

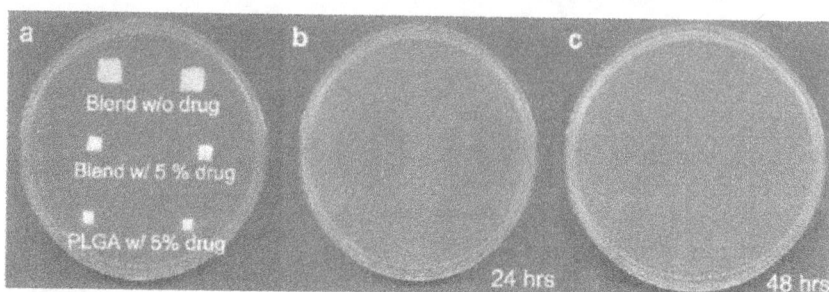

Fig. 7.10. Bacteria growth manner on agar plates. (a) Nano fibrous membranes (1cm x 1cm) with and without drugs were incubated at 37°C for 4 hours. (b) Sections were removed and S. *aureus* bacteria was plated and allowed to proliferate overnight. Regions surrounding membranes containing drugs show absence of bacteria growth both after (b) 24 hours and (c) 48 hours of incubation at 37°C [Kim et. al. (2004d)].

7.4. Tissue Scaffolds

Tissue scaffolds are utilized to repair the damaged human tissues. Seeded cells adhere and grow on tissue scaffolds immersed in the nutrient culture media and the scaffolds with cells are implanted at the human tissue defects. So far many different types of tissue scaffolds made of micron size material structure have been developed. However, the researchers realized that all of the human tissues such as blood vessel, cartilage, bone, nerve and skin, consist of nano fibrous forms from biological viewpoint. This finding brought the researcher's attention to nano structured scaffolds, and enhanced cell growth via such scaffolds at the damaged tissues has been expected. Hence, we can see much potential of the utilization of electrospun polymer nanofibers to repair human tissues. This section introduces the readers about the background and utilization of electrospun polymer nanofibers in each targeted tissue.

Blood Vessel

Blood vessels perform a very important function of carrying and transporting blood to and from the heart. As there are currently no acceptable synthetic prostheses or scaffolds for small-diameter blood vessels, surgeons have to harvest the patient's own blood vessel for the transplant. This procedure is not only time consuming, it may also lead to complications and significantly increases the recovery time for the patient. It also restricts the number of patients that are suitable for the surgery due to the limited vessels in the body appropriate for transplant. As a result, there have been many attempts to construct a viable blood vessel substitute.

In order to construct blood vessel scaffold, it is important to understand the basic structures of the natural blood vessel. The general structure of a blood vessel contains three layers, a *tunica intima* (innermost layer), a *tunica media* (middle layer), and a *tunica adventitia* (outermost layer). The *tunica intima* consists of a lining of endothelial cells (ECs) and an underlying layer of connective tissue with elastic fibers.

The *tunica media* contains concentric sheets of smooth muscle cells (SMCs) in loose connective tissue and the *tunica adventitia* consists of collagen and elastic fibers [Martini et. al. (1998)]. It is complex to construct a blood vessel scaffold to create the functional tissue engineered blood vessel substitute for the replacement of diseased organs because the cellular and humoral immune system may detect it as 'foreign', resulting in graft rejection. Sometimes, thrombosis or hyperplasia may also be a cause of vascular graft failure. Hyperplasia is caused by proliferation of smooth muscle cells in an inwards direction that blocks the lumen, thus interfering with blood flow. Whereas thrombosis is the formation of clot consisting of red blood cells, leukocytes and fibrin which was caused by an injury to the wall vessel, resulting in the obstruction of blood flow through the circulatory system.

There are many proposed methods to obtain the artificial blood vessel scaffold. These include using polymeric mesh [Hoerstrup et. al. (2001)]; filament wound tubular scaffold made from micron fibers [Gershon et. al. (1990)] and electrospinning [Kidoaki et. al. (2005)]. Artificial blood vessels can be constructed based on collagen scaffolds or biodegradable scaffolds [Shum-Tim et. al. (1999); Niklason et. al. (1999); Watanabe et. al. (2001); Hoerstrup et. al. (2001) (2002); L'Heureux et. al. (1998)]. Generally when natural collagen was used as scaffolds, poor mechanical properties were observed [Weinburg et. al. (1986); L'Heureux et. al. (1993); Hirai et. al. (1994) (1996)]. Biocompatible and biodegradable polymers such as polyglycolic acid (PGA) with porous structure are commonly used as scaffolds. In addition, it is easy to handle and fabricate into various shapes. However, PGA meshes are quickly bioabsorbed and are not capable of withstanding systemic pressures. To minimize this problem, copolymers by combining PGA with polyhydroxyalkanoate (PHA), poly(lactic acid-co-lysine), poly-4-hydroxybutyrate (P4HB), poly-L-lactic acid (PLLA) or polyethylene glycol (PEG) were used. These PGA copolymers achieved the improved mechanical properties of the scaffolds as well as regulating the cell's phenotypic characteristic via cellular interaction with the scaffold material [Kim et. al. (1999)].

Recently, electrospun polymer nanofibers have been identified as promising biodegradable matrices for supporting cell attachment and proliferation [Huang et. al. (2003)]. This process has the ability to produce nonwoven, nano fibrous structure as well as structures made of oriented fibers which makes it well suited for tissue engineering [Coombes et. al. (2002)]. So far, nanofibers obtained were of diameters from 3nm to 1000nm [Huang et. al. (2003)]. The wide variety of polymers that can be electrospun also makes electrospinning very attractive to fabricate tissue scaffolds [Huang et. al. (2003)].

Most of the natural extracellular matrix is composed of randomly oriented collagen nanofiber in nanometer scale diameters and the building components of artificial scaffolds were generally made of biodegradable PCL fibers. Attempts have been made for many years to develop viable synthetic or tissue-engineered prostheses for small blood vessels, but all had high failure rates for one reason or another. In particular, it was found that graft patency rate have been greatly reduced when synthetic grafts were used for small diameter (< 6mm) arteries, for example, coronary and infragenicular vessels [Steinthorsson and Sumpio (1999); Sapsford et. al. (1981); Veith et. al. (1986)]. The three-dimensional nanomatrix scaffolds made of collagen and cell secreting natural material can be used to seed the smooth muscle cells (SMCs) to mimic natural small diameter blood vessels. Studies suggested that muscle cells, once implanted in the scaffold developed the function, shape, morphology and cellular architecture of the normal vessel [Bowlin (2003)].

Poly(glycolide-co-ε-caprolactone) scaffolds possess elastic mechanical properties and promote SMCs adhesion and subsequent tissue formation [Lee et. al. (2003d)]. Typical materials currently used as vascular grafts include poly (ethylene terephthalate) (PET) and poly(tetrafluoroethylene) (PTFE) [Robins (1992)]. Surface modification of electrospun PET fibers improved the spreading and proliferation of endothelial cells (ECs) [Ma et. al. (2005a)]. The report of [Xu et. al. (2004a)] further suggested that not only the surface modification of nanofibers, fiber alignment of the scaffold is important. The aligned nanofiber scaffolds which mimicked the native extracellular matrix showed favorable interactions between smooth muscle cells (SMCs) and the scaffold, as shown in Fig. 7.11.

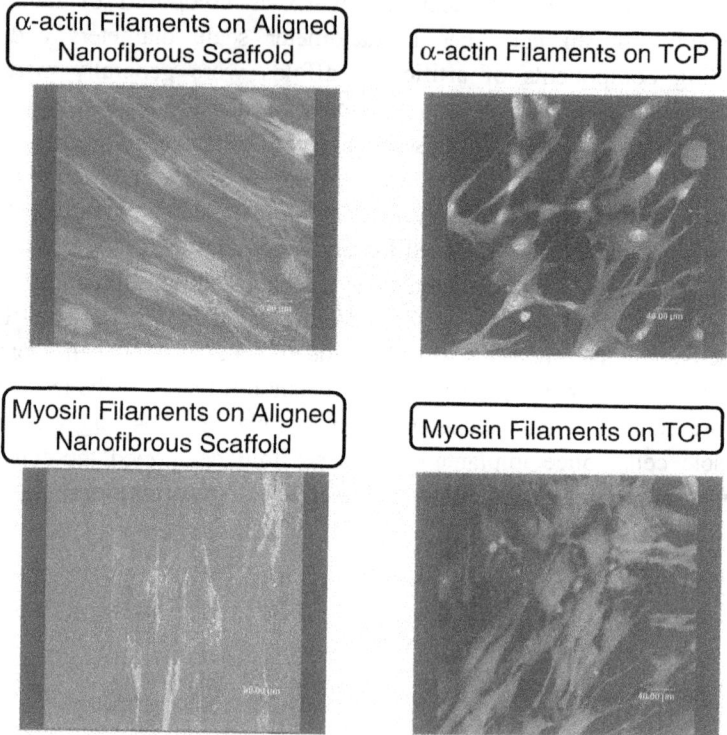

Fig. 7.11. Confocal microscope photos of immunostained a-actin and myosin filaments in SMCs after 1day of culture on aligned nanofiber scaffolds and TCP [Xu et. al. (2004a)].

The electrospinning method used to fabricate the scaffold is simple for the mass production and have huge potential for the application in blood vessel engineering. Various methods have been developed to support the seeding and growth of endothelial cells on polyurethane (PU) or other biomaterial surfaces through methods such as surface modifications by plasma treatment and photochemical grafting GRGD peptide on a modified PU surface or chitosan surface. The above mentioned surface modifications may be categorized as chemical or biological surface modifications to enhance endothelialization on biomaterial surfaces. On the other hand, modifications of physical properties of surfaces such as enhancing hydrophilic properties, changing porosity of materials and increasing roughness of surface may also enhance endothelialization of biomaterials [Chung et. al. (2003); Xu et. al. (2004b)].

Endothelialization is a promising way to minimize hyperplasia as endothelial cells secrete bioactive substances that inhibit smooth muscle cells from excessive proliferating. As the basement membrane consist of type IV collagen and laminin nanofibers and has a similar structure as non-woven polymer nanofibrous membranes, studies were conducted to facilitate viability, attachment and phenotypic maintenance of human coronary artery endothelial cells on poly(L-lactid-co-ε-caprolactone) [P(LLA-CL)] copolymer nanofibers incorporated with collagen [He et. al. (2005a) (2005b)]. It was shown that collagen coating on nanofibers improved viability and attachment of endothelial cells, as shown in Fig. 7.12. Tensile properties of fabricated nanofibrous scaffolds were equivalent to that of the human artery and were less stiff than a commonly used graft material, Dacron (PET). It was also found that endothelial cells could be spread without serum and growth factors in culture medium [He et. al. (2005b)].

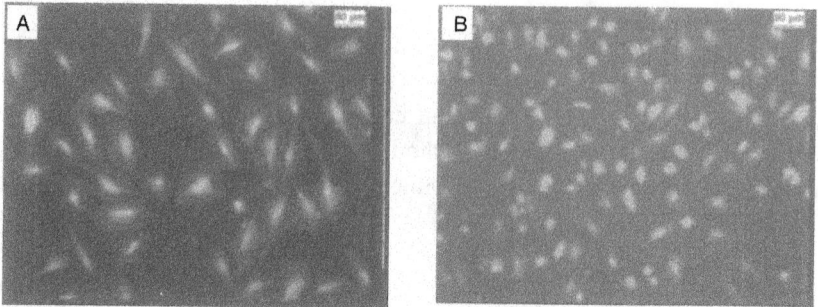

Fig. 7.12. Comparison of endothelial cells (ECs) culture between (A) tissue culture polystyrene surface and (B) collagen blended P(LLA-CL) nanofiber [Courtesy of He and Ramakrishna at National University of Singapore].

With electrospun nanofibers shown to be a good candidate as tissue scaffold through the many successes in tissue culture, the next step is to form tubular scaffold made of the electrospun nanofibers. Already, tubes of diameter as small as 1.27mm made from electrospun nanofibers has been fabricated for application as nerve guidance channel [Bini et. al. (2004)]. Tubular scaffold made of different layers of electrospun nanofibers consisting of different materials had also been made [Kidoaki et al. (2005)]. This has a great potential in creating a scaffold that offers both superior mechanical strength as well as providing a substrate suitable for cell proliferation.

As mentioned in Fig. 7.11, the orientation of the fibers encouraged cell proliferation in the direction of the fibers orientation [Xu et. al. (2004a)]. In natural blood vessel scaffold, the smooth muscle cells in the *tunica media* are arranged in a concentric manner. If SMCs are cultured on electrospun nanofibers that are oriented in a concentric manner to form a tubular scaffold, the cells is expected to grow in a concentric manner resembling that of a natural blood vessel.

Several methods have been proposed to control the orientation of the electrospun nanofibers from the use of a rotating collector [Theron et. al. (2001); Boland et. al. (2001); Matthews et. al. (2002)] to controlling the electric field during electrospinning [Li et. al. (2003b)]. The resultant mesh consisting of aligned electrospun nanofibers may then be rolled into a tubular scaffold. By using both a rotating tube and controlling the electric field, it is possible to align the electrospun nanofibers on the circumference of the small-diameter tube which can then be extracted to give a tubular scaffold [Bornat (1987)]. The pattering of nanofiber deposition is described in chapter 3 in details.

While this area remains a great challenge in the area of tissue engineering as many requirements need to be fulfilled, there is still a great potential in constructing a viable blood vessel substitute for tissue transplantations.

Cartilage

Cartilage is a dense connective tissue that is to a certain extent pliable, making it resilient. These characteristics are due to the nature of its matrix, which is rich in proteoglycans consisting of a core protein attached by the repeating units of disaccharides termed glycosaminoglycans (GAGs). Hyaluronic acid (HA), chondroitin sulfate and keratin sulfate are the most abundant GAGs found in cartilage. The matrix also has collagen fibers, but these are of a finer nature (Collagen Type II vs. Collagen Type I) than the collagen fibers in most other connective tissues. The macromolecules are bound to the thin collagen fibers by electrostatic interactions and cross-linking glycoprotein. There are 3 types of cartilage: hyaline cartilage (most abundant type), elastic cartilage and fibro cartilage.

Articular cartilages are important in load-bearing and reducing friction of the articular surfaces. Due to the limited capacity of articular cartilage to repairitself, and cartilage defect resulting from aging, joint injury and developmental disorders cause joint pain and loss of mobility [Li et. al. (2005b)], many attempts have been made to repair articular cartilage defects, which include transplantations of various tissues or cells. [Hattori et. al. (2002)], for example, used an autologous cultured chondrocyte or autologous bone marrow mesenchymal cell transplantations in repairing the articular cartilage defects. Tissue engineering approach provides a cell based therapy to repair the defect and restore joint functions [Grad et. al. (2003)]. In addition to that, many different techniques have been developed to produce nanostructured biodegradable materials such as microspheres, foams and films. It has been demonstrated that the molecular structure and morphology of PLA, PGA, PCL and their copolymers can play a major role in the degradation and mechanical properties of the final products [Lu et. al. (2001)]. Modifications of physical properties of surfaces such as enhancing hydrophilic properties, changing porosity of materials and increasing roughness of surface may also enhance cellularization of biomaterials. Polymer scaffolds are primarily used for the delivery and retention of chondrogenic cells in cartilage tissue engineering [Reinholz et. al. (2004)]. Many naturally derived and synthetic polymers are currently used as scaffolds for regeneration of articular cartilage and many others are under development. Recent advances in stem cell biology have shown that, mesenchymal stem cells can differentiate into cells of mesenchymal tissues such as bone, cartilage, muscle, tendon, ligament and fat and are expected to play an important role in the repair of skeletal defects.

Electrospinning has provided a basis for the fabrication of unique matrices and scaffolds for tissue engineering and a strong application in biomedical engineering. The structural design of artificial matrices or scaffolds that mimic the supramolecular structure and biological functions of the extracellular matrix (ECM) is a key issue in tissue engineering and the development of artificial organs. In cartilage tissue engineering, chondrocytes and mesenchymal stem cells (MSC) are commonly used for cartilage regeneration. PCL nano fibrous scaffolds structurally similar to ECM may represent promising structures for tissue engineering applications. Three dimensional nano fibrous scaffolds are characterized by high porosity with a wide distribution of pore diameter, high surface to volume ratio and morphological similarities to natural collagen fibrils. These nano fibrous scaffolds are readily fabricated in any shape and size as needed clinically and also provides sound mechanical stability to provide a carrier for MSC transplantation in tissue engineering cartilage repair. Many naturally derived and synthetic polymers are currently used as scaffolds for regeneration of articular cartilage and many others are under development. Collagen, PLLA, PLGA and PCA were shown to be suitable as a scaffold for in vitro cartilage regeneration, as demonstrated by cell densities equivalent to those found in natural tissues and by continued cellular production of type II collagen [Grande et. al. (1997); Freed and Vunjak-Novakovic (1995)].

Bone

Bone is a connective tissue distinguished by the fact that its matrix is mineralized by calcium phosphate in the form of crystals very similar to hydroxyapatite (calcium phosphate mineral). The minerals are both found in the collagen fibers (Type I) which constitute about 95% of the matrix, and in the ground substance. Bone is both resilient and hard. Its resilience is due to the organic matter (collagen), its hardness due to the inorganic minerals. Bone serves as a storage site for calcium and phosphate. There are two kinds of mature bone: compact bone (dense bone/ cortical bone) and spongy bone (cancellous bone, trabecular bone and medullary bone).

Compact bone and spongy bone are found in specific locations. In long bones, most of the thickness of the diaphysis is made of compact bone, with a small amount of spongy bone facing the marrow cavity. The ends (epiphyses) of long bones, however, consist mostly of spongy bone covered with a shell of compact bone. The flat bones of the skull have a middle layer of spongy bone sandwiched between two relatively thick layers of compact bone.

Developing natural or synthetic substances to replace the body's tissues, organs or functions is the goal of biomaterials research. In tissue engineering, synthetic compounds are used to augment patient's living tissue. Tissue-engineered bone grafts have been used to fill defects in bone caused by clinical states or trauma, such as fractures with bone loss, bone infection, or bone tumors. A bone graft is a piece of bone transplanted to another part of the skeleton where it is needed to improve function or strengthen the structure of the area. Sometimes a bone graft is taken from a cadaver, but usually it is harvested from the patient for which it will be used. Aside from blood, bone is the most frequently transplanted tissue in humans. Bone grafts are typically harvested from the patient's iliac crest (top of the hip bone), ribs, or fibula in the lower leg. They may be used in plastic surgery to reconstruct a chin or in dental surgery to provide a bony bed for a dental implant. The application and surgical site help determine how a bone graft performs, if it heals properly and offers the same durability as the original tissue.

Autologous bone is the current gold standard graft material without the risk of transfer of disease for the treatment of skeletal defects and fracture repair. In contrast, the use of allogenic bone may transmit diseases and cause an immune response, which can lead to high failure rate. One of the main goals in tissue engineering has been to develop biodegradable materials as bone graft substitutes for filling large bone defects. These materials should maintain adequate mechanical strength over critical phases of the tissue healing process and also should be oesteoconductive. Furthermore, the proper degradation at a controlled rate is required to provide space for the formation of new bone.

Engineering living tissue for reconstructive surgery requires an appropriate cell source, optimal culture conditions and a biodegradable scaffold as the basic elements. While specialized cells remain an important source, stem cells have emerged as a promising new alternative. Recent advances in stem cell biology have shown that mesenchymal stem cells (MSC) can differentiate into cells of mesenchymal tissues such as bone, cartilage, muscle, tendon, ligament and fat and are expected to play an important role in the repair of skeletal defects [Olivier et. al. (2004); Caplan (1991); Pittenger et. al. (1999)]. Osteoblasts derived from MSC of neonatal rats were cultured on poly (DL-lactide-co-glycolide) foams and mineralizations as well as three-dimensional bone formation were observed [Terai et. al. (2002)]. Cell adhesion is the most important aspects of cell interaction with a biomaterial because it is the prerequisite for further cellular activity such as spreading, proliferation and differentiation. Initial osteoblast/material interactions may be conveniently characterized by four stages: (1) protein adsorption to the surface, (2) contact of rounded cells, (3) attachment of cell to the substrate, and (4) spreading of cells. Initial cell attachment is influenced by the original surface characteristics of the materials [Leong et. al. (1985)]. In later phase of cell development, however, other factors such as polymer degradation products or bulk properties may play a more crucial role.

Biodegradable polymers often have been combined with bioceramics to produce materials for bone repair. Biodegradable polymers such as PGA, PLA and PLGA have found application as resorbable sutures [Beniceuicz and Hopper (1991)] bone fixation screws and plates, drug carriers, and they have been widely investigated as bone graft substitutes with positive results [Holland et. al. (1986); Ishaug et. al. (1994)].

Polycaprolactone (PCL) has been investigated mainly for long-term implants for drug release and support mineralized tissue formation and may be a suitable candidate for the treatment of bone defects [Yoshimoto et. al. (2003)]. An improvement in the mechanical properties of PCL has been achieved by co-polymerization with PLA, enabling its use for orthopedic applications, such as the repair of bone defects. There has been widespread use of bioceramics such as hydroxyapatite (HA) and tricalciumphosphate (TCP) for bone regeneration applications and their biocompatibility is thought to be due to their structural similarity to the mineral phase of bone. Biodegradable ceramics are used in dental and orthopedic surgery as fillers for bone defects and as coatings on metallic implants to improve implant integration in host bone. Biodegradable polymers have been used as binders for HA and TCP to overcome the problems of brittleness and the difficulty of shaping hard ceramic materials to fit bone defects [Rizzi et. al. (2001)]. At the same time, bioceramic reinforcement of polymers has been shown to improve mechanical properties and osteoconductivity [Kikuchi et. al. (1997)]. Biodegradable polymer/bioceramic composites are promising materials for bone graft replacement and this has been intensively investigated in the last decade [Piattelli et. al. (1997)]. With this respect, PCL/Calcium Carbonate ($CaCO_3$) composite nanofibers have been investigated as possible bone substrates [Fujihara et. al. (2005)]. The surface of PCL nanofibers was covered by $CaCO_3$ nano particles (see in Fig. 7.13) and osteoblast attachment was encouraged on composite nano fibrous membrane *in vitro*.

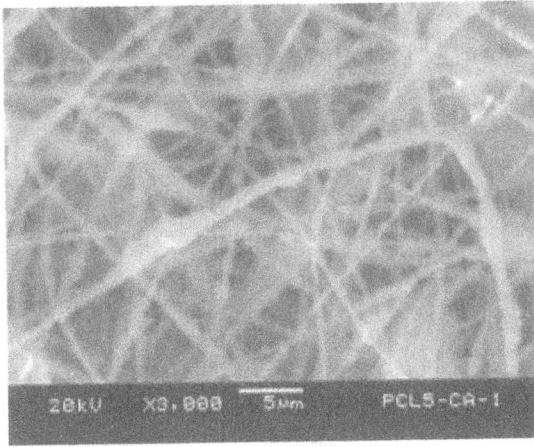

Fig. 7.13. Polycaprolactone (PCL) / CaCO$_3$ composite nanofibrous scaffolds for
bone repair therapy [Fujihara et. al. (2005)].

Carbon nanotubes and nanofibers have several properties that suggest these materials may be of value in the development of novel devices for bone reconstruction [Price et. al. (2003)]. Carbon fibers with nanometer dimensions simulate dimensions of collagen fibrils (0.1–8 μm in diameter) in bone. [Price et. al. (2003)] suggested for the first time that smaller diameter carbon nanofibers without the pyrolytic layer (i.e. PR-24 PS fibers) could be suitable for use in orthopedic/dental implant material designs due to the increase in osteoblast and decrease in osteoblast competitive cell line (fibroblasts, chondrocytes and smooth muscle cells) adhesion. More osteoblast adhesion and less competitive cell adhesion could lead to faster integration of the bone to the implant surface *in vivo*. In addition, other *in vitro* studies have shown increased osteoblast alkaline phosphatase activity and calcium deposition on nanometer carbon fibers compared to conventional fibers [Elias et. al. (2002)]. This is further evidence that nanometer carbon fibers enhance osteoblast activity for possible increased integration of bone on the orthopedic material surface.

Nerve

Nerve tissue repair is a significant treatment concept in human health care as it directly impacts on the quality of life. It is known that a properly synthesized extracellar matrix (ECM) is crucial in guiding neural outgrowth and may be relevant to the process of regeneration. The ECM of the nervous system consists of the collagen types I, II, III, IV and V, the noncollagenous glycoproteins and the glycosaminoglycans (GAGs) [Rutka et. al. (1988); Carbonetto (1984)]. From the biological point of view, a great variety of natural ECM components exist in fibrous form and structure. Examples include collagen, fibronectin and laminin. All of them are characterized by well organized hierarchical fibrous structures ranging from nanometer to millimeter scale. From this, a successful regeneration of nerves calls for the development of fibrous structure with fiber architecture for cell deposition and proliferation.

Electrospinning provides a straightforward way to fabricate fibrous polymer scaffold for neural tissue engineering. The potential of nano fibrous porous scaffold is prepared by phase separation using poly(l-lactic acid) with a similar structure of natural ECM in nerve tissue engineering [Yang et. al. (2004a)]. In a recent study [Bini et. al. (2004)], nerve guidance channel was fabricated by electrospinning poly(L-lactide-co-glycolide) biodegradable polymer nanofibers onto a Teflon mandrel. The feasibility of *in vivo* nerve regeneration was investigated through several of these conduits. The biological performance of the conduits was examined in the rat sciatic nerve model with a 10mm gap length. After implantation of the nano fibrous nerve guidance conduit to the right sciatic nerve of the rat, there was no inflammatory response. One month after implantation five out of eleven rats showed successful nerve regeneration (see in Fig. 7.14). None of the implanted conduits showed tube breakage. The nano fibrous nerve guidance conduits were flexible, permeable and showed no swelling.

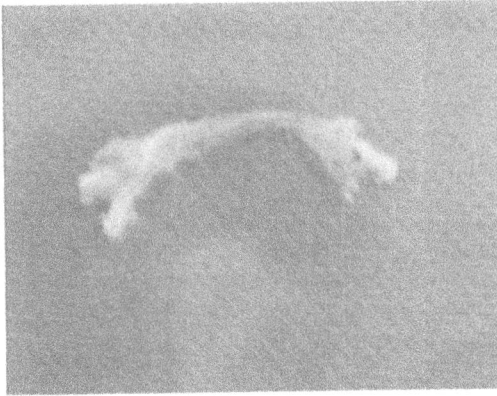

Fig. 7.14.   Macrograph of the regenerated never cable one month after implantation. [Bini et. al. (2004)].

Nerve regeneration in the central nervous system (CNS) is much more difficult than in the peripheral nervous system (PNS). One of the possible reasons is due to the glial cells in the CNS after trauma acting as a barrier to the regenerating axons and creating a hostile environment for axon regeneration [Carbonetto (1984); Schmidt and Leach (2003)]. The differences in the ECM are likely to be another reason for the disparity in regeneration between peripheral and central nervous systems [Carbonetto (1984)]. In the PNS, the linear orientation of ECM tubes in the peripheral stump of a damaged nerve provides terrain well suited for axonal regrowth, whereas there is no such framework available to assist in the process of axonal regrowth in the CNS [Rutka et. al. (1988); Bunge and Bunge (1983)]. The nano fibrous scaffolds have two unique features that make it well suited for CNS tissue engineering application. First, the morphology and architecture are similar to the natural ECM, which is the foundation of creating reproducible and biocompatible three-dimensional scaffolds for cell attachment, differentiation and proliferation; secondly, the nano fibrous scaffolds are highly porous structure with a wide variety of pore diameters, which allows for fluid transportation while inhibiting glial scars [Yang et. al. (2004b)]. Moreover, with the electrospinning technique, the aligned nano fibrous scaffold can be achieved [Yang et. al. (2005)].

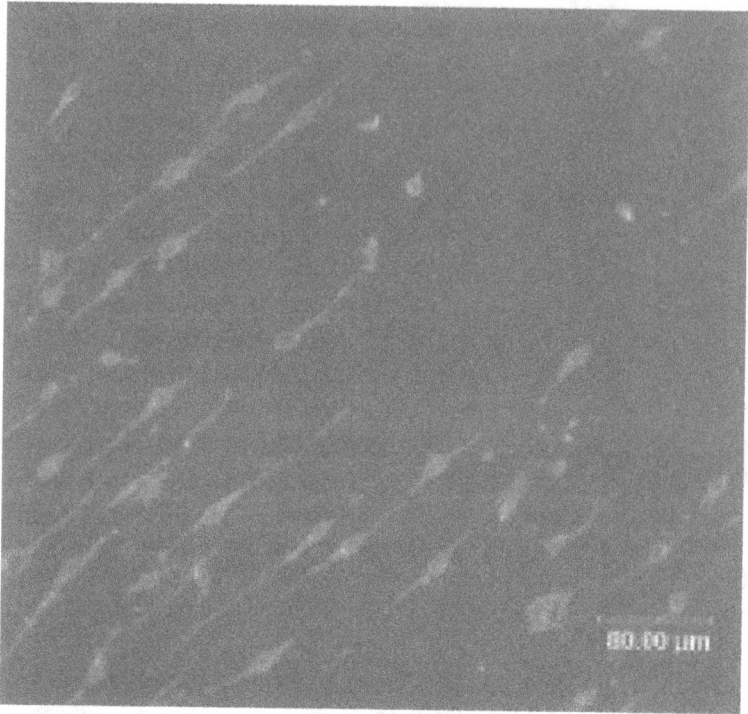

Fig. 7.15. Laser Scanning Confocal micrographs of immunostained neurofilament 200kD in C17.2 cells after 2 days of culture on aligned electrospun PLLA fibers [Yang et. al. (2005)].

The *in vitro* results showed that the fiber alignment had a strong effect on the cell phenotype: neural cells on aligned fibers grew parallel to the fiber orientation (see in Fig. 7.15). Furthermore, the aligned nanofibers improved neurite outgrowth when compared with random or micron fibrous scaffolds. The results suggest that the aligned nano fibrous scaffold may be a suitable nerve guidance channel, in particular for CNS regeneration.

7.5. Wound Dressing

Wound dressing is a therapy to repair the skin damaged by ambustion and injury. So far electrospun nanofibrous membrane exhibited the potential in wound dressing field. The membrane attained uniform adherence at wet wound surface without any fluid accumulation [Bhattarai et. al. (2004)]. Wound dressing with electrospun nanofibrous membrane can meet the requirements such as higher gas permeation and protection of wound from infection and dehydration. The goal of wound dressing is the production of an ideal structure, which gives higher porosity and good barrier. To reach this goal, wound dressing materials must be selected carefully and the structure must be controlled to confirm that it has good barrier properties and oxygen permeability. The rate of epithelialization was increased and the dermis was well organized in electrospun nanofibrous membrane and provided a good support for wound healing [Khil et. al. (2003)]. This wound dressing showed controlled evaporative water loss, excellent oxygen permeability and promoted fluid drainage ability due to the nanofibers with porosity and inherent property of polyurethane.

The materials described here is to apply physical integration of natural and synthetic polymers to provide a favorable substrate for fibroblast [Venugopal et. al. (2005)]. Biodegradable polycaprolactone (PCL) is potentially useful for the replacement of implanted material by the repair of tissues by coating collagen and improves the mechanical integrity of the matrix. The PCL and collagen nanofiber structure provides a high level of surface area for cells to attach due to its 3D feature and its high surface area to volume ratio, as shown in Fig. 7.16. This approach exploits the cell binding properties of PCL whilst avoiding the toxicological concerns associated with chemical crosslinking of collagen to impart stability. Tissue engineering scaffolds are required to exhibit a residence time but do not compromise complete space filling by new tissue at the wound site. Cell interaction study proves fibroblasts that migrated inside the collagen nano fibrous matrices showed morphologically similar to dermal substitute.

The collagen synthesized by the fibroblast enhanced the attachment of keratinocytes to the surface of artificial dermis in serum free medium. It is assumed that the presence of fibroblasts invade the wound tissue by early synthesis of new skin tissue because the fibroblasts on the artificial dermis can release biologically active substance cytokines. The dermal fibroblasts entering into the matrix through small pores in an electrospun structure by differently oriented fibers lay loosely upon each other. When cells perform amoeboid movement to migrate through the pores, they can push the surrounding fibers aside to expand the hole as small fibers offer little resistance to cell movement. This dynamic architecture of the fibers provides the cells to adjust according to the pore size and grow into the nanofiber matrices to form a dermal substitute for many types of wound healing. The nanofiber based cultured dermal fibroblast maintains the moist environment on the wound surface and thereby to promote wound healing.

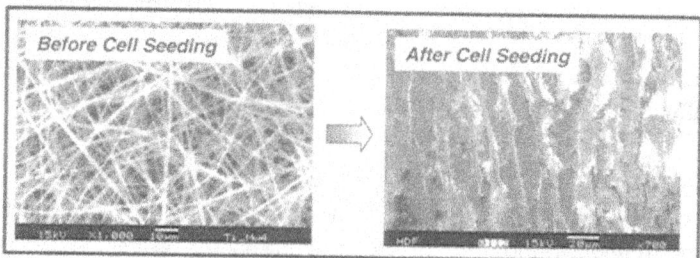

Fig. 7.16. Fibroblast cell attachment and spreading manner on polycaprolactone and collagen nanofibers [Courtesy of Venugopal and Ramakrishna at National University of Singapore].

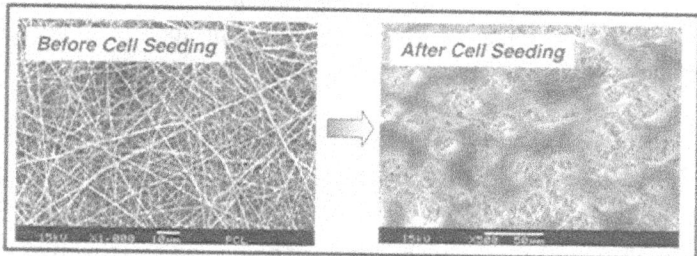

7.6. Filter Media

In industrial factory, working office and hygienic surgical operation room, air purification is essential requirement to protect people and precision equipment. Filter media is utilized to purify air which contains solid particles (virus, mine dust and anther dust etc.) and liquid particles (smog, evaporated water and chemical solvents etc.). So far, high-efficiency-particulate-air (HEPA) filter made of non-woven glass fiber mesh has been utilized to capture particles and 300nm size particles can be excluded with 99.97% filter efficiency [Maus et. al. (1997); Kemp et. al. (2001)]. However, mesh pore size must be small or thick mesh is required to remove ultra fine particles, which means that a filtration fun needs to blow the air with high pressure. In contrast, air blowing with lower pressure leads to poor ventilation through a filter media. This kind of property is called "pressure drop" and lower pressure drop is required to an excellent filter media. On such background, electrospun nanofiber membranes have gained the large potential and it was estimated that future filtration market would be up to US $700billion by the year 2020 [Suthar and Chase (2001)].

An important property of filter media is high filtration efficiency with low air-blowing resistance. [Tsai et. al. (2002)] compared the filtration efficiency of conventional meltblown fabrics (35 g/m^2 density) and electrospun polyethylene oxide (PEO) nanofiber webs (3 g/m^2 density) under NaCl aerosol which contains 100nm size particles. It was revealed that thinner PEO nanofiber webs showed similar filtration efficiency as meltblown fabrics. As PEO nanofiber webs with 16 g/m^2 density is comparable to HEPA filters, small quantity of nanofiber webs is useful to achieve the same function of HEPA filters.

Another possible benefit to use nanofibers is that nanofiber webs may be applied to remove small liquid drops in submicron range while HEPA filters are more suitable to remove solid particles not liquid drops [Hajra et. al. (2003)].

[Hajra et. al. (2003)] prepared composite filter media which contains sub-micron glass fiber substrate and polyamide nanofiber layer, as shown in Fig. 7.17. The tested filter media with small addition of nanofibers on the substrate of glass fibers effectively captured oil droplets with 210nm diameter. Similar effect was also seen in the result of [Shin and Chase (2004)] which used polyacrylonitrile, polyamide and nylon nanofibers on glass fiber substrate. Several companies have been seeking for the feasibility of commercialization of polymer nanofiber webs due to their high filtration efficiency. Firma Carl Freudenberg may be the oldest company which filed a US patent with respect to filter media made of electrospun polymer nanofiber webs [Groitzsch and Fahrbach (1986)]. Fiber Mark Gessner GmbH & Co., adopted two-layer composite nanofiber webs with supporting micron-fiber layer and targeted dust filter bags which captures the particles with below 500nm

Fig. 7.17.   SEM photograph of filter media which consists of sub-micron glass fiber substrates and electrospun polyamide nanofibers. The diameters of glass fibers and nanofibers are 4 ~ 5mm and 100 ~ 200nm, respectively [Hajra et. al. (2003)].

diameter size [Emig et. al. (2002)]. Donaldson Company, USA has already commercialized nanofiber filter media consisted of 10μm size cellulose fibers and 250nm size nanofibers, so called Ultra-Web® [Grafe and Graham (2003)]. In their trial, ISO Fine test which is commonly used for the evaluation of engine air cleaner was adopted. The test utilized particles in the size range of 0.7 ~ 70μm and investigated how much dust was captured by a filter media. It was found that micron-size cellulose filter media indicated 68% sub-micron dust reduction while Ultra-Web® indicated 92% sub-micron dust reduction.

7.7. Chemical and Biological Protective Clothing

Background

These days, the worldwide threat of chemical and biological agents by hostile military and terrorist groups has been growing. There is a grave concern in the development of reliable and stronger mechanisms of defense. Upholding the motto of "prevention is better than cure" is very essential in the field of national security, where even the smallest compromise cannot be made as it would lead to disastrous effects not only to an individual but to the community and more so to the country as a whole. Hence attention is drawn to develop protective systems to fight these chemical and biological weapons of mass destruction. With all these advancements in artillery, soldier survivability has considerably reduced. Chemical and biological weapons can be used across a wide spectrum of warfare, from acts of assassination and small-scale terrorism to various tactical and operational situations, both defensive and offensive, including strategic population attacks. The technical and economic barriers to development and weaponization have decreased.

The chemical agents can be divided into three major categories, nerve agents, blister agents/ vesicants and blood agents.
The vesicants include:

HD - sulfur mustard (Yperite)
HN - nitrogen mustard
L - lewisite (arsenical vesicants may be used in a mixture with HD)
CX - phosgene (properties and effects are very different from other vesicants)

HD and HN are the most feared mustard vesicants historically, because of their chemical stability, their persistency in the field, the insidious character of their effects by attacking skin as well as eyes and respiratory tract, and because no effective therapy is yet available for countering their effects. Hence individual protection becomes mandatory to safeguard from exposure and for these reasons, we would keep our attention on prevention of the mustard gases only.

Nerve Agents

As their name suggests, nerve agents (OP compounds) attack the nervous system of the human body. All such agents function the same way: by interrupting the breakdown of the neurotransmitters that signal muscles to contract, preventing them from relaxing. Initial symptoms following exposure to sarin (and other nerve agents) are a runny nose, tightness in the chest and dilation of the pupils. The effects of nerve agents are very long lasting and cumulative (increased successive exposures), and survivors of nerve agent poisoning almost invariably suffer chronic neurological damage. Some of the nerve agents include the volatile G type and the non-volatile V type. Some of the chemicals known to have been manufactured for this purpose include Sarin, Soman, Tabun and VX.

Blood Agents

The most important route of poisoning is through inhalation. Both gaseous and liquid hydrogen cyanide, as well as cyanide salts in solution, can also be taken up through the skin. Its high volatility probably makes hydrogen cyanide difficult to use in warfare since there are problems in achieving sufficiently high concentrations outdoors. On the other hand, the concentration of hydrogen cyanide may rapidly reach lethal levels if it is released in confined spaces. The most important toxic effect of blood agent is by inhibiting the metal-containing enzymes. One such enzyme is cytochromoxidase, containing iron. This enzyme system is responsible for the energy-providing processes in the cell where oxygen is utilized, i.e., cell respiration. When cell respiration ceases, it is no longer possible to maintain normal cell functions, which may lead to cell mortality. Symptoms of poisoning vary and depend on, for example, route of poisoning, total dose and the exposure time. If blood agent has been inhaled, the initial symptoms are restlessness and increased respiratory rate. Other early symptoms are giddiness, headache, palpitations and respiratory difficulty. These are later followed by vomiting, convulsions, respiratory failure and unconsciousness. If the poisoning occurs rapidly, e.g., as a result of extremely high concentrations in the air, there is no time for symptoms to develop and exposed persons may then suddenly collapse and die.

Bio-warfare Agents

Biological pathogens released intentionally or accidentally, or naturally occurring, can result in disease or death. Human exposure to these agents may occur through inhalation, skin (cutaneous) exposure, or ingestion of contaminated food or water. Following exposure, physical symptoms may be delayed and sometimes confused with naturally occurring illnesses. Biological warfare agents may persist in the environment and cause problems some time after their release.

The bio-warfare agents fall into four major groups, three classes of microorganisms - bacteria, rickettsia, and viruses, plus bacterial toxins, which are poisonous chemicals produced by bacteria. More than 100 years of international efforts to ban chemical weapons culminated January 13, 1993, in the signing of the Chemical Weapons Convention (CWC). Tremendous importance has been placed on the development of protective systems to survive the attack of chemical and biological warfare agents.

NBC (Nuclear Biological and Chemical) protective equipments have been in use for many years [Smith et. al. (2001); Smith and Dann (1991)]. They typically consist of fabric impregnated with activated charcoal which can adsorb the mustard gases and radioactive alpha and beta particles. These equipments were designed such that they are impermeable to air, particulate matter, bacterial contaminants and other liquids. The disadvantages associated are:

a) Heavy weight and no protection against viruses (for charcoal impregnated breathable masks)

b) Moisture retention from within (for polymer film based light weight non- breathable suits)

As such, a lightweight and breathable fabric, which is permeable to both air and water vapor, insoluble in all solvents and highly reactive with nerve gases and other deadly chemical agents are desirable. Because of their great surface area, nanofiber fabrics are capable of neutralization for chemical agents without impedance of the air and water vapor permeability to the clothing [Smith et. al. (2001)]. Electrospinning results in nanofibers laid down in a layer that has high porosity but very small pore size, providing good resistance to the penetration of chemical harm agents in aerosol form [Gibson et. al. (1999)].

Preliminary investigations have indicated that compared to conventional textiles the electrospun nanofibers present both minimal impedance to moisture vapor diffusion and extremely efficiency in trapping dust and aerosol particles [Gibson et. al. (1998); Graham et. al. (2004)]. Fig. 7.18 shows strong promises as ideal protective clothing.

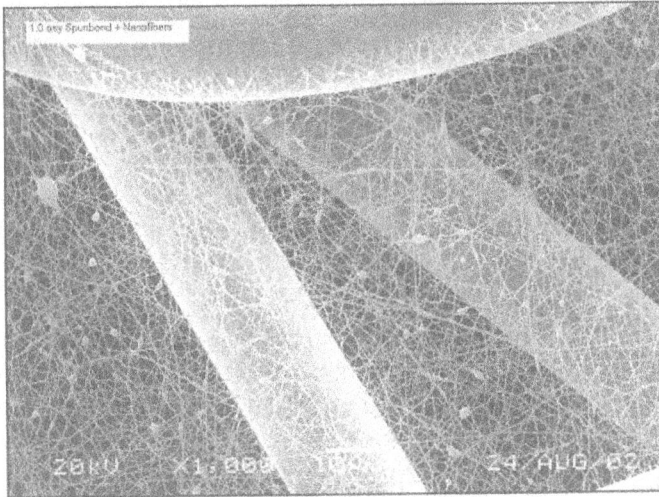

Fig. 7.18. SEM micrograph of a nanofiber filter on support for air filtration showing the dust particles trapped by electrostatic action [Graham et. al. (2004)].

Filter masks which offer protection against chemical agents are mostly impermeable and come with in-built respirators [Reneker and Chun (1996)]. Some designs allow for activated charcoal impregnated media which allows for air permeability thus eliminating the need for any self contained breathing apparatus. Such designs limit the continuous time of usage and increase weight of the equipment. Filter masks developed for protection against biological agents come with an additional HEPA filter media in addition to the activated charcoal media, which can filter out particles ranging from size 0.13-15 micrometers.

Such types are referred to as M15, M40, etc. The M17 and M40 series chemical-biological mask, when properly fitted and worn with the hood, protects against field concentrations of all known chemical and biological agents in vapor or aerosol form, but fails to protect when the air has a low-oxygen content, such as in tunnels or caves, or when the air has a high level of smoke mixtures. Nanofibers have an excellent potential to be used in the air as well as particulate filtration. These nanofibers, fabricated by electrospinning have a very large surface area and there by could be used as lining material for the face mask canisters wherein they aid in the capture and adsorption of the toxic chemicals and act as a barrier to the biological warfare agents. It was also shown that those with smaller diameter of fibers had better filtration efficiencies and could be fabricated in smaller thicknesses, as shown in Fig. 7.19 [Graham et. al. (2004)].

The Institute for Soldier Nanotechnologies (ISN) is an interdepartmental research center at Massachusetts Institute of Technology (MIT), USA. Established in 2002 by a five-year, $50 million contract from the U.S. Army, the ISN's research mission is to use nanotechnology to dramatically improve the survival rate of the soldiers. The ultimate goal is to create a 21st century battle suit that combines light weight high-tech capabilities with comfort. Nanotechnology fits into this vision in two important ways. First, it offers the potential for miniaturization, a key part of reducing the weight. Today's hefty radio worn on a harness might be reduced to a button-sized tab on the collar. A waterproof poncho could be replaced by a permanent nano-thin coating applied to everything the soldier carries.

Secondly, because nanotechnology operates at length scales whereby classical macroscopic physics no longer applies, it offers engineers the potential for creating unprecedented new material properties and devices. As such, nanotechnology can solve problems that scientists have been struggling with for decades.

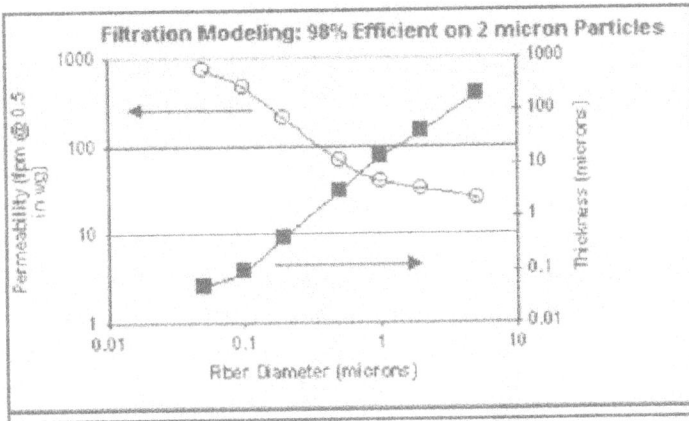

Fig. 7.19. Plot between Permeability and fiber diameter showing that smaller fibers lead to improved permeability and also reduce thickness of the filter [Graham et. al. (2004)].

Singapore realizes the importance of protective systems for chemical and biological defense. Initiative has been taken in the National University of Singapore (NUS) for the development of nanofiber based protective clothing and face mask canisters. Researchers at the NUS are trying to come up with functionalized nanofibers which can selectively bind to the chemical warfare agent on contact and decompose it into non-toxic by products. The current mechanism followed by the US Army is to only provide a barrier to the entry of toxins and eliminate them from human contact.

To encounter the above mentioned constraints, it is desired that the filter media's material in the face mask is selectively permeable. This is possible by adopting polymer nanofibers which are not only selectively permeable but are also very light weight. The proposed nanofiber media comprises of two distinct groups of fibers, one of which has the capacity to capture and decompose mustard gases and the other set of fibers is able to oxidize and deactivate nerve agents (organ phosphorus compounds). Together the mesh has an average pore size of less than 500 nm; this blocks the entry of bacterial spores of Bacillus Subtilis and Bacillus Anthracis. High surface area and light weights are the chief properties of the nanofibers which come into play to yield the increased decomposition efficiency. High surface area helps in the effective binding and deactivation of chemical and biological agents of target and the light weight is due to the increased permeability of these nanofiber webs which eliminate the need for external respirators. The proposed filter media made up of polymer nanofibers would have sufficiently high porosity and thus would be able to avoid the issues from low oxygen permeability.

Beneficial Features and Potential Impact to Defense and Security
1. Selective capture and decomposition of chemical warfare agents (Mustard and Nerve gases)

2. Removal of biological warfare agents by size separation (bacterial agents such as Bacillus Subtilis, Bacillus Anthracis)

3. Indication of contaminant levels

4. Submicron particulate filtration capability

5. Extended service lifetime of the material

6. Environmentally friendly and easy disposal

7. Cost-Effective

Hostile use of chemical and biological warfare agents is a serious threat to the men in the battlefield. This threat is amplified with the ease of availability and concealment of these agents. Hence, providing the soldiers with physical protection and other means of surviving these threats is paramount. This project focuses on developing nano materials technology that can provide significant enhancement to chemical and biological defense. Nano materials in the form of polymer nanofibers with appropriate functionalization represent a breakthrough technology that provides a means of making systems more practical and affordable for individual protective equipments.

Presently, activated carbon-based permeable media systems such as the US Joint Service Lightweight Integrated Suit Technology (JSLIST) [Smith et. al. (2001); Smith and Dunn (1991)] chemical protective over garment can effectively stop the chemical agent vapor challenge, but at a cost of heat stress, weight, and bulk. The development of lightweight and selectively permeable, breathable fabrics and filter media would be ideal and very much in conjunction with the Defense Technology Objective (DTO) "Advanced Lightweight CB Protection".

The uses of polymer nanofibers not only help in weight reduction but also help to increase the effective or available surface area for functionalization. These fibers have an estimated available surface of about $10\text{-}20 m^2/g$ compared to the $4 m^2/g$ for activated carbon adsorbent based media [Smith and Dunn (1991)]. Increase in surface area implies greater adsorption (and enhanced decomposition) of chemical and biological warfare agents and extended life of the media.

7.8. Energy and Electrical Application

Background

In this century China, which has one of the largest populations in the world, has been rapidly developing their industry and economy. Subsequently, the demand of energy consumption in China has been increasing every year. However, it is reported that estimated reserve amounts of oil and natural gas in the world are 41 years and 67 years, respectively [Source: Agency for Natural Resources and Energy, Japan, 2005]. Furthermore, the readers may be reminded that the Kyoto Protocol, a legally binding international agreement to reduce the amount of generated CO_2, was concluded in 1997 [Source: Agency for Natural Resources and Energy, Japan, 2005]. A number of countries, especially developed countries such as G7, struggle to reduce the CO_2 amount which is related to the improvement of efficient energy generating equipment. Hence, it is considered that the clean energy generation system is an urgent requirement to address Kyoto Protocol. Table 7.2 shows the typical clean energy systems without CO_2 production so far, i.e., wind generator, solar power generator, hydrogen battery and polymer battery. As described in the table, each system has advantages and disadvantages. For wind generator, although the power generating efficiency is good with low cost, the amount of generated power is greatly influenced by the weather condition. Solar power energy utilizes silicon base photovoltaic semiconductors which convert light energy to electricity.

Table 7.2. Advantage and disadvantage of typical clean energy.

Type of Clean Energy	Advantage	Disadvantage
Wind Generator	Eternal power generation Low cost, Good power generation	Non-constant power generation, Noise Landscape problem
Solar Power Generator	Eternal power generation Non-size dependent Power Efficiency	Low energy density, Non-constant power generation
Hydrogen Battery	Good Power Efficiency, Low-waste, Low-noise	Explosible gas, Gas storage problem, High cost
Polymer Battery	Light weight, Small geometry	Long charging time, Retention of Charging Low energy density

However, this system is also influenced by the weather condition and energy generation level per density (energy density) still needs to be improved. Hydrogen battery has been utilized in the power source of hybrid-engine-cars and the technology of hydrogen battery may be established in near future although the storage cost of hydrogen is still costly.

Polymer Battery

With respect to polymer batteries, conductive polymers such as polyvinylidenefluoride (PVdF) [Periasamy et. al. (1999); Singh and Sekhon (2004); Sekhon and Singh (2004); Cheng et. al. (2005)], polyacrylonitrile (PAN) [Nicotera et. al. (2004); Kim et. al. (2001); Sekhon et. al. (2000)], poly(vinyl chloride) (PVC) [Su et. al. (1998)] and poly(methyl methacrylate) (PMMA) [Kim and Oh (2002); Hashmi et. al. (2005)] have been utilized as gel condition with electrolyte solution. As geometry of polymer batteries can be compactly packed into a small package, the targeted field of polymer batteries is mobile and notebook PC batteries. In this field, usage of electrospun polymer fibers has been expected because of pore structure and sub-micron to nano level size network of fibers. A pore structure of nano fibrous membrane leads to high electrolyte uptake while large surface area of nanofiber network contributes to excellent ion conductivity [Choi et. al. (2004e)]. [Choi et. al. (2003b)] also pointed out that conventional polymer gel electrolytes need to address the handling difficulty for high-speed manufacturing of lithium polymer batteries. Hence, a few research groups have been working on the design of new polymer batteries using conductive polymer nanofibers [Choi et. al. (2003b) (2004e); Choi et. al. (2004b); Hong et. al. (2005)].

The important properties of polymer nanofiber batteries are (1) Ion Conductivity, (2) Interfacial Resistance and (3) Electrochemical Stability. The detailed discussion of each is described as follows.

[Choi et. al. (2004e)] designed the polymer batteries consist of a few different types of electrodes and electrospun polyvinilidenefluoride (PVdF) nanofibers, as shown in Fig. 7.20. A PVdF nanofiber membrane with 450nm average fiber diameter and 1μm mean pore size was soaked in lithium - base electrolyte solution (1M LiPF6 in ethylene carbonate (EC)/dimethyl carbonate (DMC)/diethylcarbonate (DEC) (1:1:1 by weight) for 1 hour under argon gas atmosphere. A nanofiber membrane which contains electrolyte solution was then sandwiched between two stainless steel electrodes and the whole assemble was sealed with a polyethylene-coated aluminum pouch under vacuuming condition. It was found that electrospun PVdF nanofiber webs indicated high uptake of electrolyte solution (320 ~ 350wt%) due to the good wettability and large surface area of nanofibers and the retention of initial absorption of electrolyte solution was 80wt%. The absorbed amount of electrolyte solution is affected by average fiber diameter of nanofibers and the thinner fibers led to higher electrolyte uptake and also higher ionic conductivity (1.0×10^{-3} S/cm at room temperature).

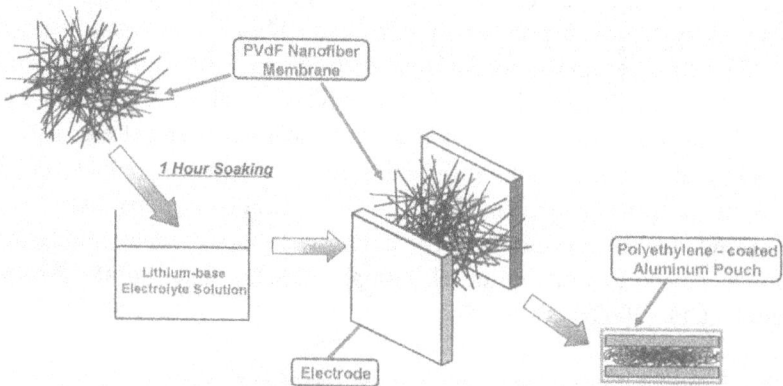

Fig. 7.20. Ionic conductivity cell made of polyvinilidenefluoride (PVdF) nanofibers [Choi et. al. (2004e)].

Another important property of the polymer nanofiber battery is interfacial resistance between a nanofiber membrane absorbed electrolyte solution and the electrodes. [Choi et. al. (2004e)] pointed out that absorption of electrolyte solution into pores of a nanofiber membrane eventually leads to fiber diameter swell during the storage. In this case, the amount of electrolyte solution is reduced between a nanofiber web and the electrode and the surface contact area of a nanofiber membrane is reduced as the fiber diameter increase through swelling. Therefore, the interfacial resistance generally increases with longer storage time. In their result, interfacial resistance of PVdF nanofiber electrolyte with lithium electrodes slightly increased with the storage time however, it is relatively lower compared to other gel polymer electrolyte. Electrochemical stability of this cell is shown in Fig. 7.21 and the good electrochemical stability was shown up to 4.3V at 60°C in the case of 0.45μm fiber diameter. Another prototype cell (MCMB anode/a PVdF nanofiber web/LiCoO2 cathode) also showed the good cycle performance which means that charge-discharge behavior was not changed with increasing the cyclic number (see in Fig. 7.22).

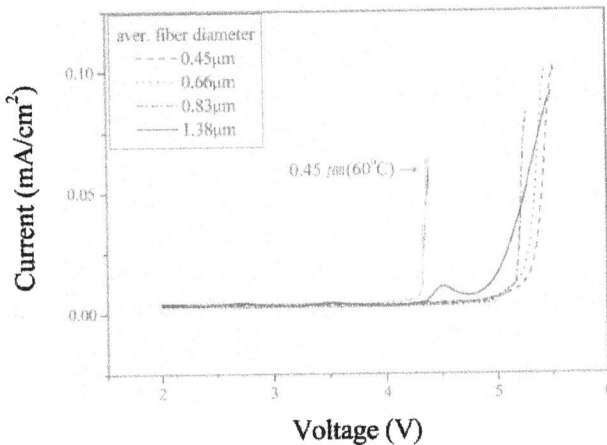

Fig. 7.21. Electrochemical stability of PVdF nanofiber electrolyte with lithium electrodes [Choi et. al. (2004e)].

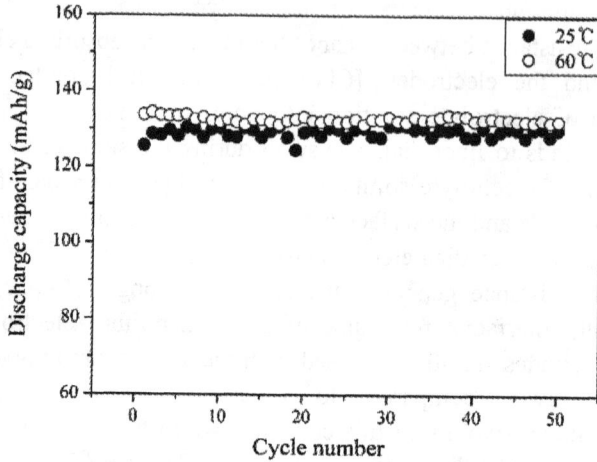

Fig. 7.22. Cycle performance of a prototype cell made of PVdF nanofiber electrolyte, MCMB anode and LiCoO$_2$ cathode [Choi et. al. (2004e)].

In order to enhance ion conductivity as well as the mechanical stability of polymer nanofiber electrolyte, the group of [Choi et. al. (2004b)] adopted the heat treatment to a PVdF nanofiber membrane to interconnect fibers together. Thermally treated PVdF nanofibers were soaked into 1M LiN(CF$_3$SO$_2$)$_2$ electrolyte solution and dried at 100°C for 30 minutes. After the thermal treatment at 160°C for 2 hours, the average fiber diameter and pore size of a membrane became 510nm and 1.23μm. The geometry of this nanofiber membrane is comparable with PVdF nanofibers of [Choi et al. (2004e)]. However, it was found that the ion conductivity of polymer electrolyte was 1.6 ~ 2.0 x 10^{-3} S/cm because of interconnection of nanofibers. [Hong et. al. (2005)] utilized the different approach to achieve high conductivity with mechanical stability by making composite nanofibers. Nylon 6 nanofibers were firstly electrospun and experienced further polymerization of aniline. Fig. 7.23 shows the grafted polyaniline on the surface of nylon 6 nanofibers and this nanofiber morphology resulted in high volume conductivity of polymer electrolyte.

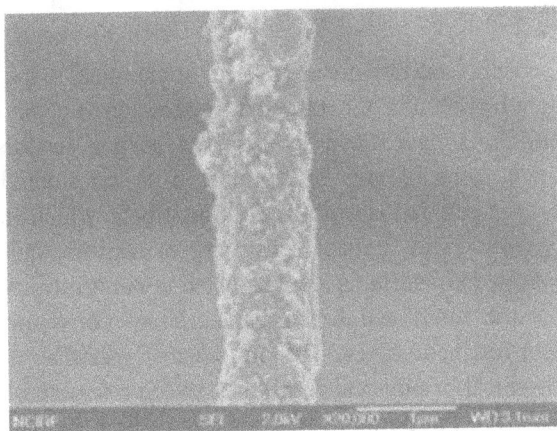

Fig. 7.23. SEM photograph of nylon 6 nanofibers coated by conductive polyaniline [Hong et. al. (2005)].

Polymer Capacitor

Except polymer batteries, another energy and electrical application of conductive nanofibers is supercapacitor reported by the group of [Kim et. al. (2004b) (2004c)]. [Kim et. al. (2004b)] electrospun polybenzimidazol (PBI) nanofiber membranes and further applied carbonization process at 700, 750, 800 and 850°C for 30 minutes under nitrogen atmosphere. The average pore size and specific surface area of activated carbon nanofibers (ACNF) are respectively 0.64 ~ 0.66nm and 500 ~ 1220m^2/g. The supercapacitor cells were assembled by ACNF nanofiber electrode, KOH aqueous solution, polypropylene separator and Ni foil of 50μm thickness. It was found that ACNF nanofibers activated at higher temperature exhibited a longer charging and discharging time to be calculated as higher specific capacitance. The specific capacitance of the designed supercapcitor cells was dependent on the activated temperature and ACNF nanofibers activated at 850°C indicated the highest value (178 F/g) at high charge current density. In another trial [Kim et. al. (2004c)], ACNF nanofibers derived from poly(amic acid) electrospun fibers exhibited high electrical conductivity (2.5 S/cm) and 175 F/g of the specific capacitance.

7.9. Sensors

The role of sensors is to transform physical or chemical responses into an electrical signal based on the targeted application. So far electrospun polymer nanofibers have been investigated as gas sensors, chemical sensors, optical sensors and biosensors. It is considered that high sensitive sensors can be assembled by nanofibers which possess high surface to volume ratio. Except sensitivity of sensors, quick response time with a targeted material is also expected to nanofiber sensors. Some cases replace the interfacial sensing material of conventional sensors with nano fibrous membranes. The principle of nanofiber sensors is to utilize the chemical or physical reaction between a targeted material and a sensing material. Furthermore, the sensors convert the result of those chemical or physical phenomena to the electrical output and finally quantitative measurement of the detected materials is conducted. Here, electrospun polymer nanofibers which aimed at gas sensors, chemical sensors, optical sensors and biosensors will be introduced in details.

Gas Sensors

Sensing toxic gases, such as carbon monoxide (CO), nitrogen dioxide (NO_2) and ammonia (NH_3) is a very serious issue to protect people in the residence, office and industrial factory. For example, CO is generated by the result of non-perfect burning of oil. In winter season, we occasionally see gas poisoned victims which wrongly used gas stoves. For NH_3 gas, it is ruled in UK that the exposure limit is 25ppm over 8 hours and 35ppm over 10 minutes [Christie et. al. (2003)]. As mentioned in introduction, for the soldier's protection from terrorist gas attacks, excellent alert system is necessary. Thus, the demand of high sensitive gas sensors has increased. For NH_3 gas detection, we can see four examples of the usage of electrospun polymer nanofibers.

[Ding et. al. (2004a)] considered to use quartz crystal microbalances (QCM) sensors with electrospun nanofibers. It is acknowledged that QCM sensors are widely used to detect the mass change in the presence of coating material which reacts with the target gas.

Although a number of materials have been used as the coating material of QCM sensors, [Ding et. al. (2004a)] applied nano fibrous membrane for the coating of QCM sensors. Poly(acrylic acid) (PAA) was chosen for nano fibrous membrane since carboxyl groups of PAA react with ammonia. However, as electrospun PAA nano fibrous membranes could not be dried well due to the strong bonding between carboxyl groups and water molecules, poly(vinyl alcohol) (PVA) was mixed with PAA solution and then electrospinning process was conducted. It was shown that sensitivity of QCM sensors coated with PVA-PAA nano fibrous membrane at 50 ppm level of NH_3 increased with increasing PAA amount. More importantly, QCM sensors coated with nanofibers demonstrated higher gas sensitivity as compared to the sensors coated with PAA-poly(allylamine hydrochloride) (PAH) film. It was revealed that large surface area of nano fibrous membrane contributed to high sensitivity to the gas. Ding et al. (2004b) further tried to improve the sensitivity of QCM sensors at lower concentration of NH_3. At this time, higher molecular weight of PAA was prepared to make nanofibers and dry process of deposited PAA nanofibers became easier. Hence, the pure PAA nano fibrous membrane could be applied to QCM sensors. As a result, sensitivity of QCM sensors was improved up to 130ppb level of NH_3.

Conducting polymer nanowires also have the potential to detect NH_3 gas [Liu et. al. (2004)]. [Liu et. al. (2004)] noticed the NH_3 gas reactivity of polyaniline (PANI). PANI exhibits the character like p-type semiconductor. It means NH_3 gas supplies electrons to PANI which results in the reduction of charge-carrier concentration on a polymer chain and consequently conductivity of PANI decreases. Because of this unique property, electrospun PANI/poly(ethylene oxide) (PEO) nanowires deposited on gold electrodes exhibited current change at 0.5ppm level of NH_3 gas. [Gouma et. al. (2003)] also utilized the change of electrical property of metal oxides under gas exposure. The behind mechanism is that metal oxides possess the property of n-type semiconductor which means electrical resistance of metal oxides increases with increasing gas existence.

Molybdenum oxide (MoO_3)-PEO nano fibrous membranes were electrospun using sol-gel technique. Through sintering process, the deposited membranes finally contained nanowires of 100nm length and 5-20nm width and MnO_3 nano particle aggregates. Rapid change of electrical resistance of nanostructured metal oxides was recognized at 100ppm level of NH_3 gas.

Chemical Sensors

The role of chemical sensors is to detect certain chemical component or a change of pH (potential of hydrogen) in a material system. [Kwoun et. al. (2000)] modified the conventional piezoelectric sensors, so called Thickness Shear Mode (TSM) sensors using electrospun nano fibrous membrane. Poly-lactic acid-co-glycolic acid (PLAGA) nanofibers were deposited on a sensing interface of TSM sensors and characterization under different liquid loading. [Kwoun et. al. (2000)] utilized the unique property of PLAGA which means that PLAGA polymer exhibits moderate hydrophobic properties against water while moderate hydrophilic properties against propanol. The response of TSM sensors coated with PLAGA nanofibers definitely indicated different manner in the water and propanol and the concept of sensors is very useful to detect certain chemical component. [Kwoun et. al. (2001)] further investigated the property of TSM sensors coated with PLAGA nanofibers and demonstrated that the sensor is useful to detect benzene gas.

Optical Sensors

Optical sensors utilize the fluorescence quenching of the sensing material against the targeted chemical molecules. [Lee et. al. (2002b)] synthesized the florescent monomer and further polymerized with methacrylate monomer. As shown in Fig. 7.24(a), the synthesized polymer possesses high photo sensitive pyrene.

Electrospun nanofibers made of this polymer had an average fiber diameter with 300 ~ 1000nm and fluorescent emission response was investigated with dinitro toluene (DNT). It is known that DNT is a toxic chemical compound and is occasionally detected in drinking water. In this case, dinitro toluene causes anemia, methemoglobonemia, cyanosis and liver damage to humans [Celin et al., 2003]. Thus, it is essential to detect DNT in the water. As shown in Fig. 7.24(b), it was found that the fluorescence intensity of the fabricated nanofibers decreased with DNT concentration.

(a)

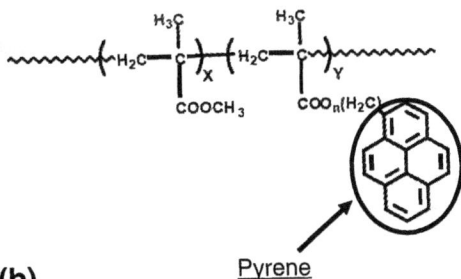

(b)

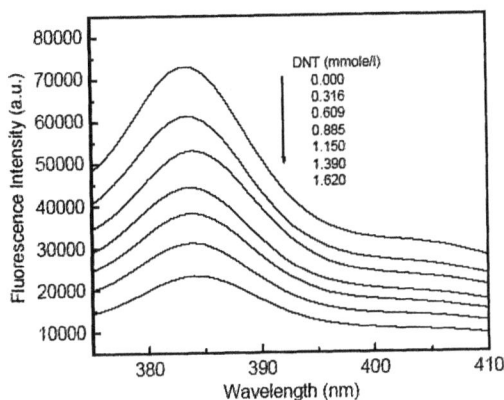

Fig. 7.24.   (a) Structure of electrospun fluorescent polymer and (b) fluorescent intensity of electrospun nanofibers against the concentration of dinitro toluene [Lee et. al. (2002b)].

[Wang et. al. (2002b)] also adopted fluorescent monomer in their synthesized polymer, i.e., poly(acrylic acid)-poly(pyrene methanol) (PAA-PM). The electrospun nanofiber membranes of this polymer had 100 ~ 400nm average fiber diameter and exhibited high fluorescent sensitivity with Hg^{2+}, Fe^{3+} and DNT. Thus, optical sensors using fluorescent polymer nanofibers could have potential.

Biosensors

Biosensors are the device which can detect or monitor the biological change of human body. Bioengineering is one of the important research fields in this century as the number of oldery people will dramatically increase in the near future. Therapy with bioengineering technology is high requirement and this background brought biosensors to be a hot topic. [Wang et. al. (2004c)] considered to detect heme-containing respiratory protein, so called cytochrome c. This protein is related to the confirmation of myocardial infarction and detection of this protein is very important to monitor the patient. Methyl viologen tethered to a ligand that is sequestered by binding to a specific and biorelevant target was also discussed. For the monitoring cytochrome c and methyl viologen, [Wang et. al. (2004c)] utilized electrospun cellulose acetate (CA) nanofibers coated by fluorescent layer, i.e., poly[2-(3-thienyl) ethanol butoxy carbonyl-methyl urethane] (H-PURET). The nanofibers possessed 100 ~ 400nm average fiber diameter. It was found that fluorescent intensity decreased with increasing concentration of cytochrome c as well as methyl viologen. It was concluded that high fluorescent sensitivity was attributed to the high surface area to volume ration of the nanofibers and efficient interaction between the fluorescent polymer and the analytes.

The group of [Sawicka et. al. (2005)] fabricated polyvinylpyrrolidone (PVP) nanofiber sensor incorporated with urease (urea decomposition enzyme) to detect urea, which can be used to check kidney and liver function problems. The behind mechanism of the nanofiber sensor is the following chemical reaction which urea is converted to ammonia under the existence of urease.

$$CO(NH_2)_2 + H_2O \xrightarrow[\text{urease}]{} CO_2 + 2NH_3$$

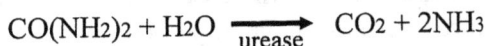

Hence, it was expected that PVP nanofiber webs with urease highly encourages the above chemical reaction. The electrospun PVP/urease nanofibers exhibited $7 \sim 100$nm average fiber diameter and the spherical aggregates of urease molecules with $10 \sim 800$nm diameter were also observed in the nanofiber network. It was found that the developed nanofiber sensors reacted with low concentration of urea up to 0.5mM. Although complete reaction time was 20 minutes, the response of urea detection was fast.

The Ramakrishna's group at National University of Singapore recently developed Nylon 6 nanofibers incorporated with gold particles as glucose sensors. Glucose detection is important for the patient who suffers from diabetes. It is known that the glucose level increases in the blood if patient has serious condition of diabetes. Non-proper treatment causes necrosis of cells and blood vessels. [Lala et. al. (2005)] noticed the glucose reaction under the enzyme of glucose oxidase (GOx) as follows.

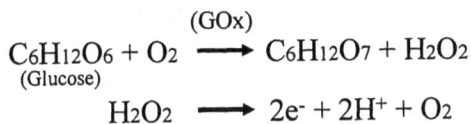

$$C_6H_{12}O_6 + O_2 \xrightarrow{\text{(GOx)}} C_6H_{12}O_7 + H_2O_2$$
$$\text{(Glucose)}$$
$$H_2O_2 \longrightarrow 2e^- + 2H^+ + O_2$$

Here, hydrogen peroxide (H_2O_2) is further decomposed to oxygen and two protons and electrons. Thus, in order to measure the amount of glucose, either change of oxygen concentration or H_2O_2 concentration may be considered. However due to the less sensitivity to monitor the oxygen and H_2O_2, [Lala et. al. (2005)] measured the output of the potentiostat by the existence of gold particles on the nanofiber surface.

Nylon 6 nanofibers were firstly electrospun and the fabricated nanofiber membranes were immersed in 1mM Chloroauric Acid (HAuCl$_4$) with pH4-5. This process resulted in the dispersion of gold nano particles on the surface of nylon nanofibers (see in Fig. 7.25) and enhanced the conductivity of nanofibers.

Fig. 7.25.   SEM micrographs of nylon 6 nanofiber membranes coated with gold particles [Courtesy of Lala and Ramakrishna at National University of Singapore].

The unreacted surface of Au/Nylon 6 nanofibers further reacted with tetrameric protein *"avidin"* (pI=10), negatively charged at pH=12 through electrostatic interaction which on further addition of biotinylated glucose oxidase (B-GO$_x$) binds specifically to avidin [Lala and Sastry (2000)]. Hence, once B-GOx reacts with glucose, electrical output generated by aforementioned glucose reaction is delivered through the conductive Au/Nylon 6 nanofibers. The developed sensor could measure glucose concentrations from solutions ranging from 1M to 1μM. The average response time was recorded as 130s with sensitivity levels of 0.113μV/mM. It is considered that the developed sensors may find applications for sensing glucose levels from human fluids such as blood and urine, as well as in the bio-industry for determination of analyte levels in fermentation and cell-culture media.

7.10. Composite Reinforcement

Fiber reinforced polymer composite materials have been utilized in aerospace, automobile, sports items and electrical products because of the superior structural properties such as high modulus and strength to weight ratios, which cannot be obtained by sole materials. As described in Fig. 7.26, polymer-based fibrous composites contain micron-size fiber reinforcements (around 7μm ~ 30μm) such as carbon, glass, ultra-high-molecular-weight polyethylene and aramid fibers and thermoset and thermoplastic matrix polymers. It is well known that there is interphase region between fibers and matrix which takes a role to deliver the stress from matrix to reinforcement fibers. Hence, polymer composites with weak interphase generally indicate poor mechanical performance. To strength the interphase adhesion between fibers and matrix, silane coupling agents and plasma-etching treatment are generally utilized. As shown in Fig. 7.27, reinforcement fibers further can take the form of several configurations, such as discontinuous chopped short fibers, non-woven fiber webs, unidirectional laminates and textile fabrics (woven, braided and knitted fabrics). The choice of reinforcement and matrix polymer is dependent on the required mechanical performance of the products and the whole productive cost. Reinforcement configuration is decided based on the geometry and the size of final products.

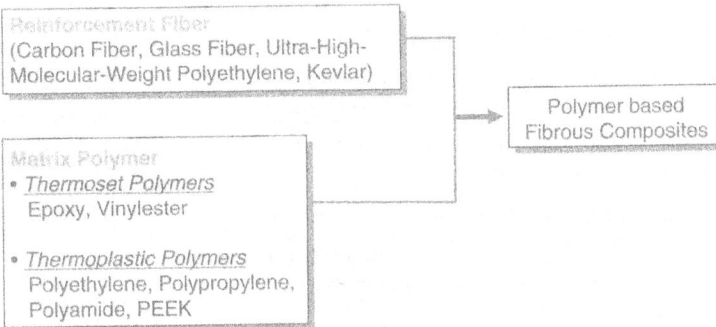

Fig. 7.26. Typical reinforcement fibers and matrix polymers of polymer based fibrous composites.

Discontinuous Chopped Fibers   **Non-Woven Fiber Web**   **Unidirectional (UD) Laminates**

Textile Reinforcement

Woven Fabric   **Braided Fabric**   **Knitted Fabric**

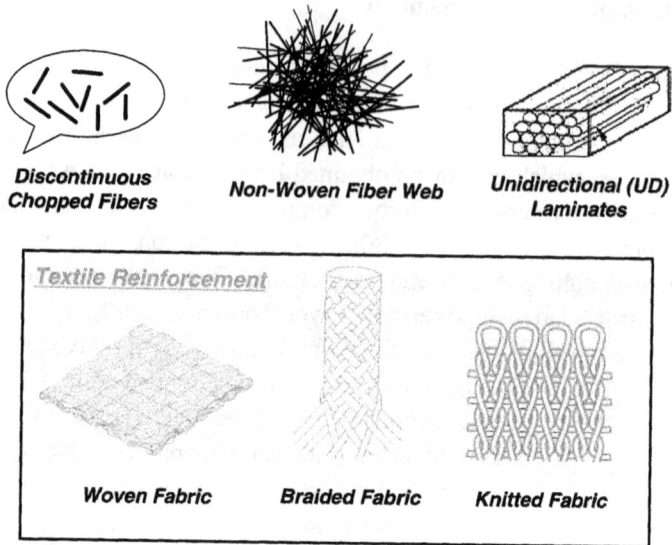

Fig. 7.27. Typical reinforcement configurations of polymer based fibrous composites.

For instance, if the final product with complicated shape needs high mechanical property, drapeable knitted fabric is suitable to reinforce along the product shape.

Since electrospun polymer nanofibers have gained the strong attention due to the high surface area to volume ratio, it is considered that nanofibers will also eventually find important applications in making nanofiber reinforced composites. For nanofiber reinforced polymer composites, it is considered that the stress which applied to the composites effectively transmitted into nanofibers due to their large surface area although there must be a hypothesis that suitable interphase adhesion is created between nanofibers and matrix polymer. Except mechanical performance improvements, polymer composites reinforced by nanofibers has very good transparency [Bergshoef and Vancso (1999)].

Accordingly nanofiber reinforced composites have much potential, although there are few examples of electrospun nanofibers as composite reinforcements. So far three different types of nanofibers, i.e., 1.Polymer Nanofibers, 2.Carbon Nanofibers and 3.Carbon Nanotube composite nanofibers have been discussed as reinforcements of polymer composites. It is noted that reinforcement configurations of nanofibers are chopped fibers or almost cases are non-woven webs.

Table 7.3 summarizes the examples of polymer nanofibers used as polymer composite reinforcements. [Kim and Reneker (1999)] electrospun polybenzimidazole (PBI) nanofiber membranes and PBI/Epoxy non-woven composites were fabricated. The fabricated composites contained 32 layers of nanofiber sheets. Chopped PBI nanofibers reinforced rubber composites were also fabricated. Both composites possessed 3-15% fiber weight fraction and tensile, 3- point bending, double torsion, and tear tests were carried out. It was shown that bending modulus and the fracture toughness of PBI/Epoxy composites slightly increased with increasing the nanofiber weight fraction, whereas the fracture energy significantly increased.

Table 7.3. Discussed mechanical properties of nanofiber reinforced polymer composites.

Nanofibers	Matrix	Discussed Mechanical Properties	Reference
Nylon	Epoxy	Tensile	Bengshoef and Vansco (1999)
	Bis-GMA/TEGDMA	Bending	Fong (2004)
Polybenzimidazole (PBI)	Epoxy, Rubber	Tensile, Bending, Double Torsion, Tear	Kim and Reneker (1999)
	Epoxy	Mode I & Mode II Fracture Toughness	Dzenis and Reneker (2001)

On the other hand, the effect of PBI nanofibers was greatly seen in rubber composites, which means tensile modulus was ten times and the tear strength was twice as large as that of the unfilled rubber material. [Bergshoef and Vancso (1999)] fabricated Nylon-4,6 nanofiber reinforced epoxy composite films. Tensile tests were conducted for the composite as well as the monolithic matrix films. Fiber weight fraction of composites was just 3.9%, however, it was reported that both tensile stiffness and strength of the composite were significantly higher than those of the reference matrix film. [Fong (2004)] recently reported the bending behavior of dental resin (BIS-GMA/TEGDMA) reinforced by electrospun nylon nanofibers (see in Fig. 7.28(a)) with 100 - 600nm fiber diameter. Bending test results revealed that the addition of nylon nanofibers with 5% weight fraction improved fracture strength by 36%, bending modulus by 26% and fracture energy by 42%. As shown in Fig. 7.28(b), the adhesion between nylon nanofibers and matrix polymer was reasonably good as nanofibers tended to break instead of pullout during the bending testing. However, it was pointed out that the interface or interphase between the nanofibers and the matrix still needs to be improved.

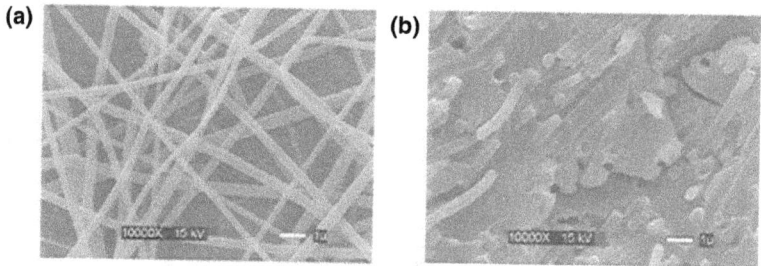

Fig. 7.28. SEM photos of (a) electrospun nylon 6 nanofibers and (b) fracture surface of nylon 6 nanofibers reinforced dental resin composites [Fong (2004)].

In addition to the stiffness and strength improvement, [Dzenis and Reneker (2001)] tried to modify interlaminar fracture resistance of unidirectional laminated composites using polybenzimidazole (PBI) nanofibers. In laminated composites, delamination fracture occasionally occurs through a main crack that propagates into resin layer, as shown in Fig. 7.29.

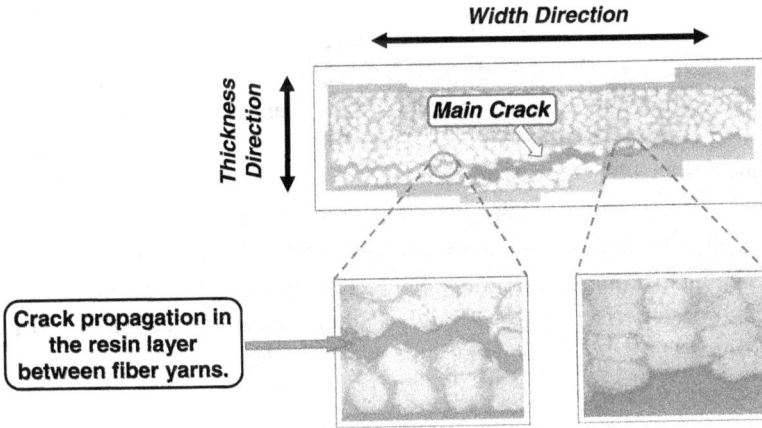

Fig. 7.29. Delaminated fracture of unidirectional Carbon/PEEK composites under bending loading. A crack propagated the resin layer between fiber yarns [Courtesy of Fujihara and Ramakrishna, National University of Singapore].

Especially, in the case of thermoset laminated composites, crack propagation occurs with very high speed due to the relatively low fracture toughness of thermoset resin. This is a very dangerous fracture mode which leads to the whole structure collapse. [Dzenis and Reneker (2001)] inserted PBI nanofiber web sheet between plies of unidirectional carbon/epoxy composites [0_{20}]. The diameter of PBI nanofibers was in the range of 300-500nm. Mode I critical energy release rate GIc increased by 15%, while an increase of 130% in the Mode II critical energy release rate GIIc was observed.

Carbon nanofibers are also a good candidate for composite applications although they have been originally used in other applications as filters, high temperature catalysts, heat management materials in aircraft and semiconductor device. Fabrication process of carbon nanofibers contains electrospinning of polyacrylonitrile (PAN) precursors, stabilization of as-spun PAN fibers at 270°C in the air and follows carbonization process in inert atmosphere at 600-1500°C [Dzenis and Wen (2002)]. AFM result showed that stiffness of carbon nanofibers is higher than that of as-spun PAN nanofibers. There is the curiosity about the mechanical property of single carbon nanofiber as compared to conventional micron-size carbon fibers. Tensile strength and modulus of micron-size carbon fibers are around 3000MPa and 230GPa. Since it is considered that micron-size fibers possess structural defect, if fiber size is reduced to nano-size level, such structural defect may be reduced which leads to ultra excellent mechanical properties. Although there is limited information about the mechanical performance of single carbon nanofiber, carbon nanofiber reinforced polymer composites have a large potential.

Another trial for composite reinforcements is carbon nanotube/polymer composite nanofibers. The idea is that polymer nanofiber itself cannot achieve high mechanical performance and hence, carbon nano tube with excellent stiffness and strength is combined with polymer nanofibers. [Ko et. al. (2003)] developed single wall carbon nanotubes (SWNT)/polyacrylonitrile (PAN) composite nanofiber yarns. Electrospun composite nanofibers possessed 50 ~ 200 nm fiber diameter and weight fraction of nanotubes was 1 ~ 4 wt%. TEM results revealed that nanotubes were dispersed in PAN nanofibers very well and were parallel to the longitudinal direction of PAN fibers. As shown in AFM result (see in Fig. 7.30), the indentation force of SWNT/PAN composite nanofibers was higher than neat PAN nanofibers. Accordingly, SWNT/PAN composite nanofibers can be expected as reinforcement in linear, planer and 3D preforms (see in Fig. 7.27) for new generation polymer composites.

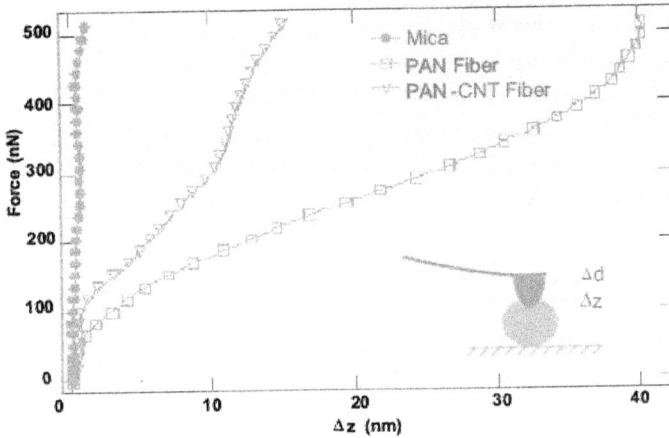

Fig. 7.30.   Load indentation curves of electrospun PAN nanofibers and PAN-CNT composite nanofibers obtained by AFM analysis [Ko et. al. (2003)].

7.11.   Conclusions

In this chapter, each potential application of electrospun polymer nanofibers has been briefly introduced. The readers may realize that the area of potential applications is very broad including bioengineering, tissue engineering, material engineering, electrical engineering and chemical engineering etc. In the field of tissue engineering and wound dressing, the cell attachment and proliferation on nano fibrous scaffolds have been investigated *in vitro* and now the research trend is heading to animal study of nano fibrous scaffolds with seeded cells. In near future, biomedical therapy using nano fibrous tissue scaffolds may be close to the practical level. The research status of electrospun polymer nanofibers has just finished analyzing the fundamental processing and, physical and chemical properties of nanofibers.

In this respect, it can be said that the application researches of nanofibers have just started. In other research fields, high research competition will be estimated due to the surrounding world affairs of this century such as terrorism attacks and the rapid economical growth of undeveloped countries. In order to encourage the application researches of electrospun nanofibers, the cooperation of researchers from various scientific fields is important. The current authors look forward to seeing the bright future of people's daily life by the novel application researches of nanofibers.

Appendix A

Glossary of Terms

Affinity membrane

Affinity is a measure of the intrinsic binding strength of the ligand binding reaction. Afinity is also known as the tendency of a molecule to associate with another. The affinity of a drug is its ability to bind to its biological target (receptor, enzyme, transport system, etc.). For pharmacological receptors it can be thought of as the frequency with which the drug, when brought into the proximity of a receptor by diffusion, will reside at a position of minimum free energy within the force field of that receptor. [IUPAC Medicinal Chemistry]. Hence affinity membranes are chromatographic membranes with capturing agent molecules (affinity ligands) on the membrane surfaces. The affinity ligands can specifically capture their corresponding target molecules (ligates) through specific interactions. Affinity membrane is different from size-exclusion membranes like microfiltration, ultrafiltration, nanofiltration and reverse osmosis membrane that base their separation functions on size discrimination of the solute. Affinity membrane is developed to permit the purification of molecules based on physical/chemical properties or biological functions rather than molecular weight/size.

Amorphous

Latin meaning without form. Non-crystalline structure

Biodegradable
The ability of a material to decompose under the action of biological agent such as bacteria, enzymes etc.

Chemical Vapor Deposition (CVD)
A coating deposition process that is characterized by a chemical reaction producing the desired film material as well as a waste stream. CVD processes frequently involve use of noxious and/or unstable materials. In hard coating technology, CVD is an older process for the deposition of TiN. It is generally only used for coating cemented carbide inserts, because of the extreme temperatures of the process. It's primary advantage is the ability to coat thick, tough layers for rough-cutting applications. Disadvantages include corner rounding, microcracking, and temperature-induced problems.

Coalescence
Aggregation of substances with affinity towards one another.

Conservation Relations
A group of relationships in engineering and physical sciences that emphasizes the conservancy of some physical quantities, which are of importance in calculation of physical phenomena. A few examples are given as follows. Kirchhoff's Current Law states that the algebraic sum of the currents in all the branches which converge in a common electrical circuit's node is equal to zero. Mass conservation states that the amount of mass in a closed-system (except nuclear reaction) remains constant. Momentum conservation states that the amount of momentum is a system of colliding particles remains the same.

Contact angle
(1) A contact angle is the angle between the solid surface (plate or vessel) and the liquid surface. (2) A means of quantifying the nonstick properties of a coating by measuring the ability of a liquid to wet its surface. (3) Also called the wetting angle.

Copolymers
A polymer resulting from addition polymerization of two different monomers. The resulting copolymers have properties different from either of the homopolymers from the two monomers or the mechanical mixture of the homopolymers.

Crystallinity
The presence of three-dimensional order on the level of atomic dimensions. In polymers, the range of order may be as small as about 2 nm in one (or more) crystallographic direction(s) and is usually below 50 nm in at least one direction. Polymer crystals frequently do not display the perfection that is usual for low-molecular mass substances. Polymer crystals that can be manipulated individually are often called polymer single crystals. Pure and Appl. chem, 1989, 61, 769 IUPAC Macromolecular Nomenclature.

Electric Field
A force field which defines what acceleration an electric charge placed at rest at any point in space will feel. Electric charges cause electric fields around them, which then apply a force to any other electric charge placed in the field. The electric field E has both a magnitude and a direction at each point in space, and the magnitude and direction of the resulting force on a charge q at that point is given by $F = qE$. When you get a shock from a door handle after scuffing your feet on a carpet you feel the effect of an electric field accelerating electrons

Extracellular Matrix
Animal and human tissues are not made up solely of cells. A substantial part of their volume is extracellular space, which is largely filled by an intricate network of macromolecules constituting the extracellular matrix (ECM) .The ECM is composed of a variety of proteins and polysaccharides that are secreted locally and assembled into an organized meshwork in close association with the surface of the cell that produced them.

Extrusion

The act squeezing out a fluid through a die or confined region.

Glass Transition Temperature

In polymer or glass chemistry, the temperature corresponding to the glass-to-liquid transition, below which the thermal expansion coefficient is low and nearly constant, and above which it is very high. Symbol: Tg.

Hydrophobic

Compounds that tend to repel water and does not dissolve or absorb water.

Instability

(1) The condition of a body if, when displaced from a state of equilibrium, it continues, or tends to continue, to depart from the original condition. (2) The condition of a body or system which responds to a specified disturbance by increasing the disturbance (perturbation) until an irreversible change has taken place. Sometimes used in context to mean convective instability. (3)

Ligand/ligate

In chromatography, the word ligand is used to describe the capturing agents which is fixed on the membrane surface or in the column packing materials like agarose gel or silica gel particles. The target molecules which is supposed to be captured by the membrane or column through the specific binding with the ligands are called ligate.

Maxwell Fluid

An idealized behavior of fluid (but not solid) which acts in a viscoelastic manner. Specifically, a Maxwell unit consists of a spring and a dashpot, representing an elastic solid and a viscous fluid respectively, in series so that upon prescribed elongation of the Maxwell unit the elongated spring contracts back to its original length whilst energy is lost via the dashpot. This allows permanent deformation, as well as a flow-like behavior, which is characteristic of a fluid.

Modeling
In a loose sense, modeling simply refers to creating a simplified representation of something else. A model can be a picture, a diagram or a mathematical formula. Modeling is an investigative technique which uses a mathematical or physical representation of a system or theory to test for effects that changes in system components may have on the overall functioning of the system. As an example, mathematical modeling using computers plays a major role in climate research, by simulating how Earth's climate will respond to changes in atmospheric concentrations of greenhouse gases. Hence modeling includes theoretical computation of natural or industrial phenomenon using raw data and basic properties.

Molecular Weight
Molecular weight of the polymer is the sum of the molecular weight of the individual monomers

Morphology
The study of form and structure of a substance without considering its function.

Nanofiber Membrane
A thin membrane that forms on the collecting plate during the electrospinning process as a result of laying of solidified or semi-solidified jet on a two-dimensional surface. This membrane is also called a nanofiber mesh or nanofiber web due to its network-like structure when viewed from an electron microscope.

Nanofiber Mesh
See nanofiber membrane

Nanofiber Web
See nanofiber membrane

Nanotechnology
(1) The science of building devices at the molecular and atomic level. For example, a single data bit might be represented by only one atom some time in the future. Beyond being used in computers and communications devices, nanotechnology could be used to build devices, change the properties of materials, and extensively in biotechnology. (2) Synonymous with molecular systems engineering. An interdisciplinary field where devices are constructed at the molecular scale and function at this scale. This technology is expected to allow the construction of very compact and high performance computing devices. (3) In recent general usage, any technology related to features of nanometer scale: thin films, fine particles, chemical synthesis, advanced microlithography, and so forth. As introduced by the author, a technology based on the ability to build structures to complex, atomic specifications by means of mechanosynthesis; this can be termed molecular nanotechnology.

Necking
A deformation phenomenon where there is a narrowing of a section of a material that is subjected to tensile stress.

Node
Conceptualized beads. In this book, we refer to nodes as idealized point mass that is equivalent to a segment of liquid jet.

Perturbation
Functionally incorporated noise to test the threshold and growth of instability.

Plasma Treatment
Plasma is a state of matter characterized by unbound negative electrons and positive ions which may conduct electrical current. It is known as the fourth state of matter, along with other states of matter: solids, liquids and gases. Scientists believe that more than 99% of matter in the universe exists as plasma, including stars, lightning, and the Aurora Borealis. Hence plasma treatment is the use of ionized-gas for performing materials treatment.

Porosity
A condition of trapped pockets of air, gas, or vacuum within a solid material, usually expressed as a percentage of the total nonsolid volume to the total volume (solid plus nonsolid) of a unit quantity of material

Relative permittivity
Indicates the ability of a material to store electrical energy when a voltage is applied to it. This may be considered a desirable property or an electrical loss depending on the application.

Relaxation time
The time constant of an exponential return of a system to equilibrium after a disturbance

Spherulite
An aggregate of ribbonlike polymer crystallites radiating from a common center with amorphous region separating the crystallites.

Substrate
A surface on which an organism grows or is attached. An underlying layer; a substratum

Surface Modification
Changing the physical properties of a surface while retaining the physical properties of the bulk by attaching molecules or altering the molecules at the surface such that these surface molecules can be made to function in a desired manner.

Surface tension
The contracting force per unit length around the perimeter of a surface is usually referred to as surface tension if the surface separates gas from liquid or solid phases, and interfacial tension if the surface separates two nongaseous phases. Surface tension can also be expressed in units of energy per unit surface area. For practical purposes surface tension is frequently taken to reflect the change in surface free energy per unit increase in surface area.

Surfactant

A molecule that lowers surface tension with a hydrophilic head that sticks to water and a hydrophobic end that sticks to organic substance.

Tissue engineering

Tissue engineering was defined by National Science Foundation (NSF) of USA in 1988 as follows "Tissue engineering is the application of the principles and methods of engineering and the life science toward the fundamental understanding of structure-function relationships in normal and pathological mammalian tissue and the development of biological substitutes to restore, maintain or improve functions." However, most researchers would think less formally of tissue engineering today as a combination of biology and engineering that aims to develop substitute tissue, to replace or otherwise restore function of damaged human tissues. Generally the tissue engineered implant will require a porous 'scaffold' to direct the cells to grow in the correct physical form, and of course, cells.

Tissue engineering scaffold

The porous scaffold used to accommodate cells in tissue engineering. Tissue engineering scaffold should have properties like high porosity and interconnectivity, proper strength and good bioactivity. If the tissue engineering scaffold is made from synthetic polymer, its surfaces usually need to be modified to improve the cell adhesion. Most of the tissue engineering scaffolds are prepared using biodegradable materials.

Transconductance

The creation of a current over two output points where a voltage is applied to two input base points, as if the conductance is transferred from one point to another.

Viscoelasticity

The dual nature of polymers, partly viscous fluid and partly elastic solid, is referred to as viscoelasticity. In flowing polymers viscoelasticity is responsible for time-dependent properties, such as stress relaxation, normal stresses, very large elongational viscosities, and numerous unusual phenomena such as extrudate swell, entry flow vortices and some flow instabilities.

Viscosity

The resistance to flow of a fluid (strictly speaking the resistance to shearing). It is defined as the ratio of shear stress (Tangential Force/Area) to shear rate (velocity/gap). The viscosity of a polymer decreases as the shear rate increases. This property is referred to as pseudoplastic behavior or shear thinning. The viscosity of a polymer at (near) zero shear for a polymer like PE might be 5,000 to 10,000 Pa.s while during flow in an extrusion channel it could be much lower (i.e. 500 Pa.s or less). Melt flow index corresponds to just one point on a viscosity curve (actually inverse). High viscosity implies low melt index and high molecular weight. Viscosity is measured in units of Pa.s or poise. 1 Pa.s = 10 poise. The viscosity of water is 10^{-3} Pa.s (1 centipoise) and for a typical polymer melt at least one million times larger (i.e. over 1000 Pa.s).

Volatility

The capability of a liquid substance to be vaporized at room temperarure and pressure.

Appendix B

Useful Websites on Electrospinning and Nanofibers

http://www.livescience.com/scienceoffiction/technovel_nanofiber_041124.html

http://www.genomenewsnetwork.org/articles/06_03/nano.shtml

http://www.donaldson.com/en/filtermedia/support/datalibrary/003317.pdf

http://www.azonano.com/SearchResults.asp?MaterialKeyWord=Nanofiber

http://webpages.sdsmt.edu/~hfong/1.1.html

http://academic.sun.ac.za/unesco/Conferences/Conference2003/ABSTRACTS/RENEKER.htm

http://www.eng.nus.edu.sg/EResnews/0302/sf/sf_3.html

http://fluid.ippt.gov.pl/sblonski/nanofibres.html

http://www.people.vcu.edu/~glbowlin/electrospinning.htm

http://tandecresearch.utk.edu/electroarticle.htm

http://www.tx.ncsu.edu/ci/nanosciences/articles_detail.cfm?article_heading=Electrospinning

http://cuhwww.upr.clu.edu/~npinto/papers/1ncur2001.PDF

http://www.azonano.com/details.asp?ArticleID=181

http://www.nanopolis.net/ES.php

http://web.mit.edu/rutledgegroup/projects/electrospinning.html

http://msewww.engin.umich.edu:81/people/milty/research/electrospinning/

http://www.zyvex.com/nanotech/nano4/renekerAbstract.html

http://www.umassd.edu/engineering/textiles/research/electrospinning/website-electrospinning.html

http://ecsmeet.peerx-press.org/ms_files/ecsmeet/2005/01/03/00022501/00/22501_0_art_file_1_1104815851.pdf

http://www.nist.gov/sigmaxi/Posters04/stephens.html

http://www.che.vt.edu/Wilkes/electrospinning/electrspinning.html

http://www.explore-technology.com/technology/E/Electrospinning.html

http://www.people.vcu.edu/~glbowlin/fibrinogen.htm

http://www.news.cornell.edu/Chronicle/03/9.11.03/electrospinning_fiber.html

http://www.asc2004.com/Manuscripts/sessionM/MP-14.pdf

http://pubs.rsc.org/ej/JM/2005/b415094e.pdf

http://www.psicorp.com/casestudies/case_electrospin.shtml

http://www.ipme.ru/e-journals/RAMS/no_3503/kim1/kim1.pdf

http://www.nsti.org/procs/Nanotech2004v3/9/T66.05

http://www.ntcresearch.org/current/year9/M98-D01.htm

http://www.jeccomposites.com/news/news_fiche.asp?id=904&

Bibliography

Aboulfaraj, M., Ulrich, B., Dahoun, A. and G'Sell, C. (1993). Spherulitic morphology of isotactic polypropylene investigated by scanning electron microscopy. Polymer 34, 23, pp. 4817-4825.

Acatay, K., Simsek, E., Akel, M., Menceloglu, Y.Z. (2003). Electrospinning of low surface energy quaternary ammonium salt containing polymers and their antibacterial activity, Proceedings of the NATO Advanced Study Institute, Belek-Antalya, Turkey. Nato Science Series II, Mathematics, Physics and Chemistry 169, pp. 97-106.

Acatay, K., Simsek, E., Ow-Yang, C. and Menceloglu, Y.Z. (2004). Tunable, superhydrophobically,stable polymeric surfaces by electrospinning, Angew. Chem. Int. Ed. 43, pp. 5210-5213.

Adam, G. and Gibbs, J. H. (1965). On the Temperature Dependence of Coorperative Relaxation Properties in Glass-Forming Liquids. .J Chem. Phys. 43, pp. 139-146.

Agency for Natural Resources and Energy, Japan. Energy in Japan 2005?(http://www.enecho.meti.go.jp).

Alfrey, T. The Influence of Solvents Composition on the Specific Viscosities of Polymer Solutions. J. Coll. Sci. 2, 1, pp. 99-114.

Alfrey, T., Bartovics, A. and Mark, H. (1942). The Effect of Temperature and Solvent Type on the Intrinsic Viscosity of High Polymer Solutions. J. Am. Chem. Soc. 64, pp. 1557-1560.

Al-Najjar, M. M., Hamid, S. H. and Hamad, E. Z. (1996). The Glass Transition Temperature of Nitrated Polystyrene/Poly(Acrylic Acid) Blends. Polym. Eng. Sci. 36, pp. 2083-2087.

Angell, C. A. (1995). Formation of Glasses from Liquids and Biopolymers. Science. 297, pp. 1924-1935.

Atala, A. and Lanza, R. P. (2002). Methods of Tissue Engineering. Academic Press USA.

Ayutsede, J., Gandhi, M., Sukigara, S., Micklus, M., Chen, H. E. and Ko, F. (2005). Regeneration of Bombyx mori silk by electrospinning. Part 3: characterization of electrospun nonwoven mat. Polymer 46, pp. 1625-1634.

Bamford, C.H., Al-Lamee, K.G., Pm-brick, M.D., Wear, T.J. (1992). Studies of a novel membrane for affinity separations: I. Functionalisation and protein coupling, J. Chromatogr. A. 606, pp.19-31.

Baumgarten, P. K. (1971). Electrostatic Spinning of Acrylic Microfibers. J. Colloid Interf. Sci. 36, pp. 75-79.

Benicewicz, B.C. and Hopper, P.K. (1991). Polymers for absorbable surgical sutures, J. Bioact. Compat. Polym. 6, pp. 64-94.

Bergshoef, M. M. and Vancso, G.J. (1999). Transparent Nanocomposites with Ultrathin Electrospun Nylon-4,6 Fiber Reinforcement. Adv. Mater. 11, pp. 1362-1365.

Berkland, C., Pack, D. W. and Kim, K. (2004). Controlling surface nano-structure using flow-limited field-injection electrostatic spraying (FFESS) of poly(D,L-lactide-co-glycolide). Biomaterials. 25, pp. 5649-5658.

Bhattarai, N., Cha, D.I., Bhattarai, S.R., Khil, M.S. and Kim, H.Y. (2003). Biodegradable electrospun mat: novel block copolymer of poly (p-dioxanone-co-L-lactide)-block-poly(ethylene glycol), J. Polym. Sci. Pt. B-Polym. Phys. 41, pp. 1955-1964.

Bhattarai, S.R., Bhattarai, N., Yi, H.K., Hwang, P.H., Cha, D.I. and Kim, H.Y. (2004). Novel biodegradable electrospun membrane: scaffold for tissue engineering, Biomaterials. 25, pp. 2595-2602.

Bigg, D. M. (1979). Mechanical and conductive properties of metal fibre-filled polymer composites, Composites. 10, pp. 95-100.

Bini, T.B., Gao, S.J., Tan, T.C., Wang, S., Lim, A., Hai, L.B. and Ramakrishna, S. (2004). Electrospun poly(L-lactide-co-glycolide) biodegradable polymer nanofibre tubes for peripheral nerve regeneration, Nanotechnology. 15, pp. 1459-1464.

Boeden, H.F., Pommerening, K., Becker, M., Rupprich, C., Holtzhauer, M. (1991). Bead cellulose derivatives as supports for immobilization and chromatographic purification of proteins, J. Chromatogr. A. 552, pp.389-414.

Bognitzki, M., Czado, W., Frese, T., Schaper, A., Hellwig, M., Steinhart, M., Greiner, A. and Wendorff, J. H. (2001). Nanostructured Fibers via Electrospinning. Adv. Mater. 13, pp. 70-72.

Bognitzki, M., Hou, H., Ishaque, M., Frese, T., Hellwig, M., Schwarte, C., Schaper, A., Wendorff, J.H. and Greiner, A. (2000). Polymer, metal, and hybrid nano- and mesotubes by coating degradable polymer template fibers (TUFT Process). Adv Mater, 12, 9, pp. 637-640.

Boland, E. D., Wnek, G. E., Simpson, D. G., Palowski, K. J. and Bowlin, G. L. (2001). Tailoring tissue engineering scaffolds using electrostatic processing techniques: A study of poly(glycolic acid) electrospinning. J. Macromol. Sci. Pur Appl. Chem. A38, 12, pp. 1231-1243.

Boland, E. D, Coleman, B. D, Barnes, C. P, Simpson, D. G, Wnek, G. E and Bowlin, G. L. (2005). Electrospinning polydioxanone for biomedical applications. Acta. Biomater. 1, pp. 115-123.

Bornat, A. (1987). Production of electrostatically spun products, US Patent No. 4,689,186.

Bowlin, G.L., (2003) Electrospinning used to create small blood vessels. http://www.futurepundit.com.

Brown, I.G. (1993). Metal-ion implantation for large-scale surface modification, J. Vac. Sci. Technol. A. 11, pp. 1480-1485.

Brunauer, S., Emmett, P. and Teller, E. (1938). Adsorption of gases in multimolecular layers, J. Amer. Chem. Soc. 60, pp. 309-319.

Buchko, C. J, Chen, L. C., Shen, Y. and Martin, D. C. (1999). Processing and microstructural characterization of porous biocompatible protein polymer thin films. Polyme.r 40, pp. 7397-7407.

Buer, A., Ugbolue, S.C. and Warner, S.B. (2001). Electrospinning and properties of some nanofibers, Tex. Res. J. 71, 4, pp. 323-328.

Bunge, R.P. and Bunge, M.B. (1983). Interrelationship between Schwann cell function and extracellular matrix production, Trends Neurosci. 6, pp. 499-505.

Bureau, E., Cabot, C., Marais, S. and Saiter, J. M. (2005). Study of the a-relaxation of PVC, EVA and 50/50 EVA70/PVC blend. Eur. Polym. J. 41, pp. 1152-1158.

Callister, W. D. (1997). Materials Science and Engineering An Introduction. John Wiley & Sons.

Cambell, D. and Pethrick, R.A., White, J.R. (2000a). Chapter 10: Scanning electron microscopy, Polymer Characterization, Physical techniques, Second Edition, pp. 293-325, Stanley Thornes (Publishers) Ltd.

Cambell, D. and Pethrick, R.A., White, J.R. (2000b). Chapter 8: X-ray diffraction, Polymer Characterization, Physical techniques, Second Edition, pp. 194-236, Stanley Thornes (Publishers) Ltd.

Cambell, D. and Pethrick, R.A., White, J.R. (2000c). Chapter 5: Vibrational spectroscopy: infra-red and raman spectroscopy, Polymer Characterization, Physical techniques, Second Edition, pp. 67-107, Stanley Thornes (Publishers) Ltd.

Campo, F. F. D., Paneque, A. , Ramirez, J. M. and Losada, M. (1963). Thermal transitions in collagen. Biochim. Biophys. Acta. 66, pp. 448-452.

Caplan, A.I. (1991). Mesenchymal stem cells, J. Orthopaed. Res. 9, pp. 641-650.

Carbonetto, S. (1984). The extracellular matrix of the nervous system. Trends Neurosci. 7, pp. 382-387.

Caruso R.A., Schattka J.H. and Greiner A. (2001). Titanium dioxide tubes from sol-gel coating of electrospun polymer fibers. Advanced Materials, 13, 20, pp. 1577-1579.

Casper, C.L., Stephens, J.S., Tassi, N.G., Chase, D.B. and Rabolt, J.F. (2004). Controlling surface morphology of electrospun polystyrene fibers: Effect of humidity and molecular weight in the electrospinning process, Macromolecules. 37, pp. 573-578.

Celin, S.M., Pandit, M., Kapoor, J.C. and Sharma, R.K. (2003). Studies on photo-degradation of 2,4-dinitro toluene in aqueous phase, Chemosphere. 53, pp. 63-69.

Chan, C.M., Ko, T.M., Hiraoka, H. (1996). Polymer surface modification by plasmas and photons, Surf. Sci. Rep. 24, pp.3-54

Chen, K. B., Lee, K. C., Ueng, T. H. and Mou, K. J. (2002). Electrical and impact properties of the hybrid knitted inlaid fabric reinforced polypropylene composites. Compos. Part A-Appl S. 33, pp. 1219-1226.

Chen, Z., Foster, M.D., Zho,u W., Fong, H., Reneker, D.H., Resendes, R. and Manners, I. (2001). Structure of poly(ferrocenyldimethylsilane) in electrospun nanofibers. Macromol., 34(18), pp. 6156-6158.

Cheng, C.L., Wan, C.C., Wang, Y.Y. and Wu, M.S. (2005). Thermal shutdown behavior of PVdF-HFP based polymer electrolytes comprising heat sensitive cross-linkable oligomers, J. Power Sources (in press).

Chipara, M., Hui, D., Notingher, V., Chipara, M. D., Lau, K. T., Sankar, J. and Panaitescu, D. (2003). On polyethylene-polyaniline composites, Compos Part B-Eng 34, pp. 637-645.

Choi, S.S., Lee, S.G., Im, S.S., Kim, S.H. and Joo, Y.L. (2003a). Silica nanofibers from electrospinning/sol-gel process. J. Mater. Sci. Lett., 22(12), pp. 891-893.

Choi, S.W., Jo, S.M., Lee, W.S., Kim, Y.R. (2003b). An electron poly(vinylidene fluoride) nanofibrous membrane and its battery applications, Adv. Mater. 15, 23, pp. 2027-2032.

Choi, J.S., Lee, S.W., Jeong, L., Bae, S.H., Min, B.C., Youk, J.H. and Park, W.H. (2004a). Effect of organosoluble salts on the nanofibrous structure of electrospun poly(3-hydroxybutyrate-co-3-hydroxyvalerate). International Journal of Biological Macromolecules, 34, 4, pp. 249-256.

Choi, S. S., Lee, Y. S., Joo, C. W., Lee, S. G., Park, J. K. and Han, K. S. (2004b). Electrospun PVDF nanofiber web as polymer electrolyte or separator. Electrochim Acta 50, pp. 338-342.

Choi, S. S., Lee, S. G., Joo, C. W., Im, S. S. and Kim, S. H. (2004c). Formation of interfiber bonding in electrospun poly(etherimide) nanofiber web. J. Mater. Sci. 39, pp. 1511-1513.

Choi, S.S., Chu, B., Lee, S.G., Lee, S.W., Im, S.S., Kim, S.H. and Park, J.K. (2004d). Titanium-doped silica fibers prepared by electrospinning and sol-gel process. J. Sol-Gel Sci. Technol., 30(3), pp. 215-221.

Choi, S.W., Kim, J.R., Jo, S.M., Lee, W.S. and Kim, B.C. (2004e). Electrospun PVdF-based fibrous polymer electrolytes for lithium ion polymer batteries, Electrochim. Acta. 50, 1, pp. 69-75.

Christanti, Y. and Walker, L. M. (2001). Surface tension driven jet break up of strain-hardening polymer solutions. J. Non-Newton. Fluid. 100, pp. 9-26.

Christie, S., Scorsone, E., Persaud, K., Kvasnik, F. (2003). Remote detection of gaseous ammonia using the near infrared transmission properties of polyaniline, Sensor Actuat B-Chem. 90, pp. 163-9.

Chu, B., Hsiao, B. S. and Fang, D. (2004) Apparatus and methods for electrospinning polymeric fibers and membranes. US Patent 6713011 B2.

Chu, P.K., Chen, J.Y., Wang, L.P., Huang, N. (2002). Plasma-surface modification of biomaterials, Mater. Sci. Eng. R-Rep. 36, pp.143-206.

Chua, K.N., Lim, W.S., Zhang, P., Lu, H., Wen, J., Ramakrishna, S., Leong, K.W., Mao, H.Q. (2005). Stable immobilization of rat hepatocyte spheroids on galactosylated nanofiber scaffold, Biomaterials 26, 2537.

Chun, Y. S., Lee, H. S. and Kim, W. N. (1996). Thermal Properties and Morphology of Blends of Poly(ether imide) and Polycarbonate. Polym. Eng. Sci. 36, pp. 2694-2972.

Chung, H.Y. (2004). Polymer, polymer microfiber, polymer nanofiber and applications including filter structures, US patent 6,743,273.

Chung, T.W., Liu, T.W., Wang, D.Z. and Wang, S.S. (2003). Enhancement of the growth of human endothelial cells by surface roughness at nanometer scale, Biomaterials. 24, pp. 4655-4661.

Clark, A.L. (1938). The Critical State of Pure Fluids. Chem Rev. 23, pp. 1-15.

Clough, R.L., Gillen, K.T., Dole, M. (1991). Chapter 3 In: Irradiation effects on polymers, ed by Clegg, D.W., Collyer, A.A., Elsevier applied science, New York.

Coombes, A.G.A., Verderio, E., Shaw, B. and Downes, D. (2002). Biocomposites of non-crosslinked natural and synthetic polymers, Biomaterials. 23, pp. 2113-2118.

Corradini, E., Mattoso, L. H. C, Guedes, C. G. F. and Rosa, D. S. (2004). Mechanical, thermal and morphological properties of poly(e-caprolactone)/zein blends. Polym. Advan. Technol. 15, pp. 340-345.

Daane, J. H. and Barker, R. E. Jr. (1964). Multiple glass transitions in cellulose 2.5 acetate. J. Polym. Sci. Pol. Lett. 2, pp. 343-347.

Dai, H., Gong, J., Kim, H. and Lee, D. (2002). A novel method for preparing ultra-fine alumina-borate oxide fibres via an electrospinning technique. Nanotechnology. 13, pp. 674-677.

Dai, N.T., Williams, M.R., Khammo, N., Adams, E.F. and Coombes, A.G.A. (2004). Composite cell support membranes based on collagen and polycaprolactone for tissue engineering of skin, Biomaterials. 25, pp. 4263-4271.

Dalton, P. D., Klee, D. and Moller, M. (2005). Electrospinning with dual collection rings. Polym. Commun. 46, 611-614.

Davis, S. A., Burkett, S. L., Mendelson, N. H. and Mann, S. (1997). Bacterial templating of ordered macrostructures in silica and silica-surfactant mesophases. Nature. 385, pp. 420-423.

de Moel, K., van Ekenstein, G.O.R.A., Nijland, H., Polushkin, E., ten Brinke, G., Maki Ontto, R. and Ikkala, O. (2001). Polymeric nanofibers prepared from self organized supramolecules. Chem. Mater., 13(12), pp. 4580-4583.

Deam, J. R. and Maddox, R. N. (1970). Interfacial Tension in Hydrocarbon Systems. J. Chem. Eng. Data. 15, pp. 216-222.

Deitzel, J. M., Kleinmeyer, J. D., Hirvonen, J. K. and Tan, N. C. B. (2001a). Controlled deposition of electrospun poly(ethylene oxide) fibers. Polymer. 42, pp. 8163-8170.

Deitzel, J. M., Kleinmeyer, J., Harris, D. and Tan, N. C. B. (2001b). The effect of processing variables on the morphology of electrospun nanofibers and textiles. Polymer. 42, pp. 261-272.

Deitzel, J.M., Kosik, W., McKnight, S.H., Tan, N.C.B., DeSimone, J.M. and Crette, S. (2002). Electrospinning of polymer nanofibers with specific surface chemistry, Polymer. 43, pp. 1025-1029.

Demir, M. M., Yilgor, I., Yilgor, E. and Erman, B. (2002). Electrospinning of polyurethane fibers. Polymer. 43, pp. 3303-3309.

Demir, M.M., Gulgun, M.A., Menceloglu, Y.Z., Erman, B., Abramchuk, S.S., Makhaeva, E.E., Khokhlov, A.R., Matveeva, V.G. and Sulman, M.G. (2004). Palladium nanoparticles by electrospinning from poly(acrylonitrile-co-acrylic acid)-PdCl2 solutions. Relations between preparation conditions, particle size, and catalytic activity, Macromolecules. 37, pp. 1787-1792.

Dersch, R., Liu, T., Schaper, A.K., Greiner, A. and Wendorff, J.H. (2003). Electrospun nanofibers: Internal structure and intrinsic orientation, J. Polym. Sci. Pt. B-Polym. Phys. 41, pp. 545-553.

Desai, S.M., Singh, R.P. (2004). Surface modification of polyethylene: long-term properties of polyolefins, Adv. Polym. Sci. 169, pp.231-293.

Dharmaraj, N., Park, H.C., Kim, C.K., Kim, H.Y. and Lee, D.R. (2004a). Nickel titanate nanofibers by electrospinning. Materials Chemistry and Physics, 87, pp. 5-9.

Dharmaraj, N., Park, H.C., Lee, B.M., Viswanathamurthi, P., Kim, H.Y. and Lee, D.R. (2004b). Preparation and morphology of magnesium titanate nanofibres via electrospinning. Inorganic Chem Commu, 7, pp. 431-433.

Ding, B., Kim, H.Y., Lee, S.C., Shao, C.L., Lee, D.R., Park, S.J., Kwag, G.B. and Choi, K.J. (2002). Preparation and characterization of a nanoscale poly(vinyl alcohol) fiber aggregate produced by an electrospinning method, J. Polym. Sci. Pt. B-Polym. Phys. 40, pp. 1261-1268.

Ding B., Kim H., Kim C., Khil M. and Park S. (2003). Morphology and crystalline phase study of electrospun TiO2-SiO2 nanofibres. Nanotechnology, 14, pp. 532-537.

Ding, B., Kim, J., Miyazaki, Y. and Shiratori, S. (2004a). Electrospun nanofibrous membranes coated quartz crystal microbalance as gas sensor for NH3 detection. Sensor Actuat B-Chem. 101, pp. 373-380.

Ding, B., Yamazaki, M. and Shiratori, S. (2004b). Electrospun fibrous polyacrylic acid membrane-based gas sensors, Sensor Actuat B-Chem. 106, pp. 477-483.

Ding, B., Kimura, E., Sato, T., Fujita, S. and Shiratori, S. (2004c). Fabrication of blend biodegradable nanofibrous nonwoven mats via multi-jet electrospinning. Polymer. 45, pp. 1895-1902.

Dong, H., Bell, T. (1999). State-of-the-art overview: ion beam surface modification of polymers towards improving tribological properties, Surf. Coat. Technol. 111, 1, pp. 29-40

Dresselhaus, M. S. (1998). New tricks with nanotubes. Nature. 391, pp. 19-20.

Drew, C., Liu, X., Ziegler, D., Wang, X., Bruno, F.F., Whitten, J., Samuelson, L.A., Kumar, J. (2003). Metal Oxide-Coated Polymer Nanofibers, Nano Lett. 3, pp. 143.

Dror, Y., Salalha, W., Khalfin, R. L., Cohen, Y., Yarin, A. L. and Zussman, E. (2003). Carbon Nanotubes Embedded in Oriented Polymer Nanofibers by Electrospinning. Langmuir. 19, pp. 7012-7020.

Duan, B., Dong, C. H., Yuan, X.Y. and Yao, K. D. (2004). Electrospinning of chitosan solutions in acetic acid with poly(ethylene oxide). J. Biomat. Sci.-Polym. E. 15, pp. 797–811.

Dzenis, Y.A., Reneker, D.H. (2001). Delamination resistant composites prepared by small diameter fiber reinforcement at ply interfaces, US Patent No. 6,265,333 B1.Dzenis, Y.A. and Wen, Y.K. (2002). Continuous carbon nanofibers for nanofiber composites, Mat Res Soc Symp Proc. 702, pp. U5.4.1-U5.4.6.

Elbert, D.L., Hubbell, J.A. (1996). Surface treatment of polymers for biocompatibility, Annu. Rev. Mater. Sci., 26, pp.365.

Elias, K.E., Price, R.L. and Webster, T.J. (2002). Enhanced functions of osteoblasts on nanometer diameter carbon nanofibers, Biomaterials. 23, pp. 3279-3287.

Ellison, C. J. and Torkelson, J. M. (2003). The distribution of glass-transition temperatures in nanoscopically confined glass formers. Nat. Mater. 2, pp. 695-700.

Emig, D., Klimmek, A., Raabe, E. (2002). Dust filter bag containing nano non-woven tissue. US Patent 6,395,046 B1.

Emmenegger, C. H., Mauron, P. H., Sudan, P., Wenger, P., Hermann, V., Gallay, R. and Zuttel, A. (2003). Investigation of electrochemical double-layer (ECDL) capacitors electrodes based on carbon nanotubes and activated carbon materials. J. Power Sources. 124, pp. 321-329.

Farag, M. M. (1989). Selection of Materials and Manufacturing Processes for Engineering Design, Prentice Hall, New York.

Feng, J.J. (2002). The stretching of an electrified non-Newtonian jet: A model for electrospinning. Phys. Fluids, 14(11), pp. 3912-3926.

Feng, J.J. (2003). Stretching of a straight electrically charged viscoelastic jet. J. Non-Newtonian Fluid. Mech., 116(1), pp. 55-70.

Feng, L., Li, S.H., Li, H.J., Zhai, J., Song, Y.L., Jiang, L. and Zhu, D.B. (2002). Super hydrophobic surface of aligned polyacrylonitrile nanofibers. Angew. Chem. Int. Ed., 41(7), pp. 1221-1223.

Fennessey, S.F. and Farris, R.J. (2004). Fabrication of aligned and molecularly oriented electrospun polyacrylonitrile nanofibers and the mechanical behavior of their twisted yarns, Polymer, 45, pp. 4217-4225.

Ferguson, J. and Zemblowski, Z. (1991). Applied Fluid Rheology. Elsevier Science Publishers Ltd, Great Britain.

Fertala, A., Han, W. B. and Ko, F. K. (2001) Mapping critical sites in collagen II for rational design of gene-engineered proteins for cell-supporting materials, J. Biomed. Mater. Res. 57, pp. 48-58.

Fong, H. (2004). Electrospun nylon 6 nanofiebr reinforced BIS-GMA/TEGDMA dental restorative composite resins, Polymer. 45, pp. 2427-2432.

Fong, H. and Reneker, D.H. (1999). Elastomeric nanofibers of styrene-butadiene-styrene triblock copolymer, J. Polym. Sci. Pt. B-Polym. Phys. 37, pp. 3488-3493.

Fong, H., Chun, I., Reneker, D. H. (1999). Beaded nanofibers formed during electrospinning. Polymer. 40, pp. 4585-4592.

Fong, H., Liu, W.D., Wang, C.S. and Vaia, R.A. (2002). Generation of electrospun fibers of nylon 6 and nylon 6-montmorillonite nanocomposite, Polymer. 43, pp. 775-780.

Formhals A. (1934). Process and apparatus for preparing artificial threads. US Patent. 1,975,504.

Fraga, A. N. and Williams, R. J. J. (1985). Thermal Properties of gelatin films. Polymer. 26, pp. 113-118.

Freed, L.E. and Vunjak-Novakovic, G. (1995). Tissue engineering of cartilage. Biomedical Engineering. Hartford, CRC Press, pp.1788-806.

Fujihara, K., Kotaki, M. and Ramakrishna, S. (2005). Guided bone regeneration membrane made of Polycaprolactone / calcium carbonate composite nano fibers, Biomaterials, 29, 16, pp. 4139-4147.

Gao, Y., Xiu, Y. Y., Pan, Z. Q., Wang, D. N., Hu, C. P. and Ying, S. K. (1994). Effect of Aromatic Diamine Extenders on the Morphology and Property of RIM Polyurethane-Urea. J. Appl. Polym. Sci. 53, 1, pp. 23-29.

Gershon, B., Marom, G. and Cohn, D. (1990). New arterial prostheses by filament winding, Clinical Materials. 5, pp. 13-27.

Ghiggino, P.K. (1989). Chapter 3 in: The effects of radiation on high-technology polymers, ed by Reichmanis, E., O'Donnell, J.H., American Chemical Society, Washington.

Gibson, H.S. and Gibson, P. (2002). Use of Electrospun Nanofibers for aerosol filtration in Textile Structures. U.S. Army Soldier Systems Center, AMSSB-RSS-MS(N) Natick, Massachusetts, 01760-5020.

Gibson, P., Schreuder-Gibson, H. and Pentheny, C. (1998). Electrospinning technology: direct application of tailorable ultrathin membranes, J. Coated Fabric. 28, pp. 63-72.

Gibson, P.W., Schreuder-Gibson, H.L. and Riven, D. (1999). Electrospun fiber mats: transport properties, AIChE J. 45, 1, pp. 190-195.

Godovsky, Y. K. and Slonimsky, G. L. (1974). Kinetics of Polymer Crystallization from the Melt (Calorimetric Approach). J. Polym. Sci. 12, pp. 1053-1080.

Gong, J., Shao, C.L., Yang, G.C., Pan, Y. and Qu, L.Y. (2003). Preparation of ultra-fine fiber mats contained $H_4SiW12O40$. Inorganic Chem Commu, 6, pp. 916-918

Gordon G. A. (1971). Glass Transition in Nylons. J. Polym. Sci. 9, pp. 1693-1702.

Gouma, P.I. (2003). Nanostructured polymorphic oxides for advanced chemosensors, Rev. Adv. Mater. Sci. 5, pp.147-154.

Grace, J.M., Gerenser, L.J. (2003). Plasma treatment of polymers, J. Dispersion Sci. Technol. 24, 3-4, pp. 305-341.

Grad, S., Kupcsik, L., Gogolewski, S. and Alini, M. (2003). The use of biodegradable polyurethane scaffolds for cartilage tissue engineering: potential and limitations, Biomaterials. 24, pp. 5163-5171.

Grafe, T.H. and Graham, K.M. (2003). Nanofiber webs from electrospinning. Nonwovens in Filtration – Fifth International Conference. pp. 1-5.

Graham, K., Gibson, H.S., Gogins, M. (2003). Incorporation of Electrospun Nanofibers into Functional Structures, Presented at INTC 2003, September 15-18, Baltimore, MD.

Graham, K., Gogins, M. and Schreuder-Gibson, H. (2004). Incorporation of electrospun nanofibers into functional structures, Int. Nonwovens J. Summer. pp. 21-27.

Grande, D.A., Halberstadt, C., Schwartz, R. and Manji, R. (1997). Evaluation of matrix scaffolds for tissue engineering of articular cartilage grafts, J. Biomed. Mater. Res. 34, pp. 211-220.

Gregg, S.J. and Sing, K.W.S. (1982). Adsorption, Surface Area and Porosity, Academic Press, London.

Griffith, L. G. (2000). Polymeric Biomaterials, Acta Mater. 48, pp. 263-277.

Griffiths, C. H., Okumura, K. and VanLaeken, A. (1977). Influence of Chemical Modification (Bromination) on Structure and Charge Transport in Poly(N-Vinylcarbazole). J. Polym. Sci.Pol. Phys.15, pp. 1627-1639.

Groiztsch, D. and Fahrbach, E. (1986). Microporous multilayer nonwoven material for medical applications, US Patent 4,618,524.

Guan, H., Shao, C., Chen, B., Gong, J. and Yang, X. (2003a). A novel method for making CuO superfine fibres via an electrospinning technique. Inorganic Chem Commu, 6, 11, pp. 1409-1411.

Guan, H., Shao, C., Wen, S., Chen, B., Gong, J. and Yang, X. (2003b). A novel method for preparing Co3O4 nanofibers by using electrospun PVA/cobalt acetate composite fibers as precursor. Mater Chem Phy, 82, 3, pp. 1002-1006.

Guan, H., Shao, C., Liu, Y., Yu, N. and Yang, X. (2004). Fabrication of NiCo2O4 nanofibers by electrospinning. Solid State Commu, 131 pp. 107-109.

Gupta, P. and Wilkes, G. L. (2003). Some investigations on the fiber formation by utilizing a side-by-side bicomponent electrospinning approach. Polymer, 44, pp. 6353-6359.

Hajra, M.G., Mehta, K. and Chase, G.G. (2003). Effects of humidity, temperature, and nanofibers on drop coalescence in glass fiber media, Sep. Purif. Technol. 30, 1, pp. 79-98.

Harris, J. (1977). Rheology and non-Newtonian flow. Longman, New York.

Hartgerink, J.D., Beniash, E. and Stupp, S.I. (2001). Self-assembly and mineralization of peptide-amphiphile nanofibers. Science, 294(5547), pp. 1684-1688.

Hashmi, S.A., Kumar, A. and Tripathi, S.K. (2005). Investigations on electrochemical supercapacitors using polypyrrole redox electrodes and PMMA based gel electrolytes, Euro. Polym. J. (in press).

Hattori, T. and Wakitani, S. (2002). Articular Cartilage regeneration, Clin Calcium. 12, 2, pp. 217-221.

He, W., Ma, Z.W., Yong, T., Teo, W.E. and Ramakrishna, S. (2005a). Fabrication of collagen-coated biodegradable copolymer nanofiber and their potential for endothelial cell growth, Biomaterials (submitted).

He, W., Yong, T., Teo, W.E., Ma, Z.W., Ramakrishna, S. (2005b). Fabrication and endothelialization of collagen-blended biodegradable polymer nanofiber: potential vascular grafts for the blood vessel tissue engineering. Tissue Eng. Accepted.

Hildebrand, J. H. and Robert, L. S. (1950). The solubility of Nonelectrolytes, 3rd edition, Reinhold, New York.

Hirai, J. and Matsuda, T. (1996). Venous reconstruction using hybrid vascular tissue composed of vascular cells and collagen: tissue regeneration process, Cell Transplant. 5, pp. 93-105.

Hirai, J., Kanda, K., Oka, T. and Matsuda, T. (1994). Highly oriented, tubular hybrid vascular tissue for a low pressure circulatory system, ASAIO J, 40, pp. M383-M388.

Hoerstrup, S.P., Kadner, A., Breymann, C., Maurus, C.F., Guenter, C.I. and Sodian, R. (2002). Living, autologous pulmonary artery conduits tissue engineered from human umbilical cord cells. Ann. Thorac. Surg. 74, pp. 46-52.

Hoerstrup, S.P., Zund, G., Sodian, R., Schnell, A.M., Grunenfelder, J. and Turina, M.I. (2001). Tissue engineering of small caliber vascular grafts, Eur. J. Cardiothorac. Surg. 20, pp. 164-169.

Hohman, M.M., Shin, M., Rutledge, G. and Brenner, M.P. (2001a) Electrospinning and electrically forced jets. I. Stability theory. Phys. Fluids, 13(8), pp. 2201-2220.

Hohman, M.M., Shin, M., Rutledge, G. and Brenner, M.P. (2001b) Electrospinning and electrically forced jets. II. Applications. Phys. Fluids, 13(8); 2221-2236.

Holland, S., Tighe, B.J. and Gould, P.L. (1986). Polymers for biodegradable medical devices. I. The potential of polyesters as controlled macromolecular release systems, J. Control. Release, 4, pp. 155-180.

Hong, K.H., Oh, K.W. and Kang, T.J. (2005). Preparation of conducting nylon-6 electrospun fiber webs by the in situ polymerization of polyaniline, J. Appl. Poly. Sci. 96, pp. 983-991.

Hou, H. and Reneker, D. H. (2004). Carbon Nanotubes on Carbon Nanofibers: A Novel Structure Based on Electrospun Polymer Nanofibers. Adv Mater. 16, pp. 69-73.

Hou, H.Q., Jun, Z., Reuning, A., Schaper, A., Wendorff, J.H. and Greiner, A. (2002). Poly(p-xylene) nanotubes by coating and removal of ultrathin polymer template fibers. Macromolecules, 35, pp. 2429-2431.

Hsu, C. M. and Shivakumar, S. (2004). N,N-Dimethylformamide Additions to the Solution for the Electrospinning of Poly(e-caprolactone) Nanofibers. Macromol. Mater. Eng. 289, pp. 334-340.

Huang, L., McMillan, R.A., Apkarian, R.P., Poudehimi, B., Conticello, V.P. and Chaikof, E.L. (2000). Generation of synthetic elastin-mimetic small diameter fibers and fiber networks, Macromolecules. 33, 8, pp. 2989-2997.

Huang, L., Nagapudi, K., Apkarian, R.P. and Chaikof, E. (2001a). Engineered collagen - PEO nanofibers and fabrics. J Biomat Sci Poly Edi, 12, 9, pp. 979-993.

Huang, L., Apkarian, R.P. and Chaikof, E.L. (2001b). High-resolution analysis of engineered Type I collagen nanofibers by electron microscopy, Scanning, 23, pp. 372-375.

Huang, Z.M., Zhang, Y.Z., Kotaki, M. and Ramakrishna, S. (2003). A review on polymer nanofibers by electrospinning and their applications in nanocomposites, Comp. Sci. Tech. 63, pp. 2223-2253.

Huang, Z. M., Zhang, Y. Z., Ramakrishna, S. and Lim, C. T. (2004). Electrospinning and mechanical characterization of gelatin nanofibers, Polymer. 45, pp. 5361-5368.

Hubbell, D. S. and Cooper, S. L. (1977). The physical properties and morphology of poly-e-caprolactone polymer blends. J. Appl. Polym. Sci. 21, pp. 3035-3061.

Ibim, S. E. M., Ambrosio, A. M. A., Kwon, M. S., El-Amin, S. F., Allcock, H. R. and Laurencin, C. T. (1997). Novel polyphosphazene/poly(lactide-co-glycolide) blends: miscibility and degradation studies. Biomaterials. 18, pp. 1565-1569.

Ikada, Y. (1994). Surface modification of polymers for medical applications, Biomaterials, 15, pp.725-736

Inai, R., Kotaki, M. and Ramakrishna, S. (2005). Structure and property of electrospun PLLA single nanofibers, Nanotechnology. 16, pp. 208-213.

Ishaug, S.L., Bizios, R. and Mikos, A.G. (1994). Osteoblast function on synthetic biodegradable polymers, J. Biomed. Mater. Res. 28, pp. 1445-1453.

Ivanov, V.S.(1992). Chapter 3 in: New concepts in polymer science: Radiation chemistry of Polymers, ed by C.R.H.I. de Jonge, Utrecht, The Netherlands.

Iwasaki, Y., Sawada, S., Ishihara, K., Khang, G. and Lee, H. B. (2002). Reduction of surface-induced inflammatory reaction on PLGA/MPC polymer blend. Biomaterials. 23, pp. 3897-3903.

Jarusuwannapoom, T., Hongrojjanawiwat, W., Jitjaicham, S., Wannatong, L., Nithitanakul, M., Pattamaprom, C., Koombhongse, P., Rangkupan, R. amd Supaphol, P. (2005). Effect of solvents on electro-spinnability of polystyrene solutions and morphological appearance of resulting electrospun polystyrene fibers. Euro. Polym. J. 41, pp. 409-421.

Jaworek, A. and Krupa, A. (1999). Classification of the modes of EHD spraying. J. Aerosol Sci., 30(7), pp. 873-893.

Jia, H., Zhu, G., Vugrinovich, B., Kataphinan, W., Reneker, D.H., Wang, P. (2002). Enzyme-Carrying Polymeric Nanofibers Prepared via Electrospinning for Use as Unique Biocatalysts, Biotechnol. Prog. 18, pp.1027-1032.

Jiang, H., Fang, D., Hsiao, B.S., Chu, B. and Chen, W. (2004). Optimization and characterization of dextran membranes prepared by electrospinning, Biomacromolecules. 5, pp. 326-333.

Jin, H. J., Fridrikh, S. V., Rutledge, G. C. and Kaplan, D. I. (2002). Electrospinning Bombyx mori silk with poly(ethylene oxide), Biomacromolecules. 3, pp. 1233-1239.

Jin, H.J., Chen, J., Karageorgiou, V., Altman, G.H. and Kaplan, D.L. (2004). Human bone marrow stromal cell responses on electrospun silk fibroin mats, Biomaterials. 25, pp. 1039-1047.

Johnson, R. W., et al. (1988). In Encyclopedia of Polymer Science and Engineering, edited by H. F. Mark et al. John Wiley and Sons, New York, vol. 11.

Jonassen, N. (2002). Electrostatics. Kluwer Academic Publishers.

Joschek, S., Nies, B., Krotz, R. and Gopferich, A. (2000). Chemical and physiochemical characterization of porous hydroxyapatite ceramics made of natural bone. Biomaterials; 21, pp. 1645-1658.

Kameoka, J., Orth, R., Yang, Y., Czaplewski, D., Mathers, R., Coates, G. and Craighead, H. G. (2003). A scanning tip electrospinning source for deposition of oriented nanofibres. Nanotechnology. 14, pp. 1124-1129.

Kameoka, J., Czaplewski, D., Liu, H. and Craighead, H. G. (2004). Polymeric nanowire architecture. J. Mater. Chem. 14, pp. 1503-1505.

Kaneko, K. (1994). Determination of pore size and pore size distribution 1. Adsorbents and catalysts, J. Memb. Sci. 96, 1-2, pp. 59-89.

Kato, K., Uchida, E., Kang, E.T., Uyama, Y., Ikada, Y. (2003). Polymer surface with graft chains, Prog. Polym. Sci. 28, 2, pp.209-259.

Katta, P., Alessandro, M., Ramsier, R. D. and Chase, G. G. (2004). Continuous Electrospinning of Aligned Polymer Nanofibers onto Wire Drum Collector. Nano. Lett. 4, pp.2215-2218.

Kaur, S., Kotaki, M., Ma, Z., Gopal, R., Ng, S.C., Ramakrishna, S. (2004). Oligosaccharide functionalized nanofibrous membrane, 1st Nano-Engineering and Nano-Science Congress, July7-9, 2004 Singapore.

Kemp, P.C., Neumeister-Kemp, H.G., Lysek, G. and Murray, F. (2001). Survival and growth of micro-organisms on air filtration media during initial loading, Atmos. Environ. 35, pp. 4739-4749.

Kenawy, E.R., Bowlin, G.L., Mansfield, K., Layman, J., Simpson, D.G., Sanders, E.H. and Wnek, G.E. (2002). Release of tetracycline hydrochloride from electrospun poly(ethylene-co-vinylacetate), poly(lactic acid), and a blend, J. Control. Release. 81, pp. 57-64.

Kenawy, E. R., Layman, J. M., Watkins, J. R., Bowlin, G. L., Matthews, J. A., Simpson, D. G., and Wnek, G. E. (2003). Electrospinning of poly (ethylene-co-vinyl alcohol) fibers. Biomaterials. 24, 907-913.

Kessick, R. and Tepper, G. (2003). Microscale electrospinning of polymer nanofiber interconnections. Appl. Phys. Lett. 83, 557-559.

Kessick, R. and Tepper, G. (2004). Microscale polymeric helical structures produced by electrospinning. Appl Phys Lett, 84, 23, pp. 4807-4809.

Kessick, R., Fenn, J., Tepper, G. (2004). The use of AC potentials in electrospraying and electrospinning processes. Polymer. 45, pp. 2981-2984.

Khil, M.S., Cha, D.I., Kim, H.Y., Kim, I.S. and Bhattarai, N. (2003). Electrospun nanofibrous polyurethane membrane as wound dressing, J. Biomed. Mater. Res. 67B, pp. 675-679.

Khil, M. S., Kim, H. Y., Kim, M. S., Park, S. Y. and Lee, D. R. (2004) Nanofibrous mats of poly(trimethylene terephthalate) via electrospinning. Polymer. 45, pp. 295-301.

Khil, M. S., Bhattarai, S. R., Kim, H. Y., Kim, S. Z. and Lee, K. H. (2005). Novel Fabricated Matrix Via Electrospinning for Tissue Engineering. J. Biomed. Res. Part B: Appl. Biomater. 72B, pp. 117-124.

Khurana, H.S., Patra, P.K. and Warner, S.B. (2003). Nanofibers from melt electrospinning. Abstr. Pap. Am. Chem. Soc., 226; U425-U425 464-POLY.

Kidoaki, S., Kwon, I.K. and Matsuda, T. (2005). Mesoscopic spatial designs of nano- and microfiber meshes for tissue-engineering matrix and scaffold based on newly devised multilayering and mixing electrospinning techniques, Biomaterials. 26, pp. 37-46.

Kikuchi, M., Cho, S.B. and Tanaka, J. (1997). In vitro tests and in vivo tests developed TCP/CPLA composites, Bioceramics. 10, pp. 407-410.

Kim, J.S. and Reneker, D.H. (1999). Mechanical properties of composites using ultrafine electrospun fibers, Polym. Composites, 20, 1, pp. 124-131.

Kim, B.S., Nikolovski, J., Bonadio, J., Smiley, E. and Mooney, D.J. (1999). Engineered smooth muscle tissues: regulating cell phenotype with the scaffold, Exp. Cell Res. 251, pp. 318-328.

Kim, Y.W., Gong, M.S. and Choi, B.K. (2001). Ionic conduction and electrochemical properties of new poly(acrylonitrile-itaconate)-based gel polymer electrolytes, J. Power Sources. 97-98, pp. 654-656.

Kim, C.S. and Oh, S.M. (2002). Spectroscopic and electrochemical studies of PMMA-based gel polymer electrolytes modified with interpenetrating networks, J. Power Sources. 109, pp. 98-104.

Kim, C., Kim, Y. J. and Kim, Y. A. (2004a). Fabrication and structural characterization of electro-spun polybenzimidazol-derived carbon nanofiber by graphitization. Solid State Commun. 132, pp. 567-571.

Kim, C., Park, S.H., Lee, W.J. and Yang, K.S. (2004b). Characteristics of supercapaitor electrodes of PBI-based carbon nanofiber web prepared by electrospinning, Electrochim. Acta. 50, 2-3, pp. 872-876.

Kim, C., Choi, Y.O., Lee, W.J. and Yang, K.S. (2004c). Supercapacitor performances of activated carbon fiber webs prepared by electrospinning of PMDA-ODA poly(amic acid) solutions, Electrochim. Acta. 50, (2-3), pp. 878-882.

Kim, K., Luuc, Y.K., Chang, C., Fang, D., Hsiao, B.S., Chua, B. and Hadjiargyrou, M. (2004d). Incorporation and controlled release of a hydrophilic antibiotic using poly(lactide-co-glycolide)-based electrospun nanofibrous scaffolds, J. Control. Release. 98, pp. 47-56.

Kim, S.J., Shin, K.M. and Kim, S.I. (2004e). The effect of electric current on the processing of nanofibers formed from poly(2-acrylamido-2-methyl-1-propane sulfonic acid), Scripta Materialia. 51, pp. 31-35.

Kim, C. (2005). Electrochemical characterization of electrospun activated carbon nanofibres as an electrode in supercapacitors. J Power Sources. Article In Press.

Kim, J. R., Choi, S. W., Jo, S. M., Lee, W. S. and Kim, B. C. (2005). Characterization and Properties of P(VdF-HFP)-Based Fibrous Polymer Electrolyte Membrane Prepared by Electrospinning. J. Electrochem. Soc. 152, pp. A295-A300.

Klein, E. (2000).Affinity membranes: a 10-year review, J. Membr. Sci., 179, pp.1-27.

Ko, F., Gogotsi, Y., Ali, A., Naguib, N., Ye, H., Yang, G., Li, C. and Willis, P. (2003). Electrospinning of continuous carbon nanotube-filled nanofiber yarns, Adv. Mater. 15, 14, pp. 1161-1165.

Koombhongse, S., Liu, W. and Reneker, D.H. (2001). Flat polymer ribbons and other shapes by electrospinning. J. Polym. Sci., Polym. Phys, 39, pp. 2598-606.

Koski, A., Yim, K. and Shivkumar, S. (2004). Effect of molecular weight on fibrous PVA produced by electrospinning. Materials Letters, 58, pp. 493-497.

Kotz, R. and Carlen, M. (2000). Principles and applications of electrochemical capacitors. Electrochim Acta. 45, pp. 2438-2498.

Krieger, I. M. and Maron, S. H. (1954). Direct Determination of the Flow Curves of Non-Newtonian Fluids. III. Standardized Treatment of Viscometric Data. J. Appl. Phys. 25, pp. 72-75.

Krishnappa, R. V. N., Sung, C. M. and Schreuder-Gibson, H. (2002). Electrospinning of Polycarbonates and their Surface Characterization using the SEM and TEM. Mat. Res. Soc. Symp. Proc. 702, pp. U6.7.1-6.7.6.

Krishnappa, R. V. N., Desai, K., Sung, C. M. (2003). Morphological study of electrospun polycarbonates as a function of the solvent and processing voltage. J. Mater. Sci. 38, pp. 2357-2365.

Kwoun, S.J., Lec, R.M., Han, B. and Ko, F.K. (2000). A novel polymer nanofiber interface for chemical sensor applications, Int. Frequency Control Symp. Exh. pp. 52-57.

Kwoun, S.J., Leo, R.M., Han, B. and Ko, F.K. (2001). Polymer nanofiber thin films for biosensor applications. Proc of the IEEE 27th Annual Northeast Bioengineering Conference. pp. 9-10.

L'Heureux, N., Germain, L., Labbe, R. and Auger, F.A. (1993). In vitro construction of a human blood vessel from cultured vascular cells: a morphologic study, J. Vasc. Surg. 17, pp. 499-509.

L'Heureux, N., Paquet, S., Labbe, R., Germain, L. and Auger, F.A. (1998). A completely biological tissue-engineered human blood vessel. FASEB, 12, pp. 47-56.

Lahann, J., Balcells, M., Rodon, T., Lee, J., Choi, I.S., Jensen, K.F., Langer, R. (2002). Reactive polymer coatings: A platform for patterning proteins and mammalian cells onto a broad range of materials, Langmuir, 18, 9, pp.3632-3638.

Lala, N.L. and Sastry, M. (2000). Nanocomposites of collidal gold particles and fatty acids formed by the high-affinity biotin-avidin interaction, Phys. Chem. Chem. Phys. 2, pp. 2461-2466.

Lala, N.L., Ramaseshan, R., Ramakrishna, S. (2005) A Novel Nanofiber Based Material System for Glucose Biosensor Application, J. Nanoengin. and Nanosyst. – Part N, submitted.

Langstaff, S., Sayer, M., Smith, T. J. N. and Pugh, S. (2001). Resorbable bioceramics based on stabilized calcium phosphates. Part II: evaluation of biological response. Biomaterials. 22, pp. 135-150.

Larrondo, L. and Manley R. S. J. (1981a). Electrostatic Fiber Spinning from Polymer Melts. I. Experimental Observations on Fiber Formation and Properties. J. Polym. Sci. Pol. Phys. 19, pp. 909-920.

Larrondo, L. and Manley R. S. J. (1981b). Electrostatic Fiber Spinning from Polymer Melts. II. Examination of the Flow Field in an Electrically Driven Jet. J. Polym. Sci. Pol. Phys. 19, pp. 921-932.

Larrondo, L. and Manley R. S. J. (1981c). Electrostatic Fiber Spinning from Polymer Melts. III. Electrostatic Deformation of a Pendant Drop of Polymer Melt. J. Polym. Sci. Pol. Phys. 19, pp. 933-940.

Larsen, G., Spretz, R. and Velarde-Ortiz, R. (2004). Use of coaxial gas jackets to stabilize taylor cones of volatile solutions and to induce particle-to-fiber transitions. Adv. Mater. 16, 2, pp. 166-169.

Laurencin, C.T. and Ko, F.K. (2004). Hybrid nanofibril matrices for use as tissue engineering devices, US Patent 6,689,166 B2.

Laurent, T. C. (1998). Chemistry, Biology and Medical Applications of Hyaluronan and Its Derivatives, London, Portland Press.

Lee, K.H., Kim, H.Y., La, Y.M., Lee, D.R. and Sung, N.H. (2002a). Influence of a mixing solvent with tetrahydrofuran and N,N-Dimethylformamide on electrospun poly(vinyl chloride) nonwoven mats, J. Polym. Sci. Pt. B-Polym. Phys. 40, pp. 2259-2268.

Lee, S.H., Ku, B.C., Wang, X., Samuelson, L.A. and Kumar, J. (2002b). Design, synthesis and electrospinning of a novel fluorescent polymer for optical sensor applications. Materials Research Society Symposium Proc. 708: pp. BB10.45.1-BB10.45.6.

Lee, K. H., Kim, H. Y., Ryu, Y. J., Kim, K. W. and Choi, S. W. (2003a). Mechanical Behavior of Electrospun Fiber Mats of Poly(vinyl chloride)/Polyurethane Polyblends.J. Polym. Sci. Pol. Phys. 41, pp. 1256-1262.

Lee, K.H., Kim, H.Y., Ra, Y.M. and Lee, D.R. (2003b). Characterization of nano-structured poly(e-caprolactone) nonwoven mats via electrospinning, Polymer. 44, pp. 1287-1294.

Lee, K.H., Kim, H.Y., Bang, H.J., Jung, Y.H. and Lee, S.G. (2003c). The change of bead morphology formed on electrospun polystyrene fibers. Polymer, 44, pp. 4029-4034.

Lee, S.H., Mooney, D.J. and Kim, Y.H. (2003d). Elastic biodegradable poly(glycolide-co-caprolactone) scaffold for tissue engineering, J. Biomed. Mater. Res. 66A, pp. 29-37.

Lee, J.S., Choi, K.H., Ghim, H.D., Kim, S.S., Chun, D.H., Kim, H.Y. and Lyoo, W.S. (2004). Role of molecular weight of atactic poly(vinyl alcohol) (PVA) in the structure and properties of PVA nanofabric prepared by electrospinning, J. Appl. Polym. Sci. 93, pp. 1638-1646.

Leong, K.W., Brott, B.C. and Langer, R. (1985). Bioerodable poly anhydrides as drug carrier matrices. I. Characterization, degradation, and release characteristics, J. Biomed. Mater. Res. 19, pp. 941-955.

Li, J., Wang, J., Liu, X. (2002a). Preparation and chromatographic behavior of acetate-fiber filter rods with dye affinity ligands, Chromatographia, 56, pp.401-406.

Li, W.J., Laurencin, C.T., Caterson, E.J., Tuan, R.S. and Ko, F.K. (2002b). Electrospun nanofibrous structure: A novel scaffold for tissue engineering, J. Biomed. Mater. Res. 60, pp. 613-621.

Li, D. and Xia, Y. (2003). Fabrication of titania nanofibers by electrospinning. Nano Letters, 3, 4, pp. 555-60.

Li, D., Herricks, T. and Xia, Y. (2003a). Magnetic nanofibers of nickel ferrite prepared by electrospinning. Appl Phy Lett, 83, 22, pp. 4586-4588.

Li, D., Wang, Y., and Xia, Y. (2003b). Electrospinning of polymeric and ceramic nanofibers as uniaxially aligned arrays. Nano Lett. 3, 8, pp. 1167-1171.

Li, D. and Xia, Y. (2004a) Direct Fabrication of Composite and Ceramic Hollow Nanofibers by Electrospinning. Nano Lett 4, pp. 933-938.

Li, D., Wang, Y. and Xia, Y. (2004b). Electrospinning nanofibers as uniaxially aligned arrays and layer-by-layer stacked films. Adv. Mater. 16, 4, pp. 361-366.

Li, D. Ouyang, G. McCann, J. T. and Xia, Y. (2005a). Collecting Electrospun Nanofibers with Patterned Electrodes. Nano. Lett. ASAP article.

Li, W.J., Tuli, R., Okafor, C., Derfoul, A., Danielson, K.G., Hall, D.J. and Tuan, R.S. (2005b). A three-dimensional nanofibrous scaffold for cartilage tissue engineering using human mesenchymal stem cells, Biomaterials. 26, pp. 599-609.

Lim, T.C., Kotaki, M., Yong, T.K.J., Fujihara, K. and Ramakrishna, S. (2004). Recent advances in tissue engineering of electrospun nanofibers. Mater. Technol., 19(1), pp. 20-27.

Lim, T.C. and Ramakrishna, S. (2005). Next-generation applications for polymeric nanofibres. Editor: Schulte, J. Nanotechnology: Global Strategies, Trends and Applications. Wiley, West Sussex, pp. 137-147.

Lin, T., Wang, H. X., Wang, H. M. and Wang, X. G. (2004). The charge effect of cationic surfactants on the elimination of fibre beads in the electrospinning of polystyrene. Nanotechnology. 15, pp. 1375-1381.

Liston, E.M., Martinu, L., Wertheimer, M.R.(1993). Plasma surface modification of polymers for improved adhesion - a critical-review, J. Adhes. Sci. Technol. 7, pp. 1091-1127.

Liu, G.J., Qiao, L.J. and Guo, A. (1996). Diblock copolymer nanofibers. Macromol., 29(16), pp. 5508-5510.

Liu, G.J., Ding, J.F., Qiao, L.J., Guo, A., Dymov, B.P., Gleeson, J.T., Hashimoto, T. and Saijo, K. (1999). Polystyrene-block-poly (2-cinnamoylethyl methacrylate) nanofibers - Preparation, characterization, and liquid crystalline properties. Chem. Eur. J., 5(9), pp. 2740-2749.

Liu, H. Q. and Hiseh, Y. L. (2002). Ultrafine Fibrous Cellulose Membranes from Electrospinning of Cellulose Acetate, J. Polym. Sci. Pol. Phys. 40, pp. 2119-2129.

Liu, H.Q. and Hsieh, Y.L. (2003). Surface methacrylation and graft copolymerization of ultrafine cellulose fibers, J. Polym. Sci. Pt. B-Polym. Phys. 41, pp. 953-964.

Liu, H., Kameoka, J., Czaplewski, D.A. and Craighead, H.G. (2004). Polymeric nanowire chemical sensor, Nano Lett. 4, 4, pp. 671-675.

Loo, J. S. C., Ooi, C. P., Tan, M. L. and Boey, F. Y. C. (2005). Isothermal annealing of poly(lactide-co-glycolide) (PLGA) and its effect on radiation degradation. Polym. Int. 54, pp. 636-643.

Lou, X. D, Detrembleur, C. and Jerome, R. (2003). Novel Aliphatic Polyesters Based on Functional Cyclic (Di)Esters. Macromol. Rapid. Comm. 24, pp. 161-172.

Lu, L., Zhu, X., Valenzuela, R.G. and Yaszemski, M.J. (2001). Biodegradable polymer scaffolds for cartilage tissue engineering, Clin. Orthop. 391, pp. S251-S270.

Luu, Y.K., Kim, K. , Hsiao, B.S. , Chu, B., Hadjiargyrou, M. (2003). Development of a nanostructured DNA delivery scaffold via electrospinning of PLGA and PLA–PEG block copolymers, J. Control. Release 89, pp.341.

Lyons, J., Ko, F.K. and Pastore, C. (2003). Developments in melt-spinning of thermoplastic polymers. Abstr. Pap. Am. Chem. Soc., 226; U400-401 300-POLY.

Lyons, J., Li, C. and Ko, F. (2004). Melt-electrospinning part I: processing parameters and geometric properties. Polymer. 45, pp. 7597-7603.

Ma, P.X. and Zhang, R. (1999). Synthetic nano-scale fibrous extracellular matrix. J. Biomed. Mater. Res., 46(1), pp. 60-72.

Ma, Z.W., Gao, C.Y., Ji, J., Shen, J.C. (2002a) Protein immobilization on the surface of poly-L-lactic acid films for improvement of cellular interactions. Europ. Polym. J. 38, pp.2279

Ma, Z.W., Gao, C.Y., Juan, J., Ji, J., Gong, Y.H., Shen, J.C. (2002b). Surface modification of poly-L-lactide by photografting of hydrophilic polymers towards improving its hydrophilicity, J. Appl. Polym. Sci. 85, pp.2163-2171.

Ma, Z.W., Kotaki, M., Yong, T., He, W. and Ramakrishna, S. (2005a). Surface engineering of electrospun polyethylene terephthalate (PET) nanofibers towards development of a new material for blood vessel engineering, Biomaterials, 26, pp. 2527-2536.

Ma, Z.W., He, W., Yong, T., Ramakrishna, S. (2005b). Grafting of gelatin on electrospun poly(caprolactone) (PCL) nanofibers to improve endothelial cell's spreading and proliferation and to control cell orientation, Tissue Eng. Accepted.

Ma, Z.W., Kotaki, M., Inai, R., Ramakrishna, S. (2005c). Potential of nanofiber matrix as tissue engineering scaffolds, Tissue Eng. 11, pp.101.

Ma, Z.W., Kotaki, M., Ramakrishna, S. (2005d) Electrospun cellulose nanofiber as affinity membrane, submitted.

Ma, Z.W., Kotaki, M., Ramakrishna, S. (2005e). Surface modified nonwoven polysulphone (PSU) fiber mesh by electrospinning: A novel affinity membrane, submitted.

Madhugiri, S., Zhou, W., Ferraris, J.P. and Balkus, K.J. (2003). Electrospun mesoporous molecular sieve fibers, Micropor Mesopor Mat. 63, pp. 75-84.

Madhugiri, S., Sun, B., Smirniotis, P. G., Ferraris, J. P. and Balkus, K. J. Jr. (2004). Electrospun mesoporous titanium dioxide fibers. Micropor Mesopor Mat. 69, pp. 77-83.

Mao, Y., Gleason, K.K. (2004). Hot filament chemical vapor deposition of poly(glycidyl methacrylate) thin films using tret-butyl peroxide as an Inititor, Lagmuir 20, pp.2484-2488.

Martini, H.F., Ober, W.C., Garrison, C.W., Welch, K. and Hutchings, R.T. (1998). Fundamentals of Anatomy and Physiology (Fourth Edition), Prentice Hall, New Jersey.

Matthews, J.A., Wnek, G.E., Simpson, D.G. and Bowlin, G.L. (2002). Electrospinning of collagen nanofibers. Biomacromolecules, 3, pp. 232-238.

Maus, R., Goppelsroder, A. and Umhauer, H. (1997). Viability of bacteria in unused air filter media, Atmos. Environ. 31, 15, pp. 2305-2310.

McConnell, D. (1965). Crystal chemistry of hydroxyapatite: Its relation to bone mineral. Arch. Oral. Biol. 10, pp. 421-431.

McCreery, Michael, J. (1995). Topical skin protectants ,United States Patent, 5607979.

McEuen, P. L. (1998). Carbon-based electronics. Nature 393, pp. 15-16.

McKee, M.G., Wilkes, G.L., Colby, R.H. and Long, T.E. (2003). Correlations of solution rheology with electrospun fiber formation of linear and branched polyesters, Macromolecules. 37, pp. 1760-1767.

McKee, M.G., Elkins, C.L. and Long, T.E. (2004). Influence of self-complementary hydrogen bonding on solution rheology/electrospinning relationships. Polymer, 45, 26, pp. 8705-8715.

McKee, M. G., Park, T., Unal, S., Yilgor, I. and Long, T. E. (2005). Electrospinning of linear and highly branched segmented poly(urethane urea)s. Polym, 47, pp. 2011-2015.

Megelski, S., Stephens, J.S., Chase, D.B. and Rabolt, J.F. (2002). Micro- and nanostructured surface morphology on electrospun polymer fibers, Macromolecules. 35, pp. 8456-8466.

Min, B. M., Lee, G., Kim, S. H., Nam, Y. S., Lee, T. S. and Park, W. H. (2004). Electrospinning of silk fibroin nanofibers and its effect on the adhesion and spreading of normal human keratinocytes and fibroblasts in vitro. Biomaterials. 25, pp. 1289-1297.

Mit-uppatham, C., Nithitanakul, M. and Supaphol, P. (2004). Ultrafine Electrospun Polyamide-6 Fibers: Effect of Solution Conditions on Morphology and Average Fiber Diameter. Macromol. Chem. Physic. 205, pp. 2327-2338.

Mo, X. M., Xu, C. Y., Kotaki, M., Ramakrishna, S. (2004). Electrospun P(LLA-CL) nanofiber: a biomimetic extracellular matrix for smooth muscle cell and endothelial cell proliferation. Biomaterials. 25, pp. 1883-1890.

Morozov, V. N., Morozova, T. Y. and Kallenbach N. R. (1998). Atomic force microscopy of structures produced by electrospraying polymer solutions. Int. J. Mass. Spectrom. 178, pp. 143-159.

Mottola, H.A. (1992). Chemical immobilization in chemistry. In: Chemically modified surfaces, ed by Mottola, H.A., Steinmetz, J.R. , Elsevier, New York, pp. 1-14.

Mucha, M. and Pawlak, A. (2005). Thermal analysis of chitosan and its blends. Thermochim. Acta. 427, pp. 69-76.

Nagapudi, K., Brinkman, W.T., Leisen, J.E., Huang, L., McMillan, R.A., Apkarian, R.P., Conticello, V.P. and Chaikof, E.L. (2002). Photomediated solid-state cross-linking of an elastin-mimetic recombinant protein polymer, Macromolecules. 35, 5, pp. 1730-1737.

Nah, C., Han, S.H., Lee, M.H., Kim, J.S. and Lee, D.S. (2003). Characterization of polyimide ultrafine fibers prepared through electrospinning, Polym. Int. 52, pp. 429-432.

Nicotera, I., Coppola, L., Oliviero, C., Russo, A. and Ranieri, G.A. (2004). Some physicochemical properties of PAN-based electrolytes: solution and gel microstructures, Solid State Ionics. 167, pp. 213-220.

Nieh, S. and Nguyen, T. (1988). Effects of Humidity, Conveying Velocity, and Particle Size on Electrostatic Charges of Glass Beads in a Gaseous Suspension Flow. J. Electrostat. 21, 99-114.

Niklason, L.E., Gao, J., Abott, W.M., Hirschi, K.K., Houser, S. and Marini, R. (1999). Functional arteries grown in vitro, Science. 284, pp. 489-493.

Niklason, L. E. (2000). Engineering of bone grafts. Nature Biotechnol. 18, pp. 929-930.

Nuno-Donlucas, Puig J, Katime I. (2001) Hydorgen Bonding and Miscibility Behavior in Poly[ethylene-co-(acrylic acid)] and Poly(N-vinylpyrrolidone) Mixtures. Macromol. Chem. Phys., 202, pp. 3106-3111.

O'Donnell, J.H. (1989). Chapter 1 in: The effects of radiation on high-technology polymers, ed by E. Reichmanis, J.H. O'Donnell, American Chemical Society, Washington.

Odom, T. W., Huang, J. L., Kim, P. and Lieber, C. M. (1998). Atomic structure and electronic properties of single-walled carbon nanotubes. Nature. 391, pp. 62-64.

Ohgo, K., Zhao, C. H., Kobayashi, M. and Asakura, T. (2003). Preparation of non-woven nanofibers of Bombyx mori silk, Samia cynthia ricini silk and recombinant hybrid silk with electrospinning method, Polymer 44, pp. 841-846.

Ohkawa, K., Cha, K., Kim, H. Y., Nishida, A. and Yamamoto, H. (2004a). Electrospinning of Chitosan. Macromol Rapid Comm. 25, pp. 1600-1605.

Ohkawa, K., Kim, H. Y. and Lee, K. H. (2004b). Biodegradation of Electrospun Poly(e-caprolactone) Non-woven Fabrics by Pure-Cultured Soil Filamentous Fungi. J. Polym. Environ. 12, pp. 211-218.

Olivier, V., Faucheux, N. and Hardouin, P. (2004). Biomaterial challenges and approaches to stem cell use in bone reconstructive surgery, Drug Discov. Today. 9, pp. 803-811.

Ondarcuhu, T. and Joachim, C. (1998). Drawing a single nanofibre over hundreds of microns. Europhys. Lett., 42(2), pp. 215-220.

Oren, Y., Tobias, H. and Soffer, A. (1983). Removal of Bacteria from water by electroadsorption on porous carbon electrodes. Biochemistry and Bioenergetics. 11, pp. 347-351.

Ovchinnikov, Y. K., Kuz'min, N. N., Markova, G. S. and Bakeyev, N. F. (1979). A Study of the Amorphous Constituent in Partially Crystalline Oriented Polyethylene. Polym. Sci U.S.S.R. 20, pp. 1959-1973.

Pakalns, T., Haverstick, K.L., Fields, G.B., McCarthy, J.B., Mooradian, D.L., Tirrell, M. (1999). Cellular recongnition of synthetic peptide amphiphiles in self-assembled monolayer films, Biomaterials 20, pp.2265.

Pampuch, R. (1991). Constitution and properties of ceramic materials. Elsevier Science Publishing Co., Inc.

Park, C., Ounaies, Z., Watson, K. A., Pawlowski, K., Lowther, S. E,, Connell, J.W., Siochi, E. J., Harrison, J. S., Clair T. L. S. (2002). Polymer-single Wall Carbon Nanotube Composites for Potential Spacecraft Applications. Mat. Research. Society. Symp. Proceedings. 706, pp. Z3.30.1-Z3.30.6.

Pawlowski, K. J., Belvin, H. L., Raney, D. L., Su, J., Harrison, J. S. and Siochi, E. J. (2003). Electrospinning of a micro-air vehicle wing skin. Polymer. 44, pp. 1309-1314.

Pedicini, A. and Farris, R.J. (2003). Mechanical behavior of electrospun polyurethane, Polymer. 44, pp. 6857-6862.

Penning, J. P., Dijkstra, H. and Pennings, A. J. (1993). Preparation and properties of absorbable fibres from L-lactide copolymers. Polymer. 34, pp. 942-951.

Periasamy, P., Tatsumi, K., Shikano, M., Fujieda, T., Sakai, T., Saito, Y., Mizuhata, M., Kajinami, A. and Deki, S. (1999). An electrochemical investigation on polyvinylidene fluoride-based gel polymer electrolytes, Solid State Ionics. 126, pp. 285-292.

Piattelli, A., Santello, M.T. and Scarano, A. (1997). Resorption of composite polymer-hydroxyapatite membranes: A time-course study in rabbit, Biomaterials. 18, PP. 629-633.

Pittenger, M.F., Douglas, R. and Marshak, D.R. (1999). Multilineage potential of adult human mesenchymal stem cells, Science. 284, pp. 143-147.

Prego, M., Cabeza, O., Carballo, E., Franjo, C. F. and Jimenez, E. (2000). Measeurements and interpretation of the electrical conductivity of 1-alcohols from 273K to 333K. J. Mol. Liq. 89, pp. 233-238.

Price, R.L., Haberstroh, K.M., Webster, T.J. (2003). Selective bone cell adhesion on formulations containing carbon nanofibers, Biomaterials. 24, pp. 1877-1887.

Progelhof, R. C. and Throne, J. L. (1993). Polymer Engineering Principles. Hanser Publishers. Germany.

Pu, H. T. (2003). Studies on polybenzimidazole/poly(4-vinylpyridine) blends and their proton conductivity after doping with acid. Polym. Int. 52, pp. 1540-1545.

Rabinowitsch, B. (1929). Z. Phys. Chem-Leipzig. A145, pp. 1-26.

Ramanathan, K., Bangar, M.A., Yun, M., Chen, W., Myung, N.V., Mulchandani, A. (2005). Bioaffinity sensing using biologically functionalized conducting-polymer nanowire, J. Am. Chem. Soc. 127, 2, pp. 496-497.

Ratner, B.D. (1995). Surface modification of polymers - chemical, biological and surface analytical challenges, Biosens. Bioelectron. 10, pp. 797-804.

Reid, R. C., Prausnitz, J. M. and Poling B. E. (1998). The Properties of Gases and Liquids, McGraw-Hill Book Company, Singapore, Fourth Edition.

Reinholz, G.G., Lu, L. and Driscoll, S.W. (2004). Animal models for cartilage reconstruction, Biomaterials. 25, pp. 1511-1521.

Reneker, D.H. and Chun, I. (1996). Nanometre diameter fibres of polymer, produced by electrospinning, Nanotechnology. 7, pp. 216-223.

Reneker, D. H., Yarin, A. L., Fong, H. and Koombhongse, S. (2000). Bending instability of electrically charged liquid jets of polymer solutions in electrospinning. J. Appl. Phys. 87, pp. 4531-4547.

Riboldi, S. A., Sampaolesi, M., Neuenschwander, P., Cossu, G. and Mantero, S. (2005). Electrospun degradable polyesterurethane membranes: potential scaffolds for skeletal muscle tissue engineering. Biomaterials. Article in Press.

Rizzi, S.C., Bock, N. and Downes, S. (2001). Biodegradable polymer/hydroxyapatite composites: Surface analysis and initial attachment of human osteoblasts, J. Biomed. Mater. Res. 55, pp. 475-486.

Robins, B.D. (1992). The history of vascular grafts, Br. J. Theatre. Nurs. 12, pp. 9-12.

Roper, D.K. and Lightfoot, E.N. (1995). Separation of biomolecules using adsorptive membranes, J Chromatogra. A, 702, pp. 3-26.

Russell, J. and Kerpel, R. G. V. (1957). Transitions in Plasticized and Unplasticizied Cellulose Acetates. J. Polym. Sci. 25, pp. 77-96.

Rutka, J.T., Apodaca, G., Stern, R. and Rosenblum, M. (1988). The extracellular matrix of the central and peripheral nervous systems: structure and function, J. Neurosurg. 69, pp. 155-170.

Rutledge, G. C., Li, Y., Fridrikh, S., Warner, S. B., Kalayci, V. E. and Patra, P. (2000). Electrostatic Spinning and Properties of Ultrafine Fibers, National Textile Center, 2000 Annual Report (M98-D01), National Textile Center. pp. 1-10

Sakabe, H., Ito, H., Miyamoto, T., Noishiki, Y. and Ha, W.S. (1989). In vivo blood compatibility of regenerated silk fibroin, Sen-i Gakkaishi. 45, pp. 487–490.

Sakurai, K., Maegawa, T. and Takahashi, T. (2000). Glass transition temperature of chitosan and miscibility of chitosan/poly(N-vinyl pyrrolidone) blends. Polymer. 41, pp. 7051-7056.

Sangster, D.F. (1989). Chapter 2 in: The effects of radiation on high-technology polymers, ed by E. Reichmanis, J.H. O'Donnell, American Chemical Society, Washington.

Sapsford, R.N., Oakley, G.D. and Talbot, S. (1981). Early and late patency of expanded polytetrafluoroethylene vascular grafts in aorta-coronary bypass. J. Thorac. Cardiovasc. Surg. 81, pp. 860-864.

Saville, D. A. (1997). Electrohydrodynamics: The Talyor-Melcher Leaky Dielectric Model. Annu. Rev. Fluid Mech. 29, pp. 27-64.

Sawicka, K., Gouma, P. and Simon, S. (2005). Electrospun biocomposite nanofibers for urea biosensing, Sensor Actuat B-Chem (in press).

Scardino, F. L. and Balonis, R. J. (2001). Fibrous structures containing nanofibrils and other textile fibers. US Patent 6,308,509 B1.

Schaffer, J. P., Saxena, A., Antolovich, S. D., Sanders, T. H. Jr. and Warner, S. B. (1999). The Science and Design of Engineering Materials. McGraw-Hill Companies, Inc. Singapore

Schmidt, C.E. and Leach, J.B. (2003). Neural tissue engineering: strategies for repair and regeneration, Annals Biomed Eng, 5, pp. 293-347.

Schreuder-Gibson, H., Gibson, P., Senecal, K., Sennett, M., Walker, J., Yeomans, W., Ziegler, D., Tsai, P.P. (2002). Protective textile materials based on electrospun nanofibers, J. Adv. Mater. 34, 3, pp. 44-55.

Sekhon, S.S., Arora, N. and Agnihotry, S.A. (2000). PAN-based gel electrolyte with lithium salts, Solid State Ionics, 136-137, pp. 1201-1204.

Sekhon, S.S. and Singh, H.P. (2004). Proton conduction in polymer gel electrolytes containing chloroacetic acids, Solid State Ionics, 175, pp. 545-548.

Senador Jr, A.E., Shaw, M.T. and Mather, P.T. (2001). Electrospinning of polymeric nanofibers: Analysis of jet formation. Mater. Res. Soc. Symp. Proc., 661; KK5.9.1-KK5.9.6.

Shao, C., Guan, H., Liu, Y., Gong, J., Yu, N. and Yang, X. (2004a). A novel method for making ZrO2 nanofibres via an electrospinning technique. J Crystal Growth, 267, 1-2, pp. 380-384.

Shao, C., Yang, X., Guan, H., Liu, Y. and Gong, J. (2004b). Electrospun nanofibers of NiO/ZnO composite. Inorganic Chem Commu, 7, pp. 625-627.

Shao, C., Guan, H., Liu, Y., Li, X. and Yang, X. (2004c). Preparation of Mn2O3 and Mn3O4 nanofibers via an electrospinning technique. J. Solid State Chem. 177, pp. 2628-2631.

Shenoy, S. L., Bates, W. D., Frisch, H. L. and Wnek, G. E. (2005). Role of chain entanglements on fiber formation during electrospinning of polymer solutions: good solvent, non-specific polymer-polymer interaction limit. Polymer. 46, 3372-3384.

Shields, K. J., Beckman, M. J., Bowlin, G. L., Wayne, J. S. (2004). Mechanical Properties and Cellular Proliferation of Electrospun Collagen Type II. Tissue Eng. 10, pp. 1510-1517.

Shin, C. and Chase, G.G. (2004). Water-in-oil coalescence in micro-nanofiber composite fibers, AIChE J. 50, 2, pp. 343-350.

Shummer, P. and Tebel, K. H. (1983). A New Elongational Rheometer For Polymer Solutions. J. Non-Newton. Fluid. 12, 331-347.

Shum-Tim, D., Stock, U., Hrkach, J., Shinoka, T., Lien, T.J. and Moses, M.A. (1999). Tissue engineering of autologous aorta using a new biodegradable polymer, Ann. Thorac. Surg. 68, pp. 2298-2305.

Singh, H.P. and Sekhon, S.S. (2004). Non-aqueous proton conducting polymer gel electrolytes, Electrochim. Acta. 50, pp. 621-625.

Smith, D., Reneker, D.H., Schreuder, G.H., Mello, C., Sennett, M. and Gibson, P. (2001), PCT/US00/27776.

Smith, W.J. and Dunn, M.A. (1991). Medical defense against blistering chemical warfare agents. Arch. Dermatol. 127, 8, pp. 1207-1213.

Soffer, A. and Folman, M. (1972). The electrical double layer of high surface porous carbon electrode. J. Electroanal. Chem. 38, pp. 25-43.

Son, W. K., Youk, J. H., Lee, T. S. and Park, W. H. (2004a).The effects of solution properties and polyelectrolyte on electrospinning of ultrafine poly(ethylene oxide) fibers. Polymer. 45, pp. 2959-2966.

Son, W. K., Youk, J. H. and Park, W. H. (2004b). Preparation of Ultrafine Oxidized Cellulose Mats via Electrospinning. Biomacromolecules. 5, pp. 197-201.

Son, W.K., Youk, J.H., Lee, T.S. and Park, W.H. (2004c). Electrospinning of ultrafine cellulose acetate fibers: studies of a new solvent system and deacetylation of ultrafine cellulose acetate fibers, J. Polym. Sci. Pt. B-Polym. Phys. 42, pp. 5-11.

Soraru, G. D., Babonneau, F. and MacKenzie, J. D. (1988). Structural Concepts on New Amorphous Covalent Solids. J. Non-cryst. Solids. 106, pp. 256-261.

Spivak, A.F. and Dzenis, Y.A. (1998). Asymptotic decay of radius of a weakly conductive viscous jet in an external electric field. Appl. Phys. Lett., 73(21), pp. 3067-3069.

Spivak, A.F., Dzenis, Y.A. and Reneker, D.H. (2000). A model for steady state jet in the electrospinning process. Mech. Res. Commun., 27(1), pp. 37-42.

Srnivasarao, M., Collings, D., Philips, A. and Patel. (2001). Three-Dimensionally Ordered Array of Air Bubbles in a Polymer Film. Science. 292, pp. 79-83.

Steffensa, G.C.M., Nothdurfta, L., Busea, G., Thissenb, H., Ockerb, H.H., Klee, D. (2002). High density binding of proteins and peptides to poly(d,l-lactide) grafted with polyacrylic acid, Biomaterials, 23, pp. 3523.

Steinthorosson, G. and Sumpio, B. (1999). Clinical and biological relevance of vein cuff anastomosis, Acta. Chir. Belg. 99, pp. 282-288.

Stephens, J. S., Chase, D. B. and Rabolt, J. F. (2004). Effect of the Electrospinning Process on Polymer Crystallization Chain Conformation in Nylon-6 and Nylon-12. Macromolecules. 37, pp. 877-881.

Su, L., Xiao, Z. and Lu, Z. (1998). All solid-state electrochromic window of electrodeposited WO3 and Prussian blue film with PVC gel electrolyte, Thin Solid Films. 320, pp. 285-289.

Sukigara, S., Gandhi, M., Ayutsede, J., Micklus, M. and Ko, F.K. (2003). Regeneration of bombyx mori silk by electrospinning - part 1: processing parameters and geometric properties, Polymer, 44, pp. 5721-5727.

Sun, Z., Zussman, E., Yarin, A.L., Wendorff, J.H. and Greiner, A. (2003). Compound core-shell polymer nanofibers by co-electrospinning, Adv. Mater. 15, 22, pp.1929-1932.

Sundaray, B., Subramanian, V., Natarajan, T. S., Xiang, R. Z., Chang, C. C. and Fann, W. S. (2004). Electrospinning of continuous aligned polymer fibers. Appl Phy Lett. 84, 7, pp. 1222-1224.

Suthar, A. and Chase, G. (2001), Nanofiber in filter media. Chemical Engineer, December, pp. 26-28.

Tan, E.P.S. and Lim, C.T. (2004). Physical properties of a single polymeric nanofiber. Appl. Phy. Lett. 84, 9, pp. 1603-1605.

Tan, E.P.S., Ng, S.Y. and Lim, C.T. (2005). Tensile testing of a single ultrafine polymeric fiber, Biomaterials. 26, pp. 1453-1456.

Tan, S. J., Verschueren, A. R. M. and Dekker, C. (1998). Room-temperature transistor based on a single carbon nanotube. Nature. 393, pp. 49-52.

Taylor, G. (1964). Disintegration of Water Drops in an Electric Field. Proc. R. Soc. Lond. A. 280, pp. 383-397.

Teo, W. E., Kotaki, M., Mo, X.M. and Ramakrishna, S. (2005). Porous tubular structures with controlled fibre orientation using a modified electrospinning method. Nanotechnology. 16, pp. 918-924.

Terai, H., Yamano, Y. and Vacanti, J.P. (2002). In vitro engineering of bone using a rotational oxygen-permeable bioreactor system, Mater. Sci. Eng. C. 20, pp. 3-8.

Theron, A., Zussman, E., Yarin, A. L. (2001). Electrostatic field-assisted alignment of electrospun nanofibers. Nanotechnology. 12, pp. 384-390.

Theron, S. A., Zussman, E. And Yarin, A. L. (2004). Experimental investigation of the governing parameters in the electrospinning of polymer solutions. Polymer, 45, pp. 2017-2030.

Theron, S. A., Yarin, A. L., Zussman, E. and Kroll, E. (2005). Multiple jets in electrospinning: experiment and modeling. Polymer, 46, pp. 2889-2899.

Tomaszewicz, W. (2001). Surface-potential decay of disordered solids. J. Electrostat. 51-52, pp. 340-344.

Tressler, J. F., Alkoy, S., Dogan, A. and Newnham, R. E. (1999). Functional composites for sensors, actuators and transducers, Compos Part A-Appl S. 30, pp. 477-482.

Tsai, P.P., Schreuder-Gibson, H. and Gibson, P. (2002). Different electrostatic methods for making electret filters, J. Electrostatics. 54, pp. 333-341.

Um, I. C., Fang, D. F., Hsiao, B. S., Okamoto, A. and Chu, B. (2004). Electro-Spinning and Electro-Blowing of Hyaluronic Acid, Biomacromolecules. 5, pp. 1428-1436.

Uyama, Y., Kato, K., Ikada, Y. (1998). Surface modification of polymers by grafting, Adv. Polym. Sci.137, pp. 1-39.

Veith, F.J., Gupta, S.K., Ascer, E., White-Flores, S., Samson, R.H. and Scher, L.A. (1986). Six-year prospective multicenter randomized comparison of autologous saphenous vein and expanded polytetrefluoroethylene grafts in infrainguinal arterial reconstructions, J. Vasc. Surg. 3, pp. 104-114.

Venugopal, J., Ma, L.L. and Ramakrishna, S. (2005). Biocompatible nanofiber matrices for engineering dermal substitute for skin regeneration, Tissue Eng. (in press).

Verreck, G., Chun, I., Rosenblatt, J., Peeters, J., Van Dijck, A., Mensch, J., Noppe, M. and Brewster, M.E. (2003a). Incorporation of drugs in an amorphous state into electrospun nanofibers composed of a water-insoluble, nonbiodegradable polymer. J. Control. Release. 92, pp. 349-360.

Verreck, G., Chun, I., Peeters, J., Rosenblatt, J. and Brewster, M.E. (2003b). Preparation and characterization of nanofibers containing amorphous drug dispersions generated by electrostatic spinning, Pharm. Res. 20, 5, pp. 810-817.

Viswanathamurthi, P., Bhattarai, N., Kim, H.Y., Lee, D.R., Kim, S.R. and Morris, M.A. (2003a). Preparation and morphology of niobium oxide fibres by electrospinning. Chem Phy Lett, 374, pp. 79-84.

Viswanathamurthi, P., Bhattarai, N., Kim, H.Y. and Lee, D.R. (2003b). Vanadium pentoxide nanofibers by electrospinning. Scripta Materialia, 49, pp. 577-81.

Viswanathamurthia, P., Bhattarai, N., Kim, H.Y., Cha, D.I. and Lee, D.R. (2004a). Preparation and morphology of palladium oxide fibers via electrospinning. Mater Lett, 58, pp. 3368-72.

Viswanathamurthi, P., Bhattarai, N., Kim, C.K., Kim, H.Y. and Lee, D.R. (2004b). Ruthenium doped TiO2 fibers by electrospinning. Inorganic Chem Commu, 7, pp. 679-82.

Viswanathamurthi, P., Bhattarai, N., Kim, H. Y., Lee, D. R. (2004c). The photoluminescence properties of zinc oxide nanofibres prepared by electrospinning. Nanotechnology. 15, pp. 320-323.

Wang, D.A., Ji, J., Feng, L.X. (2000). Surface analysis of poly(ether urethane) blending stearyl poly(ethylene oxide) coupling polymer, Macromolecules 33, 22, pp. 8472-8478.

Wang, C.E., Zhen, Y.L., Li, D., Yang, Q.B. and Hong, Y.L. (2002a). Preparation and stability of the nanochains consisting of copper nanoparticles and PVA nanofiber. Int J Nanosci, 1, 5-6, pp. 471-476.

Wang, X., Drew, C., Lee, S.H., Senecal, K.J., Kumar, J. and Samuelson, L.A. (2002b). Electrospun nanofibrous membranes for highly sensitive optical sensors, Nano Lett. 2, 1, pp. 1273-1275.

Wang, Y., Serrano, S. and Santiago-Aviles, J.J. (2003). Raman characterization of carbon nanofibers prepared using electrospinning. Synthetic Metals, 138, 3, pp. 423-427.

Wang, Y. H. and Hsieh, Y. L. (2004). Enzyme Immobilization to Ultra-Fine Cellulose Fibers via Amphiphilic Polyethylene Glycol Spacers. J. Polym. Sci. Part A. 42, pp. 4289-4299.

Wang, Y. and Santiago-Aviles, J.J. (2004). Synthesis of lead zirconate titanate nanofibres and the Fourier-transform infrared characterization of their metallo-organic decomposition process. Nanotechnology, 15, pp. 32-36.

Wang, J. X., Wen, L. X., Wang, Z. H., Wang, M., Shao, L. and Chen, J. F. (2004a). Facile synthesis of hollow silica nanotubes and their application as supports for immobilization of silver nanoparticles. Scripta Mater. 51, pp. 1035-1039.

Wang, M., Jin, H. J., Kaplan, D. L. and Rutledge, G. C. (2004b). Mechanical Properties of Electrospun Silk Fibers. Macromolecules. 37, pp. 6856-6864.

Wang, X., Kim, Y.G., Drew, C., Ku, B.C., Kumar, J. and Samuelson, L.A. (2004c). Electrostatic assembly of conjugated polymer thin layers on electrospun nanofibrous membranes for biosensors, Nano Lett. 4, 2, pp. 331-334.

Wang, Y., Furlan, R., Ramos, I. and Santiago-Aviles, J.J. (2004d). Synthesis and characterization of micro/nanoscopic Pb(Zr0.52Ti0.48)O3 fibers by electrospinning. Appl. Phys. A, 78, pp. 1043-1047.

Wang, Y., Aponte, M., Leon, N., Ramos, I., Furlan, R., Evoy, S. and Santiago-Aviles, J.J. (2004e). Synthesis and characterization of tin oxide microfibres electrospun from a simple precursor solution. Semicond. Sci. Technol, 19, pp. 1057-1060.

Wannatong, L., Sirivat, A. and Supaphol, P. (2004). Effects of solvents on electrospun polymeric fibers: preliminary study on polystyrene. Polym. Int. 53, 1851-1859.

Watanabe, M., Shinoka, T., Tohyama, S., Hibino, N., Konuma, T. and Matsumura, G. (2001). Tissue-engineered vascular autograft: inferior vena cava replacement in a dog model, Tissue Eng. 7, pp. 429-439.

Weinberg, C.B. and Bell, E. (1986). Blood vessel model constructed from collagen and cultured vascular cells, Science, 231, pp. 397-400.

Wertheimer, M.R., Fozza, A.C., Holländer, A. (1999). Industrial processing of polymers by low-pressure plasmas: the role of VUV radiation, Nucl. Instrum. Methods Phys. Res. Sect. B-Beam Interact. Mater. Atoms 151, pp. 65-75.

Wind, S. J., Appenzeller, J., Martel, R., Derycke, V. and Avouris, P. (2002). Vertical scaling of carbon nanotube field-effect transistors using top gate electrodes. Appl. Phys. Lett. 80, pp. 3817-3819.

Wnek, G.E., Carr, M.E., Simpson, D.G. and Bowlin, G.L. (2003). Electrospinning of nanofiber fibrinogen structures, Nano Lett. 3, 2, pp. 213-216.

Woerdeman, D. L., Ye, P., Shenoy, S., Parnas, R. S., Wnek, G. E. and Trofimova, O. (2005). Electrospun Fibers from Wheat Protein: Investigation of the Interplay between Molecular Structure and the Fluid Dynamics of the Electrospinning Process. Biomacromolecules. Article in Press.

Xie, J.B., Hsieh, Y.L. (2003). Ultra-high surface fibrous membranes from electrospinning of natural proteins: casein and lipase enzyme, J Mater. Sci. 38, pp.2125 – 2133.

Xu, C.Y., Inai, R., Kotaki, M. and Ramakrishna, S. (2004a). Aligned biodegradable nanofibrous structure: a potential scaffold for blood vessel engineering, Biomaterials. 25, pp. 877-886.

Xu, C.Y., Yang, F., Wang, S. and Ramakrishna, S. (2004b). In vitro study of human vascular endothelial cell function on materials with various surface roughness, J. Biomed. Mater. Res. 71A, pp. 154-161.

Xu, J., Pan, Q., Shun, Y. and Tian, Z. (2000) Grain size control and gas sensing properties of ZnO gas sensor. Sensor. Actuat. B-Chem. 66, pp. 277-279.

Yan, X.H., Liu, G.J., Liu, F.T., Tang, B.Z., Peng, H., Pakhomov, A.B. and Wong, C.Y. (2001). Superparamagnetic triblock copolymer/Fe2O3 hybrid nanofibers. Angew. Chem. Int. Ed., 40(19), pp. 3593-3596.

Yang, F., Murugan, R., Ramakrishna, S., Wang, X., Ma, Y.X. and Wang, S. (2004a). Fabrication of nano-structured porous PLLA scaffold intended for nerve tissue engineering, Biomaterials. 25, pp. 1891-1900.

Yang, F., Xu, C.Y., Kotaki, M., Wang, S. and Ramakrishna, S. (2004b) Characterization of neural stem cells on electrospun poly(L-lactic acid) nanofibrous scaffold. J. Biomat. Sci- Polym. E. 15, pp. 1483-1497.

Yang, Q. B., Li, Z. Y., Hong, Y. L., Zhao, Y. Y., Qiu, S. L., Wang, C. and Wei, Y. (2004c). Influence of Solvents on the Formation of Ultrathin Uniform Poly(vinyl pyrrolidone) Nanofibers with Electrospinning. J. Polym. Sci. Pol. Phys. 42, pp. 3721-3726.

Yang, X., Shao, C., Guan, H., Li, X. and Gong, J. (2004d). Preparation and characterization of ZnO nanofibers by using electrospun PVA/zinc acetate composite fiber as precursor. Inorganic Chem Commu, 7, pp. 176-178.

Yang, F., Murugan, R., Wang, S. and Ramakrishna, S. (2005). Electrospinning of nano/micro scale poly(L-lactic acid) aligned fibers and their potential in neural tissue engineering. Biomaterials. 26, pp. 2603-2610.

Yarin, A.L. (1993). Free liquid jets and films: Hydrodynamics and rheology. Longman, Wiley, 1993.

Yarin, A.L., Koombhongse, S. and Reneker, D.H. (2001a). Bending instability in electrospining of nanofibers. J. Appl. Phys., 89(5), pp. 3018-3026.

Yarin, A.L., Koombhongse, S. and Reneker, D.H. (2001b). Taylor cone and jetting from liquid droplets in electrospinning of nanofibers. J. Appl. Phys., 90(9), pp. 4836-4846.

Yarin, A. L. and Zussman, E. (2004). Upward needleless electrospinning of multiple nanofibers. Polymer. 45, 9, pp. 2977-2980.

Ye, H., Lam, H., Titchenal, N., Gogotsi, Y. and Ko, F. (2004). Reinforcement and rupture behavior of carbon nanotubes-polymer nanofibers. Appl. Phys. Lett. 85, pp. 1775-1777.

Yentov, V. M., Kordonskii, V. I., Prokhorov, I. V., Rozhkov, A. N., Toropov, A. I., Shul'man, Z. P. and Yarin, A. L. (1988). Intense stretching of moderately concentrated polymer solutions. Polym. Sci. U.S.S.R 30, pp. 2663-2669.

Ying, L., Yu, W.H., Kang, E.T., Neoh, K.G. (2004). Functional and surface-active membranes from poly(vinylidene fluoride)-graft-poly(acrylic acid) prepared via RAFT-mediated graft copolymerization, Langmuir 20, 14, pp. 6032-6040.

Yoshimoto, H., Shin, Y. M., Terai, H. and Vacanti, J. P. (2003). A biodegradable nanofiber scaffold by electrospinning and its potential for bone tissue engineering. Biomaterials. 24, pp. 2077-2082.

You, Y., Min, B. M., Lee, S. J., Lee, T. S. and Park, W. H. (2005). In Vitro Degradation Behavior of Electrospun Polyglycolide, Polylactide, and Poly(lactide-co-glycolide). J. Appl. Polym. Sci. 95, pp. 193-200.

Yuan, X., Zhang, Y., Dong, C. and Sheng, J. (2005). Morphology of ultrafine polysulfone fibers prepared by electrospinning. Polym. Int. In Press.

Zarkoob, S., Eby, R.K., Reneker, D.H., Hudson, S.D., Ertley, D. and Adams, W.W. (2004). Structure and morphology of electrospun silk nanofibers, Polymer, 45, pp. 3973-3977.

Zeng, J., Xu, X., Chen, X., Liang, Q., Bian, X., Yang, L. and Jing, X. (2003). Biodegradable electrospun fibers for drug delivery, J. Control. Release, 92, pp. 227-231.

Zhang, S., Yan, L., Altman, M., LaKssle, M., Nugent, H., Frankel, F., Lauffenburger, D.A., Whitesides, G.M., Rich, A. (1999). Biological surface engineering: a simple system for cell pattern formation. Biomaterials 20, pp.1213

Zhang, X. Q. and Wang, C. H. (1994). Physical aging and dye diffusion in polysulfone below the glass transition temperature. J. Polym. Sci. Pol. Phys. 32, pp. 569-572.

Zhang, Y. Z., Huang, Z. M., Xu, X. J., Lim, C. T. and Ramakrishna, S. (2004). Preparation of Core-Shell Structured PCL-r-Gelation Bi-Component Nanofibers by Coaxial Electrospinning. Chem. Mater., 16, 3406-3409.

Zhang, Y., Ouyang, H., Lim, C. T., Ramakrishna, S. and Huang, Z. M. (2005). Electrospinning of Gelatin Fibers and Gelatin/PCL Composite Fibrous Scaffolds. J. Biomed. Mater. Res. Part B. 72B, pp. 156-165.

Zhao, S., Wu, X., Wang, L. and Huang, Y. (2003). Electrostatically generated fibers of ethyl-cyanoethyl cellulose, Cellulose. 10, pp. 405-409.

Zhao, S. L., Wu, X. H., Wang, L. G. and Huang, Y. (2004). Electrospinning of Ethyl-Cyanoethyl Cellulose/Tetrahydrofuran Solutions. J. Appl. Polym. Sci. 91, pp. 242-246.

Zhong, X. H., Kim, K. S., Fang, D. F., Ran, S. F., Hsiao, B. S. and Chu, B. (2002). Structure and process relationship of Electrospun bioabsorbable nanofiber membranes. Polymer. 43, pp. 4403-4412.

Zhu, Y.B., Gao, C.Y., Liu, X.Y., Shen, J.C. (2002). Surface modification of polycaprolactone membrane via aminolysis and biomacromolecule immobilization for promoting cytocompatibility of human endothelial cells, Biomacromolecules 3, 6, pp. 1312-1319.

Zhu, Y.B., Gao, C.Y., He, T., Liu, X.Y., Shen, J.C. (2003) Layer-by-layer assembly to modify poly(L-lactic acid) surface toward improving its cytocompatibility to human endothelial cells, Biomacromolecules 4, pp.446

Zong, X., Kim, K., Fang, D., Ran, S., Hsiao, B. S. and Chu, B. (2002). Structure and process relationship of electrospun bioabsorbable nanofiber membranes. Polymer. 43, pp. 4403-4412.

Zong, X., Ran, S., Fang, D., Hsiao, B.S. and Chu, B. (2003). Control of structure, morphology and property in electrospun poly(glycolide-co-lactide) non-woven membranes via post-draw treatments, Polymer, 44, pp.4959-4967.

Index

www.ingramcontent.com/pod-product-compliance
Lightning Source LLC
Chambersburg PA
CBHW061616220326
41598CB00026BA/3787